Proceedings
of the
STEKLOV INSTITUTE
OF MATHEMATICS

1981, ISSUE 2

Boundary Value Problems of Mathematical Physics. X

Edited by

O. A. Ladyženskaja

Translation of
ТРУДЫ
ордена Ленина
МАТЕМАТИЧЕСКОГО ИНСТИТУТА
имени В. А. СТЕКЛОВА

Том 147 (1980)

АКАДЕМИЯ НАУК

СОЮЗА СОВЕТСКИХ СОЦИАЛИСТИЧЕСКИХ РЕСПУБЛИК

ТРУДЫ

ордена Ленина

МАТЕМАТИЧЕСКОГО ИНСТИТУТА

имени В. А. СТЕКЛОВА

CXLVII

КРАЕВЫЕ ЗАДАЧИ МАТЕМАТИЧЕСКОЙ ФИЗИКИ. X

Ответственный редактор (Editor-in-chief)
академик С. М. НИКОЛЬСКИЙ (S. M. Nikol'skiĭ)

Заместитель ответственного редактора (Assistant to the editor-in-chief)
доктор физ.-матем. наук Е. А. ВОЛКОВ (E. A. Volkov)

Ленинград "Наука"
Ленинградское отделение 1980

Library of Congress Cataloging in Publication Data

Main entry under title:

Boundary value problems of mathematical physics.

(Proceedings of the Steklov Institute of Mathematics, ISSN 0081-5438; 1981, issue 2)

Translation of Kraevye zadachi matematicheskoĭ fiziki.

Includes bibliographies.

1. Boundary value problems. 2. Differential equations. 3. Mathematical physics. I. Ladyzhenskaia͡, O. A. (Ol'ga Aleksandrovna), ed. II. Series.

QA1.A413 no. 83, etc. 515.3'53 67-6187

ISBN 0-8218-3002-3 (v. 5) AACR2

1981 subscription consists of 4 issues, published quarterly by American Mathematical Society. Subscription price of 1981 is list $210.00, member $105.00. Previously translated Russian Numbers are sold as separate books. Write to the Society for prices.

Subscription and orders should be addressed to American Mathematical Society, P. O. Box 1571, Annex Station, Providence, R. I. 02901. *All orders must be accompanied by payment.* Other correspondence should be addressed to P. O. Box 6248, Providence, R. I. 02940.

Proceedings of the Steklov Institute of Mathematics (ISSN 0081–5438) is published quarterly by the American Mathematical Society at 201 Charles Street, Providence, Rhode Island 02904. Controlled Circulation postage paid at Providence, Rhode Island 02940. Postmaster: Send address changes to Proceedings of the Steklov Institute of Mathematics, American Mathematical Society, P. O. Box 6248, Providence, RI 02940.

June 1981

Translation authorized by the All-Union Agency for Author's Rights, Moscow

PROCEEDINGS OF THE STEKLOV INSTITUTE OF MATHEMATICS IN THE ACADEMY OF SCIENCES OF THE USSR

TABLE OF CONTENTS

RUSSIAN TABLE OF CONTENTS*

Стр.

* The American Mathematical Society scheme for transliteration of Cyrillic may be found at the end of index issues of *Mathematical Reviews.*

1980 *Mathematics Subject Classification.* Primary 34E15, 35A05, 35B45, 35D05, 35J60, 35J65, 35J67, 35J70, 35K50, 35P25, 35Q10, 35Q99, 43A80, 43A90, 47A40, 47A45, 47A55, 58F30, 76W05; Secondary 35B45, 35J60, 35P05, 43A30, 46E40, 47A55, 70F15, 70H99, 76W05.

ABSTRACT. This collection is devoted to the study of nonlinear partial differential equations, group analysis of the Navier-Stokes equations and boundary layer theory, the construction of functional models of perturbation theory and their application in scattering theory, and the solvability of problems in magnetohydrodynamics.

Editor's note. In order to economize on production costs, displayed formulas in this translation are photographically reproduced from the Russian original. In addition to the intrinsic appearance of the formulas themselves, this means that symbols implemented in the text are not always exact replicas of their counterparts in displays.

Труды Матем. Инст. Стеклов.
147(1980)

Proceedings of the Steklov
Institute of Mathematics
1981, Issue 2

SPECTRAL THEORY AND SCATTERING FOR THE D'ALEMBERT OPERATOR WITH A VECTOR POTENTIAL

UDC 517.9

V. G. DEIČ AND D. R. JAFAEV

ABSTRACT. The selfadjoint operator H_1 in $L_2(\mathbb{R}^{m+1})$ corresponding to the differential expression $\sum_{\alpha=0}^m \gamma^\alpha(\partial_\alpha + \kappa A_\alpha(x))^2 + v(x)$, $\gamma^0 = 1$, $\gamma^\alpha = -1$, $\alpha = 1, \dots, m$, is studied. Under particular conditions for the decay of the coefficients $A_\alpha(x)$ and $v(x)$ at infinity and a certain supplementary condition (which is satisfied, in particular, for small κ) the existence and completeness of the wave operators for the pair $H_0 = \Box$, H_1 is established. Under the same conditions the absolute continuity of the positive spectrum is proved. Conditions for the negative singular spectrum to be discrete are indicated.

Bibliography: 10 titles.

1. Elliptic differential operators (or closely related operators) with coefficients which stabilize at infinity are the traditional object of investigation in mathematical scattering theory [1]. Until recently there were only isolated results for essentially nonelliptic operators [2], [1]. A systematic study of potential scattering for differential (and pseudodifferential) operators of general form was undertaken in [3] and [4]. Apparently, there are at present no results on scattering when essentially nonelliptic operators are perturbed by differential operators. Difficulties in such problems arise even in the attempt to correctly define the perturbing operator.

In the present paper we consider the operator generated in the space $\mathfrak{H} = L_2(\mathbb{R}^{m+1})$ by the differential expression

$$H_1 f = \sum_{\alpha=0}^{m} \gamma^\alpha (\partial_\alpha + \varkappa A_\alpha(x))^2 f + v(x) f. \tag{1}$$

Here $\gamma^0 = 1$, $\gamma^\alpha = -1$ for $\alpha = 1, \dots, m$, $x = (x_0, \mathrm{x}) \in \mathbb{R}^{m+1}$, $\partial_\alpha = -i\partial/\partial x_\alpha$, κ is a real parameter, and v and A_α are real-valued, measurable functions. We show that under natural conditions the operator H_1 defined on the Sobolev space $\mathfrak{W}_2^2(\mathbb{R}^{m+1})$ is essentially selfadjoint in $\mathfrak{H}$.

The main result consists in proving the existence and completeness of the wave operators (w. o.) for the pair H_0, H_1, where $H_0 = \sum_0^m \gamma^\alpha \partial_\alpha^2 \equiv \Box$. As in [3] and [4], it is found natural to formulate the conditions for the existence and completeness of the w. o. separately for the positive ($\mathbb{R}^+ = (0, \infty)$) and negative ($\mathbb{R}^- = (-\infty, 0)$) spectral semiaxes. Moreover, it is established under our assumptions that the positive spectrum of H_1 is absolutely continuous. In the case $m = 1$ the negative spectrum of H_1 is also absolutely continuous. For $m > 1$ conditions are given which ensure that the negative singular spectrum is discrete.

1980 *Mathematics Subject Classification.* Primary 35P25; Secondary 35P05.

Study of the operator H_1 is of particular physical interest: the scattering operator for the problem of the interaction of a quantized scalar field of mass μ with external electromagnetic (A_α) and scalar (v) fields is constructed (see, for example, [5]) on the basis of the S matrix for the pair H_0, H_1 at the point $\lambda = \mu^2$.

2. We construct a selfadjoint operator in $\mathfrak{H}$ related to the differential expression (1).

THEOREM 1. *Suppose that the function v is bounded, while the functions A_α are bounded and have bounded derivatives $B_\alpha = \partial_\alpha A_\alpha$. Then the operator H_1 is essentially selfadjoint on $\mathfrak{W}_2^2(\mathbb{R}^{m+1})$.*

PROOF. We shall show that if

$$(H_1 w,\ f) = (w,\ h) \tag{2}$$

for some $h \in \mathfrak{H}$ and any $w \in \mathfrak{W}_2^2$, then f belongs to the domain $\mathfrak{D}$ of the closure of H_1. We set $f_\delta(x) = (J_\delta f)(x) = \int j_\delta(x - x')f(x')\,dx'$, where j_δ is Sobolev's smoothing kernel, and it is understood that an integral with no limits indicated extends over the whole space $\mathbb{R}^{m+1}$. Then $f_\delta \in \mathfrak{W}_2^2$ and $f_\delta \to f$ in $\mathfrak{H}$ as $\delta \to 0$. We shall show that $H_1 f_\delta \to h$ in $\mathfrak{H}$ as $\delta \to 0$. It follows from (1) that

$$\begin{aligned}(H_1 f_\delta)(x) = \Big\{\int \Box j_\delta(x-x')f(x')\,dx' - \varkappa \sum_{\alpha=0}^{m} \gamma^\alpha \int ((\partial'_\alpha A'_\alpha + A'_\alpha \partial'_\alpha)\, j_\delta(x-x'))\\ \times f(x')\,dx' + r' f_\delta(x)\Big\} + \Big[\varkappa \sum_{\alpha=0}^{m} \gamma^\alpha \int ((\partial_\alpha A_\alpha + A_\alpha \partial_\alpha\\ + \partial'_\alpha A'_\alpha + A'_\alpha \partial'_\alpha)\, j_\delta(x-x'))f(x')\,dx' + (r - r')f_\delta(x)\Big],\end{aligned} \tag{3}$$

where $r = \kappa^2 \Sigma_0^n \gamma^\alpha A_\alpha^2 + v$, ∂' denotes differentiation with respect to the variables x', and functions with primes depend only on the argument x'. The expression in braces on the right side of (3) is equal to $\overline{(H_1 j_\delta(x - \cdot), f)}$. After elementary transformations the expression in square brackets in (3) can be rewritten in the form

$$\begin{aligned}\varkappa \sum_{\alpha=0}^{m} \gamma^\alpha \Big\{2 \int (\partial'_\alpha (A'_\alpha - A_\alpha)\, j_\delta(x-x'))f(x')\,dx'\\ + \int (B_\alpha - B'_\alpha)\, j_\delta(x-x')f(x')\,dx'\Big\} + \int (r - r')\, j_\delta(x-x')f(x')\,dx'.\end{aligned}$$

It is not hard to show (see Lemmas 2.7.3 and 2.7.4 of [6]) that all the terms in this expression tend to zero in $\mathfrak{H}$ as $\delta \to 0$. It now follows from (3) and (2) that

$$\lim_{\delta\to 0} H_1 f_\delta = \lim_{\delta\to 0} (f,\ H_1 j_\delta(x - .)) = \lim_{\delta\to 0} (h,\ j_\delta(x - .)) = h,$$

and hence $f \in \mathfrak{D}$. The proof of the theorem is complete.

The hypotheses of Theorem 1 are henceforth assumed to be satisfied.

3. We present an abstract criterion for the existence and completeness of wave operators (w. o.) which is used below. Let H_i, $i = 0, 1$, be selfadjoint operators in a Hilbert space $\mathfrak{H}$; let $\Delta \subseteq \mathbb{R}$ be an interval of the spectral axis. We denote by $R_i(z)$,

$E_i(\cdot)$ and $E_i^{ac}(\cdot)$ respectively the resolvent, spectral measure, and absolutely continuous part of the spectral measure of the operator H_i. The w. o. for the pair H_0, H_1 on the interval Δ (if they exist) are the strong limits

$$W_\pm(H_1, H_0; \Delta) = s - \lim_{t\to\pm\infty} \exp(itH_1)\exp(-itH_0)E_0^{ac}(\Delta).$$

A w. o. is called *complete* if its range coincides with $E_1^{ac}(\Delta)\mathfrak{H}$.

THEOREM 2. *Suppose that the set $\mathfrak{D}(H_0) \cap \mathfrak{D}(H_1)$ is dense in $\mathfrak{H}$ and the operator $V = H_1 - H_0$ admits a factorization $V = V_1V_0$, where V_1 is a bounded selfadjoint operator with dense range, the operators $V_0R_j(z)$ $(j = 0, 1)$ are bounded in $\mathfrak{H}$ for* $\operatorname{Im} z \neq 0$, *and the operators $V_iR_j(z)V_1$ $(i, j = 0, 1)$ have strong limits as $z \to \lambda \pm i0$ for almost all $\lambda \in \Delta$. Then there exist complete w. o. for the pair H_0, H_1.*

Theorem 2 is a modification of the criterion for the existence and completeness of w. o. of [7]. The difference between the hypotheses of the Theorem 2 and those of [7] is that the operators $V_0E_0(\tilde{\Delta})$, where $\tilde{\Delta} \subset \Delta$ is a compact interval, are not assumed to be bounded. However, the assumption on the boundedness of the operators $V_0R_j(z)V_1$ makes it possible to carry over verbatim the proof in [7]. We denote by Π^+ (Π^-) the closed upper (lower) half plane; $\Pi_\Delta^\pm$ is the intersection of $\Pi^\pm$ with the strip $\{z\colon \operatorname{Re} z \in \Delta\}$.

REMARK 1. *If the operator-valued function $V_1R_i(z)V_1$ is uniformly bounded on compact subsets of $\Pi_\Delta^\pm$, then the spectrum of the operator H_i is absolutely continuous on Δ.*

In applications the condition of Theorem 2 regarding the "unperturbed" resolvent is usually verified directly, while verification of the assumptions regarding the "perturbed" resolvent $R_1(z)$ is realized on the basis of the following analytic assertion [8].

LEMMA 1. *Let $N(z)$ be an operator-valued function which is continuous in norm for $z \in \Pi_\Delta^\pm$, and suppose that the operators $N(z)$ are compact and, for* $\operatorname{Im} z \neq 0$, *$[I - N(z)]^{-1}$ exists as a bounded operator. Denote by e the set of those $\lambda \in \Delta$ for which one of the equations $f = N(\lambda \pm i0)f$ has a nontrivial solution. Then e is a closed set of measure zero, and the operator-valued function $[I - N(z)]^{-1}$ is continuous in z in the uniform operator topology for $z \in \Pi_\Delta^\pm \setminus e$.*

4. To verify the conditions of Theorem 2 regarding the operator H_0 we shall need some auxiliary assertions. Let $t(u) = t_\varepsilon(u) = (1 + |u|)^{-(1+\varepsilon)/2}$, $u \in \mathbb{R}$, $\varepsilon > 0$; we denote by T_α, $\mathbf{T}$ and T the operators in $\mathfrak{H} = L_2(\mathbb{R}^{m+1})$ of multiplication by the functions $t(x_\alpha)$, $t(|\mathbf{x}|)$, and $t(|x|)$ respectively. We denote by $Q = Q_q$ the operator of multiplication by any bounded function $q(x)$ which tends to zero as $|x| \to \infty$. The following assertion is borrowed from [4].

LEMMA 2. *The operator-valued functions $T_0R_0(z)T_0$ and $\mathbf{T}R_0\mathbf{T}$ are continuous in the uniform operator topology for $z \in \Pi_{\mathbb{R}^+}^\pm$ and $z \in \Pi_{\mathbb{R}^-}^\pm$ respectively. The operator $QR_0(z)$ is compact for* $\operatorname{Im} z \neq 0$.

We now consider the operators $G_{\alpha,\beta} = T_\beta\partial_\alpha R_0(z)T_\beta$. *Everywhere below the estimates are carried out for* $\operatorname{Im} z \geqslant 0$. Suppose first that $\beta = 0$. It is convenient to consider the operator $G_{\alpha,0}$ in the "mixed" representation $(x_0, \mathbf{p})$ where Fourier transformation has been carried out on the variable $\mathbf{x}$. The action of the operator $G_{\alpha,0}$ is then given by

$$(G_{\alpha,0}(z)f)(x_0, \mathbf{p}) = \int_{-\infty}^{\infty} s_\alpha(x_0, y_0, \mathbf{p}; z) f(y_0, \mathbf{p})\, dy_0, \tag{4}$$

where

$$s_0(x_0, y_0, \mathbf{p}; z) = (i/2)\, t(x_0) \operatorname{sgn}(x_0 - y_0)\, t(y_0) \exp[i(\mathbf{p}^2 + z)^{1/2} |x_0 - y_0|], \tag{5}$$

$$s_\alpha(x_0, y_0, \mathbf{p}; z) = (i/2)\, t(x_0)\, t(y_0)\, p_\alpha (\mathbf{p}^2 + z)^{-1/2} \times \exp[i(\mathbf{p}^2 + z)^{1/2} |x_0 - y_0|], \quad \alpha \neq 0, \tag{6}$$

and $\operatorname{Im}(\mathbf{p}^2 + z)^{1/2} \geqslant 0$. To obtain (4)–(6) it is simplest to use the fact that in the "momentum" representation the operator $\partial_\alpha R_0(z)$ is the operator of multiplication by the function $p_\alpha(p_0^2 - \mathbf{p}^2 - z)^{-1}$ and to take the inverse Fourier transform on the variable p_0.

For fixed $\mathbf{p}$ the kernel $s_\alpha(x_0, y_0, \mathbf{p}; z)$ generates an integral operator $g_{\alpha,0}(\mathbf{p}; z)$ in $L_2(\mathbb{R})$. The following inequalities obviously hold:

$$\|G_{\alpha,0}(z)\| \leqslant \sup_{\mathbf{p}} \|g_{\alpha,0}(\mathbf{p}; z)\|,$$
$$\|G_{\alpha,0}(z) - G_{\alpha,0}(z')\| \leqslant \sup_{\mathbf{p}} \|g_{\alpha,0}(\mathbf{p}; z) - g_{\alpha,0}(\mathbf{p}; z')\|. \tag{7}$$

It can be verified directly that the Hilbert-Schmidt norms of the operators $g_{\alpha,0}(\mathbf{p}; z)$ in $L_2(\mathbb{R})$ are bounded uniformly with respect to $\mathbf{p}$; moreover, for $g_{0,0}(\mathbf{p}; z)$ this estimate is uniform with respect to $z \in \Pi^\pm$, while for $g_{\alpha,0}(\mathbf{p}; z)$, $\alpha \neq 0$, it is uniform with respect to $z \in \Pi^\pm_{\mathbb{R}^+}$.

To study the continuity of the operators $g_{\alpha,0}(\mathbf{p}; z)$ in the parameter z we make use of the obvious inequalities

$$|\exp[i(\mathbf{p}^2 + z)^{1/2} a] - \exp[i(\mathbf{p}^2 + z')^{1/2} a]|$$
$$\leqslant C_\varepsilon |(\mathbf{p}^2 + z)^{1/2} - (\mathbf{p}^2 + z')^{1/2}|^\varepsilon |a|^\varepsilon, \quad a \geqslant 0, \quad \varepsilon \in [0, 1],$$
$$|(a + z)^{1/2} - (a + z')^{1/2}| \leqslant C |z - z'|^{1/2}, \quad a \in \mathbb{R}. \tag{8}$$

By means of these inequalities it is easy to deduce from (5) and (6) that the operators $g_{\alpha,0}(\mathbf{p}; z)$ are Hölder continuous in z for complex z uniformly with respect to $\mathbf{p}$. Moreover, for $g_{0,0}(\mathbf{p}; z)$ the continuous dependence on z is preserved up to the real axis, while for $g_{\alpha,0}(\mathbf{p}; z)$ there is continuity for $z \in \Pi^\pm_{\mathbb{R}^+}$.

Application of (7) now leads to the following assertion.

LEMMA 3. *The operators $G_{\alpha,0}(z)$ are bounded for* $\operatorname{Im} z \neq 0$ *and depend continuously on z. Moreover, the operator-valued function $G_{0,0}(z)$ is continuous for $z \in \Pi^\pm$, while $G_{\alpha,0}(z)$, $\alpha \neq 0$, are continuous for $z \in \Pi^\pm_{\mathbb{R}^+}$.*

We shall also need the following lemma.

LEMMA 4. *For any pair of points z and z' of the upper (or lower) half plane the operators $Q[G_{\alpha,0}(z) - G_{\alpha,0}(z')]$ are compact.*

PROOF. We denote by Γ_δ the operator in $\mathfrak{H}$ which in the momentum representation acts as the operator of multiplication by $\chi_b(p^2)$, where χ_b is the characteristic function of the interval $[0, b]$. It obviously suffices to check that the operators $Q\partial_\alpha R_0(z)\Gamma_b$ are compact and

$$\lim_{b\to\infty} \| T_0\partial_\alpha [R_0(z) - R_0(z')](I - \Gamma_b) T_0 \| = 0. \tag{9}$$

The operator $Q\partial_\alpha R_0(z)\Gamma_b$ is a pseudodifferential operator with symbol $q(x)\varphi(p)$, where $\varphi(p) = p_\alpha(p_0^2 - \mathbf{p}^2 - z)^{-1}\chi_b(\mathbf{p}^2) \to 0$ as $|p| \to \infty$. Moreover, $q(x) \to 0$ as $|x| \to \infty$ by hypothesis. Therefore the operator $Q\partial_\alpha R_0(z)\Gamma_b$ is compact.

According to the second inequality of (7), to prove (9) it suffices to verify that

$$\lim_{b\to\infty} \sup_{\mathbf{p}^2\geqslant b} \| g_{\alpha,0}(\mathbf{p};\ z) - g_{\alpha,0}(\mathbf{p};\ z') \| = 0. \tag{10}$$

The relation (10) follows from combining (5) and (6) with (8) and the equality

$$\sup_{\mathbf{p}^2\geqslant b} |(\mathbf{p}^2 + z)^{1/2} - (\mathbf{p}^2 + z')^{1/2}| = O(b^{-1}),\ b \to \infty.$$

The proof of the lemma is complete.

We now consider the operators $G_{\alpha,\alpha}(z)$ for $\alpha \neq 0$.

LEMMA 5. *The operators* $G_{\alpha,\alpha}(z)$, $\alpha \neq 0$, *are bounded and continuous in* z *for* $z \in \Pi^\pm$.

PROOF. We consider the operators $G_{\alpha,\alpha}(z)$ in the mixed representation where the Fourier transform has been taken on all variables except x_α. We assume that $\mathbf{p} = (p_\alpha, \hat{p}_\alpha)$, $\mathbf{p}^2 = p_\alpha^2 + \hat{p}_\alpha^2$. In correspondence with (4) and (5) the action of $G_{\alpha,\alpha}(z)$ is given by

$$(G_{\alpha,\alpha}(z) f)(p_0,\ x_\alpha,\ \hat{p}_\alpha) = \int_{-\infty}^{\infty} l_\alpha(p_0,\ x_\alpha,\ y_\alpha,\ \hat{p}_\alpha;\ z) f(p_0,\ y_\alpha,\ \hat{p}_\alpha)\, dy_\alpha,$$

where

$$l_\alpha(p_0,\ x_\alpha,\ y_\alpha,\ \hat{p}_\alpha;\ z) = (-i/2)\, t(x_\alpha) \operatorname{sgn}(x_\alpha - y_\alpha)\, t(y_\alpha) \times \exp[i(p_0^2 - \hat{p}_\alpha^2 - z)^{1/2} |x_\alpha - y_\alpha|]$$

and $\operatorname{Im}(p_0^2 - \hat{p}_\alpha^2 - z)^{1/2} \geqslant 0$. The assertion of the lemma regarding the operators $G_{\alpha,\alpha}(z)$, $\alpha \neq 0$, is now established in analogy to the case $\alpha = 0$ considered previously.

5. We proceed to verify the hypotheses of Theorem 2. According to (1), the operator $V = H_1 - H_0$ is given by the differential expression

$$V = \sum_{\alpha=0}^{m} 2\varkappa\gamma^\alpha A_\alpha(x)\, \partial_\alpha + u(x), \tag{11}$$

where

$$u(x) = \sum_{\alpha=0}^{m} [\varkappa\gamma^\alpha B_\alpha(x) + \varkappa^2\gamma^\alpha A_\alpha^2(x)] + v(x).$$

The operator V is defined on $\mathfrak{W}_2^2(\mathbb{R}^{m+1})$ and admits closure; we retain the notation V for the corresponding closed operator. We set

$$\Omega(x)=\sum_{\alpha=0}^{m}|A_\alpha(x)|+|u(x)|.$$

As we noted in §1, the existence and completeness of the w. o. $W_\pm(H_1, H_0; \mathbb{R}^+)$ and $W_\pm(H_1, H_0; \mathbb{R}^-)$ are established under different assumptions. In considering the positive part of the spectrum we apply the following condition:

$$\lim_{|x|\to\infty}(1+|x_0|)^{1+\varepsilon}\Omega(x)=0,\quad \varepsilon>0; \tag{12}$$

in studying the negative part of the spectrum we assume that

$$\begin{gathered}\Omega(x)\leqslant C(1+|x|)^{-1-\varepsilon},\quad \varepsilon>0,\ m>1,\\ \lim_{|x|\to\infty}(1+|x_1|)^{1+\varepsilon}\Omega(x)=0,\quad \varepsilon>0,\ m=1.\end{gathered} \tag{13}$$

Below we consider the positive and negative parts of the spectrum in parallel. We represent V as the product $V=V_1^{(\pm)}V_0^{(\pm)}$, where $V_1^{(+)}=T_0$, $V_1^{(-)}=T$ $(m>1)$ and $V_1^{(-)}=T_1$ $(m=1)$. The indices $\pm$ here refer respectively to the positive and negative semiaxes.

We set $L_i^{(\pm)}=V_i^{(\pm)}R_0(z)V_1^{(\pm)}$. The validity of the hypothesis of Theorem 2 regarding $L_1^{(\pm)}$ follows from Lemma 2. We consider $L_0^{(\pm)}$. In correspondence with (11) and (12) the continuity of the operator-valued function $L_0^{(+)}$ for $z\in\Pi^\pm_{\mathbb{R}^+}$ follows from the continuity of the operator-valued functions $G_{\alpha,0}(z)$ established in Lemma 3. Similarly, by (13), to prove the continuity for $z\in\Pi^\pm_{\mathbb{R}^-}$ of the function $L_0^{(-)}(z)$ it suffices to check the continuity on this set of the functions $G_{\alpha,\alpha}(z)$ for $m>1$ and the continuity of $G_{\alpha,1}(z)$, $\alpha=0,1$, for $m=1$. The continuity of $G_{\alpha,\alpha}(z)$ is established in Lemma 5. Since for $m=1$ the variables x_0 and x_1 are equivalent, the continuity of $G_{0,1}(z)$ for $m=1$ and $z\in\Pi^\pm_{\mathbb{R}^-}$ follows from the continuity of $G_{1,0}(z)$ for $z\in\Pi^\pm_{\mathbb{R}^+}$.

We proceed to verify the conditions of Theorem 2 regarding $K_i^{(\pm)}(z)=V_i^{(\pm)}R_1(z)V_1^{(\pm)}$. It formally follows from the resolvent identity that

$$K_i^{(\pm)}(z)=L_i^{(\pm)}(z)[I+L_0^{(\pm)}(z)]^{-1}. \tag{14}$$

Since H_1 is selfadjoint, for $\operatorname{Im} z\neq 0$ the kernel of $I+L_0^{(\pm)}(z)$ is trivial. We suppose that there exist points z_1, $\operatorname{Im} z_1>0$, and z_2, $\operatorname{Im} z_2<0$, such that the operators $I+L_0^{(\pm)}(z_i)$ are boundedly invertible. We shall show that the functions $K_i^{(\pm)}(\lambda\pm i\varepsilon)$ then have limits as $\varepsilon\to 0$ for almost all $\lambda\in\mathbb{R}^\pm$. To be specific, we consider as before the case $\operatorname{Im} z>0$ and drop the indices $\pm$.

We set

$$N(z)=[I+L_0(z_1)]^{-1}[L_0(z)-L_0(z_1)]; \tag{15}$$

then

$$I+L_0(z)=[I+L_0(z_1)][I+N(z)]. \tag{16}$$

Conditions (12) or (13) of Lemmas 2 and 4 imply the compactness of the operator $V_0[R_0(z)-R_0(z_1)]V_1$, which in conjunction with (15) implies the compactness of

$N(z)$ for $\operatorname{Im} z \neq 0$. Equation (16) implies that the kernel of $I + N(z)$ is trivial, and therefore $I + N(z)$ is boundedly invertible for $\operatorname{Im} z \neq 0$. In particular, this implies that the invertibility of $I + L_0(z_1)$ at any point z_1, $\operatorname{Im} z_1 \geqslant 0$, implies the invertibility of $I + L_0(z)$ for all z of the upper half plane. Combining (14) and (16) gives us

$$K_i(z) = L_i(z)[I + N(z)]^{-1}[I + L_0(z_1)]^{-1}.$$

It was established above that the functions $L_i^{(+)}(z)$ ($L_i^{(-)}(z)$) and hence $N^+(z)$ ($N^-(z)$) are continuous for $z \in \Pi^{\pm}_{\mathbb{R}^+}$ ($z \in \Pi^{\pm}_{\mathbb{R}^-}$). It follows from Lemma 1 that the operators $[I + N^{(\pm)}(\lambda \pm i\varepsilon)]^{-1}$ and hence also $K_i^{(\pm)}(\lambda \pm i\varepsilon)$ have limits in norm as $\varepsilon \to 0$ for almost all $\lambda \in \mathbb{R}^{\pm}$. Application of Theorem 2 now leads to the main assertion of the present paper.

THEOREM 3. 1) *Suppose that condition* (12) *is satisfied and for some* z_1, $\operatorname{Im} z_1 \geqslant 0$, *and* z_2, $\operatorname{Im} z_2 \leqslant 0$, *the operators* $I + T_0^{-1}VR_0(z_i)T_0$ *are boundedly invertible. Then the complete w. o.* $W_{\pm}(H_1, H_0; \mathbb{R}^+)$ *exist for the pair* H_0, H_1.

2) *Suppose that* (13) *is satisfied and for some* z_1, $\operatorname{Im} z_1 \geqslant 0$, *and* z_2, $\operatorname{Im} z_2 \leqslant 0$, *the operators* $I + T^{-1}VR_0(z_i)T$ *are boundedly invertible. Then the complete w. o.* $W_{\pm}(H_1, H_0; \mathbb{R}^-)$ *exist for the pair* H_0, H_1.

In correspondence with (11) the operators $I + L_0^{(\pm)}(z)$ can be represented in the form $I + \kappa M_1^{(\pm)}(z) + M_2^{(\pm)}(z)$, where the operators $M_1^{(\pm)}(z)$ are uniformly bounded, while the norm of $M_2^{(\pm)}(z)$ ($M_2^{(+)} = T_0^{-1}UR_0T_0$, $M_2^{(-)} = T^{-1}UR_0T$, and U is the operator of multiplication by $u(x)$) tends to zero as $\operatorname{Im} z \to \pm\infty$. This implies that for sufficiently small κ the operators $I + L_0^{(\pm)}(z)$ are boundedly invertible for $\operatorname{Im} z \neq 0$. The next theorem now follows directly from Theorem 3.

THEOREM 4. *Let* κ *be sufficiently small. If condition* (12) *is satisfied, then there exist the complete w. o.* $W_{\pm}(H_1, H_0; \mathbb{R}^+)$, *while if* (13) *is satisfied there exist the complete w. o.* $W_{\pm}(H_1, H_0; \mathbb{R}^-)$.

6. We now establish the absolute continuity of the positive spectrum of the operator H_1 and the discreteness of its negative singular spectrum. In correspondence with Remark 1 and Lemma 1 only those points λ can belong to the singular spectrum of H_1 for which one of the equations

$$f + V_0R_0(\lambda \pm i0)V_1f = 0 \tag{17}$$

has nontrivial solutions.

LEMMA 6. *Suppose that* f *satisfies* (17) *for* $\lambda \neq 0$. *Then the Fourier transform of the function* V_1f *vanishes on the hyperboloid* $p_0^2 - \mathbf{p}^2 = \lambda$.

The proof is a repetition of the corresponding argument in [9].

We first establish that there are no nontrivial solutions of (17) for $\lambda > 0$. Suppose that (17) has a nontrivial solution for $z = \lambda + i0$. We introduce the operator $D(\lambda)$ which in the momentum representation of the space $\mathfrak{H}$ is the operator of multiplication by the generalized function $[(p_0 + i0)^2 - \mathbf{p}^2 - \lambda]^{-1}$. It is not hard to show (see, for example, [10]) that in the momentum representation the difference $R_0(\lambda + i0) - D(\lambda)$ is given by the operator of multiplication by the function

$2\pi i\theta(p_0)\delta(p_0^2 - \mathbf{p}^2 - \lambda)$. From (17) and Lemma 6 it now follows that

$$f + V_0 D(\lambda) V_1 f = 0. \tag{18}$$

Taking the (inverse) Fourier transform of $[(p_0 + i0)^2 - \mathbf{p}^2 - \lambda]^{-1}$ on the variable p_0, we find that in the mixed representation $(x_0, \mathbf{p})$ the operator $D(\lambda)$ is given by a formula of type (4), where the role of s_α is played by the function

$$d(x_0, y_0, \mathbf{p}; \lambda) = (\mathbf{p}^2 + \lambda)^{-1/2}\theta(y_0 - x_0)\sin[(\mathbf{p}^2 + \lambda)^{1/2}(x_0 - y_0)].$$

This implies that $V_0 D(\lambda) V_1$ is a Volterra operator in the variable x_0. It is thus possible to show by standard arguments that under condition (12) we have $\|(V_0 D(\lambda) V_1)^n\| \leqslant C^n (n!)^{-1}$. Equation (18) therefore has only trivial solutions. The following theorem has thus been established.

THEOREM 5. *Suppose that the hypotheses of the first part of Theorem 3 are satisfied. Then the positive spectrum of the operator H_1 is absolutely continuous.*

We now consider the negative part of the spectrum. In the case $m = 1$, because of the equivalence of the variables x_0 and x_1, it is possible to assert as before that under the conditions of the second part of Theorem 3 there is no negative singular spectrum. In the case $m > 1$ we assume that $\Omega(x) \leqslant C(1 + |x|)^{-3/2-\varepsilon}$, $\varepsilon > 0$, and for V_1 we take the operator of multiplication by $(1 + |x|)^{-1-\varepsilon/2}$. To prove that H_1 has no negative singular continuous spectrum, we show that if f satisfies (17) for $z = \lambda \pm i0$, $\lambda < 0$, then $\psi = R_0(\lambda \pm i0) V_1 f$ is an eigenfunction of H_1 corresponding to the eigenvalue λ. For this it suffices to verify that

$$\int_{-\infty}^{\infty} (\mu - \lambda)^{-2} \frac{d}{d\mu} (E_0(\mu) V_1 f, V_1 f)\, d\mu < \infty. \tag{19}$$

In accordance with the results of [4] the function $[(d/d\mu)(E_0(\mu)V_1 f, V_1 f)]^{1/2}$ is locally Hölder continuous (uniformly with respect to f for $\|f\| \leqslant 1$) with exponent $(1 + \varepsilon)/2$; moreover, according to Lemma 6, it vanishes for $\mu = \lambda$. This proves (19).

Applying the technique of [9], we further establish that the eigenvalues of H_1 have no negative accumulation points. Assuming otherwise, we find that for some λ_n, $\lambda_n \to \lambda_0$, $\lambda_0 < 0$, equation (17) has nontrivial solutions f_n, $\|f_n\| = 1$. It follows from (16) that $f_n + N(\lambda_n \pm i0)f_n = 0$. Since the $N(z)$ are compact operators, the sequence f_n is compact. It may be assumed that $f_n \to f_0$, $\|f_0\| = 1$. Since $\psi_n = R_0(\lambda_n \pm i0) V_1 f_n$ are eigenfunctions of H_1, for $n \neq m$ we have $(\psi_n, \psi_m) = 0$. Using Lemma 6 and the fact, noted above, that the function $[(d/d\mu)(E_0(\mu)V_1 f_n, V_1 f_n)]^{1/2}$ is locally Hölder continuous, it is not hard to justify passing to the limit in the equation $(\psi_n, \psi_m) = 0$ first as $m \to \infty$ and then as $n \to \infty$. This implies that $f_0 = 0$. The assumption that the eigenvalues of H_1 have a negative accumulation point has thus led to a contradiction.

THEOREM 6. *Suppose that for $m > 1$ the conditions of the second part of Theorem 3 are satisfied and $\Omega(x) \leqslant C(1 + |x|)^{-3/2-\varepsilon}$, $\varepsilon > 0$. Then the negative singular spectrum of the operator H_1 consists only of eigenvalues of finite multiplicity having no negative finite accumulation points. In the case $m = 1$ under the hypotheses of the second part of Theorem 3 the negative singular spectrum of H_1 is empty.*

BIBLIOGRAPHY

1. M. Š. Birman, *Scattering problems for differential operators with constant coefficients*, Funkcional Anal. i Priložen. **3** (1969), no. 3, 1–16; English transl. in Functional Anal. Appl. **3** (1969).

2. L. P. Nižnik, *The spectral properties of selfadjoint partial differential operators which are close to the operators with constant coefficients*, Outlines Joint Soviet-American Sympos. Partial Differential Equations (Novosibirsk, 1963), Sibirsk. Otdel. Akad. Nauk SSSR, Novosibirsk, 1963, pp. 195–199.

3. V. G. Deič, E. L. Korotjaev and D. R. Jafaev, *Potential scattering taking account of spatial anisotropy*, Dokl. Akad. Nauk SSSR **235** (1977), 749–752; English transl. in Soviet Math. Dokl. **18** (1977).

4. ______, *Theory of potential scattering taking account of spatial anisotropy*, Zap. Naučn. Sem. Leningrad. Otdel. Mat. Inst. Steklov. (LOMI) **73** (1977), 35–51; English transl., to appear in J. Soviet Math.

5. J. Bellissard, *Quantized fields in interaction with external fields*. I, Comm. Math. Phys. **41** (1975), 235–266.

6. M. Nagumo, *Lectures on the contemporary theory of partial differential equations*, Kyorītsu Shuppan, Tokyo, 1957 (Japanese); Russian transl., "Mir", Moscow, 1967.

7. V. G. Deič, *A stationary local method in scattering theory for a pair of spaces*, Problemy Mat. Fiz., vyp. 6, Izdat. Leningrad. Gos. Univ., Leningrad, 1973, pp. 76–90. (Russian)

8. S. T. Kuroda, *An abstract stationary approach to perturbation of continuous spectra and scattering theory*, J. Analyse Math. **20** (1967), 57–117.

9. L. D. Fadeev, *On the Friedrichs model in the theory of perturbations of a continuous spectrum*, Trudy Mat. Inst. Steklov. **73** (1964), 292–313; English transl. in Amer. Math. Soc. Transl. (2) **62** (1967).

10. N. N. Bogoljubov and D. V. Širkov, *Introduction of the theory of quantized fields*, 2nd ed., "Nauka", Moscow, 1973; English transl. of 1st ed., Interscience, 1959.

Translated by J. R. SCHULENBERGER

Труды Матем. Инст. Стеклов.
147(1980)

Proceedings of the Steklov
Institute of Mathematics
1981, Issue 2

ON THE SOLVABILITY OF THE FIRST BOUNDARY VALUE PROBLEM FOR CERTAIN CLASSES OF DEGENERATING QUASILINEAR ELLIPTIC EQUATIONS OF THE SECOND ORDER

UDC 514.946

A. V. IVANOV AND P. Z. MKRTYČJAN

ABSTRACT. Certain classes of second order quasilinear elliptic equations are specified, admitting degeneracies of the ellipticity on a set of measure zero. The existence of a generalized solution of the Dirichlet problem is established. It is proved that under certain conditions the elliptic operators considered effect a homeomorphism between suitable function spaces.

Bibliography: 12 titles.

In this paper certain classes of second order quasilinear elliptic equations are specified, admitting degeneracies of a certain type. For equations in these classes, the existence of a (generalized) solution of the Dirichlet problem is established. In the nondegenerate case, the result on generalized solvability of the Dirichlet problem for certain classes of quasilinear elliptic equations was obtained by Ladyženskaja and Ural'ceva [1]. In the present paper classes of degenerating elliptic operators are also specified which effect a homeomorphism between suitable function spaces. The proof of our results is based on a general theory of operator equations in Banach spaces, possessing known properties of coercivity and monotonicity, which is due to Browder and Minty (see, for example, [2] and [3]). The further development of this theory may be learned from the surveys [4]–[6]. In this paper we introduce a certain variant of the condition ensuring the solvability of an operator equation. Matters relating to the general theory of operator equations in a Banach space are covered in §1. In §2 we examine differential operators acting on the space $\overset{\circ}{H}{}^1_m(A, \Omega)$. In §3, generalizations of the results of §2 are given.

§1. Operator equations in a Banach space

Let X be a separable reflexive Banach space,(¹) and X^* the dual space. The values of the functional $f \in X^*$ at an element $x \in X$ will be denoted by $\langle f, x\rangle$.

We consider a (nonlinear, in general) operator $A: X \to X^*$, acting from X into X^*. We recall the definition of certain terms which will be used in the paper.

The operator $A: X \to X^*$ is called *semicontinuous* if for any $u, v, w \in X$

$$\lim_{\varepsilon\to 0} \langle A(u + \varepsilon v), w\rangle = \langle Au, w\rangle. \tag{1.1}$$

1980 *Mathematics Subject Classification*. Primary 35J70; Secondary 35J60

(¹) We limit ourselves to the case of a real Banach space.

The operator $A: X \to X^*$ is called *demicontinuous* if it is continuous from X with the strong topology into X^* with the weak topology.

The operator $A: X \to X^*$ will be called *coercive*(²) if there exists a number $\rho_0 > 0$ such that for any $u \in X$, with $\|u\| = \rho_0$,

$$\langle Au, u\rangle \geqslant 0. \tag{1.2}$$

We say that the operator $A: X \to X^*$ has *semibounded variation* if, for any $\rho > 0$ and any $u, v \in X$ such that $\|u\|, \|v\| \leqslant \rho$, the inequality

$$\langle Au - Av, u - v\rangle \geqslant -\gamma(\rho, \|u - v\|') \tag{1.3}$$

holds, where the function $\gamma(\rho, \sigma)$ is nonnegative, continuous, and such that

$$\lim_{\sigma\to+0} \frac{\gamma(\rho, \sigma)}{\sigma} = 0, \qquad \forall\rho \geqslant 0, \qquad \forall\tau \geqslant 0, \tag{1.4}$$

and $\|\cdot\|'$ is a norm compact with respect to $\|\cdot\|$.

In the case when $\gamma(\rho, \sigma) \equiv 0$, the condition of semiboundedness of the operator $A: X \to X^*$ coincides with the condition of monotonicity. We shall prove the following theorem.

THEOREM 1.1. *Let the operator $A: X \to X^*$ be demicontinuous, coercive, and have semibounded variation. Then the equation*

$$Au = 0 \tag{1.5}$$

has at least one solution $u \in X$.

Results like Theorem 1.1 are given, for example, in [4]–[6]. In [4] the proof of solvability of (1.5) is given under the assumptions that the operator A is demicontinuous, coercive and monotone. It is clear that this result is a particular case of Theorem 1.1. In [5] the solvability of (1.5) is proved under the conditions of boundedness,(³) semicontinuity, coercivity, and monotonicity of A. In [6] (Chapter II, §1, Theorem 1.1) a result more general than that in [5] was proved; namely

THEOREM 1.1′. *Let the operator $A: X \to X^*$ be bounded, semicontinuous, coercive, and have semibounded variation. Then* (1.5) *has at least one solution $u \in X$.*

Theorem 1.1′ is a particular case of Theorem 1.1 (i.e., the conditions of Theorem 1.1′ imply those of Theorem 1.1). This assertion follows from the following proposition.

PROPOSITION 1.1. *If the operator $A: X \to X^*$ is bounded, semicontinuous, and has semibounded variation, then it is demicontinuous.*

PROOF. We use the construction which was applied in [5] to prove that every bounded semicontinuous monotone operator is demicontinuous. We assume that under the conditions of Proposition 1.1 the operator A is not demicontinuous. Then there exists a sequence $\{u_n\}$ such that $u_n \to u$ strongly in X, and Au_n, $n = 1, 2, \ldots,$ is bounded in X^*. Then a subsequence $\{u_\nu\}$ may be exhibited such that

(²) The somewhat unusual form used here for the coercivity condition is explained by the fact that in this paper equations of the form $Au = 0$ are considered.

(³) The operator $A: X \to X^*$ is called *bounded* if it takes bounded sets in X into bounded sets in X^*. We note that if a monotone operator is semicontinuous and locally bounded, then it is demicontinuous (see [4]).

$Au_\nu \to f$ weakly in X^*, and $f \neq Au$. Then

$$\lim_{\nu\to\infty} \langle Au_\nu, u_\nu - u\rangle = 0. \tag{1.6}$$

We set $w = (1-\theta)u + \theta v$, where $\theta \in (0, 1)$ and v is a fixed element in X. Taking account of condition (1.3) for semiboundedness of variation, we write the inequality

$$\langle Au_\nu - Aw, u_\nu - w\rangle \geqslant -\gamma(\rho, \|u_\nu - w\|'), \tag{1.7}$$

where $\rho = \sup_\nu \|u_\nu\| + \|v\|$. From (1.7) we obtain

$$\theta\langle Au_\nu, u - v\rangle \geqslant -\langle Au_\nu, u_\nu - u\rangle + \langle Aw, u - v\rangle - \gamma(\rho, \|u_\nu - w\|'). \tag{1.8}$$

Letting ν tend to ∞ in (1.8) and considering (1.6) and the properties of the function γ, we obtain

$$\langle f, u - v\rangle \geqslant \langle Aw, u - v\rangle - \frac{\gamma(\rho, \|u - w\|')}{\theta}. \tag{1.9}$$

Noting that $\|u - w\| = \theta\|u - v\|$, we take the limit in (1.9) as $\theta \to +0$. Using (1.1) and (1.4), we thus obtain

$$\langle f, u - v\rangle \geqslant \langle Au, u - v\rangle,$$

from which, by the arbitrariness of v, it immediately follows that $f = Au$. This contradiction establishes Proposition 1.1.

The proof of Theorem 1.1 uses the following known lemma.

LEMMA 1.1. *Let $\mathfrak{D}$ be a convex bounded neighborhood of the origin in $\mathbb{R}^N$, and let the mapping F: $\mathbb{R}^N \to \mathbb{R}^N$ be continuous on $\overline{\mathfrak{D}}$. If on the boundary $\partial\mathfrak{D}$ of $\mathfrak{D}$ the inner product* $(F\mathbf{c}, \mathbf{c})$ *is nonnegative, i.e.*

$$(F\mathbf{c}, \mathbf{c}) \geqslant 0 \quad \forall \mathbf{c} \in \partial\mathscr{D},$$

then the equation $F\mathbf{c} = 0$ has at least one solution $\mathbf{c} \in \mathfrak{D}$.

The proof is given, for example, in [5], Chapter 1, §4.3 (see the proof of Lemma 4.3).

PROOF OF THEOREM 1.1. Let $v_1, v_2, \ldots$ be a fundamental system of linearly independent elements in the space X (i.e., the linear span of this system is dense in X). We set

$$u^N = \sum_{k=1}^N c_k^N v_k, \tag{1.10}$$

where c_k^N, $k = 1, \ldots, N$, are defined by the system of N nonlinear equations

$$\langle Au^N, v_k\rangle = 0, \qquad k = 1, \ldots, N. \tag{1.11}$$

We first prove that for each $N = 1, 2, \ldots,$ the system (1.11) has at least one solution. We denote by F^N: $\mathbb{R}^N \to \mathbb{R}^N$ the mapping defined by

$$F^N = (F_1^N, \ldots, F_N^N), \qquad F_k^N(\mathbf{c}) = \left\langle A\left(\sum_{i=1}^N c_i v_i\right), v_k\right\rangle,$$
$$k = 1, \ldots, N, \qquad \mathbf{c} = (\mathbf{c}_1, \ldots, \mathbf{c}_N) \in \mathbb{R}^N. \tag{1.12}$$

We denote by P_N the linear span of the first N elements of the given fundamental system $\{v_i\}_1^\infty$.

Let ω be a real isomorphism taking $\mathbb{R}^N$ onto P_N, i.e. $\omega(\mathbf{c}) = \sum_1^N c_i v_i$, $\mathbf{c} \in \mathbb{R}^N$. Since in a finite-dimensional space all norms are equivalent (in particular, in $\mathbb{R}^N$ the Euclidean norm $|\mathbf{c}|$ and the norm $|\mathbf{c}|' \equiv \|\sum_1^n c_i v_i\|$ are equivalent) it follows that ω is also a homeomorphism. Therefore P_N is closed in X. We write

$$F_k^N(\mathbf{c}) = \langle A\omega(\mathbf{c}), v_k \rangle, \qquad k = 1, \dots, N.$$

We see, by virtue of the demicontinuity of A, that the mapping F^N is continuous in $\mathbb{R}^N$ (as the superposition of the continuous mappings $\mathbf{c} \to \omega(\mathbf{c})$ and $v \to (\langle Av, v_1 \rangle, \dots, \langle Av, v_n \rangle)$). Moreover, let $K_{\rho_0} \equiv \{v \in P_N : \|v\| < \rho_0\}$ be the ball of radius ρ_0 in P_N. It is easy to see that $\omega^{-1}(K_{\rho_0}) \equiv D^N$ is a convex bounded closed neighborhood of the origin in $\mathbb{R}^N$, and the boundary ∂K_{ρ_0} of K_{ρ_0} corresponds to the boundary $\partial \mathfrak{D}^N$ of the neighborhood $\mathfrak{D}^N$. Thus, because of the coercivity of the operator A, for all $\mathbf{c} \in \partial \mathfrak{D}^N$ we obtain

$$(F^N(\mathbf{c}), \mathbf{c}) = \langle Av, v \rangle \geqslant 0,$$

since $v = \omega(\mathbf{c}) = \sum_1^n c_i v_i \in \partial K_{\rho_0}$ and $\|v\|_{P_N} = \|v\|_X = \rho_0$. Applying Lemma 1.1, we conclude that there is a $\mathbf{c} = c^N \in \mathfrak{D}^N$ such that $F^N(\mathbf{c}^N) = 0$. Therefore (1.11) is solvable for each $N = 1, 2, \dots,$ and the solutions $u^N = \omega(\mathbf{c}^N)$ satisfy

$$\|u^N\| \leqslant \rho_0. \tag{1.13}$$

Let $\{u^N\}$ be any sequence of such approximate solutions. Because of the reflexivity of X, there is a subsequence of it (which we again denote by $\{u^N\}$) which converges weakly in X to some element $u \in X$. We fix some element $\xi \in p_{N'}$. Then for any $N \geqslant N'$, by virtue of the semiboundedness of A we have

$$\langle Au^N, u^N - \xi \rangle - \langle A\xi, u^N - \xi \rangle \geqslant -\gamma(\rho, \|u^N - \xi\|'), \tag{1.14}$$

where $\rho = \sup_N \|u^N\| + \|\xi\|$. Subtracting from (1.14) the equation $\langle Au^N, u^N - \xi \rangle = 0$, which follows from (1.11), we obtain

$$-\langle A\xi, u^N - \xi \rangle \geqslant -\mu(\rho, \|u^N - \xi\|'). \tag{1.15}$$

Letting N tend to ∞ in (1.15) and considering the compactness of the norm $\|\cdot\|'$ relative to $\|\cdot\|$ (from the weak convergence of u^N to u in X it follows that $\lim_{N\to\infty} \|u^N - u\|' = 0$), and also the continuity of the function γ, we obtain for any $\xi \in \bigcup_1^\infty P_N$

$$-\langle A\xi, u - \xi \rangle \geqslant -\gamma(\rho, \|u - \xi\|'). \tag{1.16}$$

Since $\bigcup_1^\infty P_N$ is dense in X, we conclude, using the demicontinuity of A and the properties of the function γ, that (1.16) is in fact satisfied for all $\xi \in X$. Setting $\xi = u + \varepsilon\eta$, where η is an arbitrary element in X, we obtain

$$\langle A(u + \varepsilon\eta), \eta \rangle - \frac{\gamma(\rho, \varepsilon\|\eta\|')}{\varepsilon}. \tag{1.17}$$

Since $u + \varepsilon\eta \to u$ as $\varepsilon \to +0$, on passing to the limit as $\varepsilon \to +0$ in (1.17), by virtue of the demicontinuity of A and property (1.4) of the function γ we obtain

$$\langle Au, \eta \rangle \geqslant 0, \qquad \forall \eta \in X.$$

From this, since η is arbitrary, we clearly obtain that $Au = 0$. Theorem 1.1 is proved.

The proof of Theorem 1.1 used approaches typical for the theory of monotone operators (see [2]–[6]).

§2. Equations in the space $\mathring{H}_m^1(A, \Omega)$

1. *The space* $\mathring{H}_m^1(A, \Omega)$. Let Ω be a bounded domain in $\mathbb{R}^N$ and $A = \|a^{ij}\|$ a symmetric matrix with elements $a^{ij} = a^{ij}(x)$, $i, j = 1, \dots, n$, defined and measurable in Ω. We denote by $\lambda = \lambda(x)$ and $\Lambda = \Lambda(x)$, respectively, the least and greatest eigenvalues of the matrix A, and we assume that

$$\lambda > 0 \text{ almost everywhere (a.e.) in } \Omega, \quad \lambda^{-1} \in L^r(\Omega), \quad r \in (0, +\infty],$$
$$\Lambda^{m/2} \in L^1(\Omega), \qquad m > 1. \tag{2.1}$$

LEMMA 2.1. *Assume that conditions* (2.1) *are fulfilled, and let r, m, q, and l satisfy the conditions*

$$r > 0,\ m > 1,\ 1/q = 1/2r + 1/m \leqslant 1,\ 1/l = 1/q - 1/n \ \text{for}\ q < n,$$
$$l \in [m, +\infty) \ \text{for}\ q = n,\ l = +\infty \ \text{for}\ q > n. \tag{2.2}$$

Then every function $u \in \mathring{C}^1(\Omega)$ satisfies

$$\|u\|_{l,\Omega} \leqslant c\|u\|_{\mathring{W}_q(\Omega)} \leqslant c\|\lambda^{-1}\|_{r,\Omega}^{1/2}\left(\int_\Omega |A^{1/2}u_x|^m\,dx\right)^{1/m}, \tag{2.3}$$

where $\|u\|_{l,\Omega} = \|u\|_{L^l(\Omega)}$, $c = c(n, q)$ *for* $q < n$, $c = c(n, l)$ *for* $q = n$, *and* $c = c(n)$ *for* $q > n$.

PROOF. By Sobolev's imbedding theorem, for any $u \in \mathring{C}^1(\Omega)$ we have

$$\|u\|_{l,\Omega} \leqslant c\|u\|_{\mathring{W}_q^1(\Omega)}, \tag{2.4}$$

where $1/l = 1/q - 1/n$ for $q < n$, $l \in [1, +\infty)$ for $q = n$, $l = +\infty$ for $q > n$, and $c = c(n, q)$ for $q < n$, $c = c(n, l)$ for $q = n$, and $c = c(n)$ for $q > n$. Let q satisfy the conditions $1/q = 1/2r + 1/m \leqslant 1$, $r \in (0, +\infty)$. Then $q \in [1, m]$. Applying Hölder's inequality, we obtain the estimate

$$\|u\|_{\mathring{W}_q^1}(\Omega) = \left(\int_\Omega \lambda^{-q/2}\lambda^{q/2}|u_x|^q\,dx\right)^{1/q} \leqslant \|\lambda^{-1}\|_{r,\Omega}^{1/2}\left(\int_\Omega |A^{1/2}u_x|^m\,dx\right)^{1/m}. \tag{2.5}$$

From (2.4) and (2.5) we obtain (2.3). Lemma 2.1 is proved.

We denote by $\mathring{H}_m^1(A, \Omega)$ the closure of the set $\mathring{C}^1(\Omega)$ with respect to the norm

$$\|u\|_{\mathring{H}_m^1(A,\Omega)} = \left(\int_\Omega |A^{1/2}u_x|^m\,dx\right)^{1/m}. \tag{2.6}$$

We note that by virtue of (2.3), the equation $\|u\|_{\mathring{H}_m^1(A,\Omega)} = 0$ for $u \in \mathring{C}^1(\Omega)$ implies the equation $u \equiv 0$ in Ω. Therefore (2.6) is a legitimate norm. Clearly $\mathring{H}_m^1(A, \Omega)$ is a separable Banach space.

LEMMA 2.2. *Assume conditions* (2.1) *and* (2.2) *hold. Then*[4] $\mathring{H}_m^1(A, \Omega) \hookrightarrow \mathring{W}_q^1(\Omega) \hookrightarrow L^l(\Omega)$, *and for any* $u \in \mathring{H}_m^1(A, \Omega)$ *the inequality*

([4]) Following [7], we denote the imbedding of a Banach space E in a Banach space G by the symbol $E \hookrightarrow G$. This means that all elements of E are contained in G and that there exists a constant K, independent of u, such that $\|u\|_E < K\|u\|_G$ $\forall u \in E$.

$$\|u\|_{l,\Omega}\leqslant c\|u\|_{\mathring{W}^1_q(\Omega)}\leqslant c\|\lambda^{-1}\|^{1/2}_{r,\Omega}\|u\|_{\mathring{H}^1_m(A,\Omega)}, \tag{2.7}$$

holds, where the constant c is the same as in (2.3).

PROOF. The assertions of Lemma 2.2 clearly follow from (2.3), the definition of $\mathring{H}^1_m(A, \Omega)$ and known properties of the spaces $L^l(\Omega)$ and $\mathring{W}^1_q(\Omega)$.

LEMMA 2.3. *The space $\mathring{H}^1_m(A, \Omega)$ is reflexive. Each linear functional F on $\mathring{H}^1_m(A, \Omega)$ may be represented in the form*([5])

$$\langle F, \eta\rangle=\int_\Omega f^i\eta_{x_i}dx,\quad \eta\in\mathring{H}^1_m(A,\Omega), \tag{2.8}$$

where $|B^{1/2}f|\in L^{m/(m-1)}(\Omega)$, $B = A^{-1}$, $\mathbf{f} = (f^1, \dots, f^n)$, with

$$\|F\|_{\mathring{H}^1_m(A,\Omega)^*}=\|B^{1/2}\mathbf{f}\|_{m/(m-1),\Omega}. \tag{2.9}$$

PROOF. Since for each $u\in\mathring{H}^1_m(A, \Omega)$ the vector-valued function $A^{1/2}u_x$ belongs to $L^m(\Omega)\times\dots\times L^m(\Omega)$, and

$$\|u\|_{\mathring{H}^1_m(A,\Omega)}=\|A^{1/2}u_x\|_{L^m(\Omega)\times\dots\times L^m(\Omega)},$$

the mapping $\pi\colon \mathring{H}^1_m(A, \Omega)\to L^m(\Omega)\times\dots\times L^m(\Omega)$ defined by $\pi(u) = A^{1/2}u_x$ is a linear isometry of $\mathring{H}^1_m(A, \Omega)$ onto a closed subspace of the reflexive Banach space $L^m(\Omega)\times\dots\times L^m(\Omega)$. But any closed subspace of a reflexive Banach space is reflexive.

Let F be a linear functional in $\mathring{H}^1_m(A, \Omega)$. Then $F\circ\pi^{-1}$ is a linear functional on the subspace $\pi(\mathring{H}^1_m(A, \Omega))\subset L^m(\Omega)\times\dots\times L^m(\Omega)$. By the Hahn-Banach theorem, such a functional may be extended, with retention of the norm, onto the entire space $L^m(\Omega)\times\dots\times L^m(\Omega)$. It is known that the general form of a linear functional on the latter space is given by the formula

$$\Phi(\mathbf{v})=\int_\Omega h^iv_idx,\quad \mathbf{v}=(v_1,\dots,v_n)\in L^m(\Omega)\times\dots\times L^m(\Omega),$$

where $\mathbf{h} = (h^1, \dots, h^n)\in L^{m/(m-1)}(\Omega)$. Thus

$$\langle F,\eta\rangle=\Phi(A^{1/2}\eta_x)=\int_\Omega(\mathbf{h},A^{1/2}\eta_x)\,dx=\int_\Omega(A^{1/2}\mathbf{h},\eta_x)\,dx,\quad \eta\in\mathring{H}^1_m(A,\Omega).$$

If we let $A^{1/2}\mathbf{h} = \mathbf{f}$, then $|B^{1/2}\mathbf{f}|\in L^{m/(m-1)}(\Omega)$. Now, by virtue of the Hahn-Banach theorem,

$$\|F\|_{(\mathring{H}^1_m(A,\Omega))^*}=\|\mathbf{h}\|_{m/(m-1),\Omega}=\|B^{1/2}\mathbf{f}\|_{m/(m-1),\Omega}.$$

Lemma 2.3 is proved.

REMARK 2.1. If $f_0\in L^{l/(l-1)}(\Omega)$, where l satisfies condition (2.2), then the integral

$$\int_\Omega f_0\eta dx,\quad \eta\in\mathring{H}^1_m(A,\Omega), \tag{2.10}$$

([5]) Here and below, summation over repeated indices is assumed.

defines a linear functional on $\mathring{H}^1_m(A, \Omega)$ with norm not exceeding $c\|\lambda^{-1}\|^{1/2}_{r,\Omega}\|f_0\|_{l/(l-1),\Omega}$, where c is the constant in (2.7).

Actually, in view of Lemma 2.2, $\mathring{H}^1_m(A, \Omega) \hookrightarrow L^l(\Omega)$, and $\mathring{H}^1_m(A, \Omega)$ is dense in $L^l(\Omega)$. Therefore the space of linear functionals on $L^l(\Omega)$ may be identified with some subspace of $(\mathring{H}^1_m(A, \Omega))^*$. Since (2.10) (under the condition $f_0 \in L^{l/(l-1)}(\Omega)$) yields the general form for a linear functional on $l^l(\Omega)$, the first part of Remark 2.1 is proved. The second part follows from the inequality

$$\left|\int_\Omega f_0\eta\,dx\right| \leqslant \|f_0\|_{l/(l-1),\Omega}\|\eta\|_{l,\Omega} \leqslant c\|\lambda^{-1}\|^{1/2}_{r,\Omega}\|f_0\|_{l/(l-1),\Omega}\|\eta\|_{\mathring{H}^1_m(A,\Omega)}.$$

2. *The solvability theorem.* We consider in Ω the quasilinear differential operator formally written as

$$\mathfrak{L}u \equiv -\frac{d}{dx_i}\mathfrak{L}^i(x, u, u_x) + \mathfrak{L}_0(x, u, u_x), \tag{2.11}$$

where $u_x = (u_{x_1}, \ldots, u_{x_n})$. We shall formulate a condition under which an exact definition of this operator, as a mapping from $\mathring{H}^1_m(A, \Omega)$ into $(\mathring{H}^1_m(A, \Omega))^*$, may be given.

CONDITION I. *We assume that in $\Omega \times \mathbb{R} \times \mathbb{R}^n$ the functions $\mathfrak{L}^i(x, u, p)$, $i = 1, \ldots, n$, and $\mathfrak{L}_0(x, u, p)$ satisfy the Carathéodory condition (i.e. they are measurable with respect to x in Ω for any $u \in \mathbb{R}$ and $p \in \mathbb{R}^n$, and continuous in u and p for almost all $x \in \Omega$). Suppose, moreover, that for any $u(x) \in L^l(\Omega)$ and $p(x)$ such that $|A^{1/2}p| \in L^m(\Omega)$ the condition*

$$|B^{1/2}\mathfrak{L}(x, u(x), p(x))| \in L^{m/(m-1)}(\Omega), \qquad \mathfrak{L}_0(x, u(x), p(x)) \in L^{l/(l-1)}(\Omega) \tag{2.12}$$

holds, where $B = A^{-1}$, A is a symmetric matrix satisfying (2.1), $\mathfrak{L} = (\mathfrak{L}^1, \ldots, \mathfrak{L}^n)$, *and the index l satisfies* (2.2).

It is clear that, under Condition I, for any $u, \eta \in \mathring{H}^1_m(A, \Omega)$ the integral

$$\langle \mathfrak{L}u, \eta\rangle \equiv \int_\Omega \left[\mathfrak{L}^i(x, u, u_x)\eta_{x_i} + \mathfrak{L}_0(x, u, u_x)\eta\right] dx \tag{2.13}$$

converges. In fact, because of (2.12), Lemma 2.3 and Remark 2.1, the integral (2.13) for any $u \in \mathring{H}^1_m(A, \Omega)$ defines a linear functional with respect to η in $\mathring{H}^1_m(A, \Omega)$ (and therefore converges for any $u, \eta \in \mathring{H}^1_m(A, \Omega)$).

Suppose Condition I is fulfilled. Then $\mathfrak{L}$ is a mapping of $\mathring{H}^1_m(A, \Omega)$ into $(\mathring{H}^1_m(A, \Omega))^*$ defined by (2.13), in which $u, \eta \in \mathring{H}^1_m(A, \Omega)$. (The expression $\langle \mathfrak{L}u, \eta\rangle$ is the value of the functional $\mathfrak{L}u \in (\mathring{H}^1_m(A, \Omega))^*$ on the function $\eta \in \mathring{H}^1_m(A, \Omega)$.)

We consider, under Condition I, the equation

$$\mathfrak{L}u = F, \tag{2.14}$$

where $F \in (\mathring{H}^1_m(A, \Omega))^*$. (2.14) may always be written in the form

$$\tilde{\mathfrak{L}}u = 0, \tag{2.15}$$

where

$$\tilde{\mathfrak{L}}u = -\frac{d}{dx_i}\tilde{\mathfrak{L}}^i(x, u, u_x) + \mathfrak{L}_0(x, u, u_x), \qquad \tilde{\mathfrak{L}}(x, u, p) = \mathfrak{L}(x, u, p) + \mathbf{f},$$

$$\mathbf{f} = (f^1, \dots, f^n), \qquad |B^{1/2}\mathbf{f}| \in L^{m/(m-1)}(\Omega), \qquad \int_\Omega f^i\eta_{x_i}\,dx = \langle F, \eta\rangle.$$

Here Condition I clearly holds for $\tilde{\mathfrak{L}}(x, u, p)$ and $\mathfrak{L}_0(x, u, p)$ (see Lemma 2.3). Therefore in the following we limit consideration to the equation $\mathfrak{L}u = 0$, where the symbol 0 on the right of (2.15) denotes the zero functional in $\mathring{H}^1_m(A, \Omega)$.

THEOREM 2.1. *Let Condition* I *be fulfilled for the operator* $\mathfrak{L}$: $\mathring{H}^1_m(A, \Omega) \to (\mathring{H}^1_m(A, \Omega))^*$ *(defined by* (2.13)) *and let the operator* $\mathfrak{L}$ *be coercive and have semibounded variation. Then there exists at least one solution of* (2.15).

PROOF. Consider the operator $\mathfrak{M}$: $L^l(\Omega) \times L^m(\Omega) \times \dots \times L^m(\Omega) \to L^{m/(m-1)}(\Omega)$ defined by

$$(\mathfrak{M}\mathbf{u})(x) = |B^{1/2}\mathfrak{L}(x, u(x), p(x))|, \qquad \mathbf{u} = (u, p),$$

and also the operator $\mathfrak{M}_1$: $L^l(\Omega) \times L^m(\Omega) \times \dots \times L^m(\Omega) \to L^{l(l-1)}(\Omega)$, defined by

$$(\mathfrak{M}_1\mathbf{u})(x) = \mathfrak{L}_0(x, u(x), p(x)), \qquad \mathbf{u} = (u, p).$$

From Condition I it follows that $\mathfrak{M}$ and $\mathfrak{M}_1$ transform the space $L^l(\Omega) \times L^m(\Omega) \times \dots \times L^m(\Omega)$ into $L^{m/(m-1)}(\Omega)$ and $L^{l/(l-1)}(\Omega)$ respectively. Moreover, from Condition I it follows that the functions $(x, u, p) \to |B^{1/2}\mathfrak{L}(x, u, p)$ and $(x, u, p) \to \mathfrak{L}_0(x, u, p)$ satisfy the Carathéodory condition. Hence, adjusting in a suitable manner the reasoning given in [8] (see Chapter I, §2) in connection with the operator N: $L^{p_1}(\Omega) \to L^{p_2}(\Omega)$ for $p_1 \geq 1$ and $p_2 \in [1, \infty)$, defined by

$$(Nu)(x) = \Phi(x, u(x)),$$

where $\Phi(x, u)$ is a function satisfying the Carathéodory condition, we obtain the following facts:

1°) The operators $\mathfrak{M}$ and $\mathfrak{M}_1$ are continuous.

2°) They are bounded.

From 1° it follows that the operator $\mathfrak{L}$: $\mathring{H}^1_m(A, \Omega) \to (\mathring{H}^1_m(A, \Omega))^*$ defined by (2.13) is continuous. In fact, we write $\mathfrak{L}$ in the form $\mathfrak{L} = \mathfrak{L}^{(1)} + \mathfrak{L}^{(2)}$, where $\mathfrak{L}^{(1)}$ and $\mathfrak{L}^{(2)}$ act from $\mathring{H}^1_m(A, \Omega)$ into $(\mathring{H}^1_m(A, \Omega))^*$ and are defined by

$$\langle \mathfrak{L}^{(1)}u, \eta\rangle = \int_\Omega \mathfrak{L}^i(x, u, u_x)\eta_{x_i}\,dx, \qquad \langle \mathfrak{L}^{(2)}u, \eta\rangle = \int_\Omega \mathfrak{L}_0(x, u, u_x)\,\eta dx.$$

From Lemma 2.3, for any $u, u_n \in \mathring{H}^1_m(A, \Omega)$ with

$$\lim_{n\to\infty} \|u_n - u\|_{\mathring{H}^1_m(A, \Omega)} = 0,$$

it follows that

$$\|\mathfrak{L}^{(1)}_{u_n} - \mathfrak{L}^{(1)}_u\|_{(\mathring{H}^1_m(A, \Omega))^*} \leq \|B^{1/2}\mathfrak{L}(x, u_n, \nabla u_n) - B^{1/2}\mathfrak{L}(x, u, \nabla u)\|_{m/(m-1),\Omega};$$

therefore 1° implies the continuity of $\mathfrak{L}^{(1)}$. In a similar way, the continuity of $\mathfrak{L}^{(2)}$ follows from 1°, it being considered as an operator from $\mathring{H}^1_m(A, \Omega)$ into $L^{l/(l-1)}(\Omega) = (L^l(\Omega))^*$. But then from Remark 2.1 it follows that $\mathfrak{L}^{(2)}$ is continuous also as an operator from $\mathring{H}^1_m(A, \Omega)$ onto $(\mathring{H}^1_m(A, \Omega))^*$. Thus $\mathfrak{L}$ is continuous as the sum of two continuous operators. And hence $\mathfrak{L}$ is continuous (and, all the more, demicontinuous), coercive, and of semibounded variation. Applying Theorem 1.1, we obtain Theorem 2.1.

We next introduce different sufficient conditions ensuring the satisfaction of the conditions of Theorem 2.1.

3. *Sufficient conditions for Condition* I. Let the functions $\mathfrak{L}^i(x, u, p)$, $i = 1, \dots, n$, and $\mathfrak{L}_0(x, u, p)$ satisfy the Carathéodory condition in $\Omega \times \mathbb{R} \times \mathbb{R}^n$, and suppose for almost all $x \in \Omega$ and any $u \in \mathbb{R}$ and $p \in \mathbb{R}^n$ the inequalities

$$|B^{1/2}\mathfrak{L}(x, u, p) \leqslant \mu_1|A^{1/2}p|^{m-1} + \mu_2|u|^{l(m-1)/m} + \varphi_1(x),$$

$$|\mathfrak{L}_0(x, u, p) \leqslant \mu_3|A^{1/2}p|^{m(l-1)/l} + \mu_4|u|^{l-1} + \varphi_2(x) \tag{2.16}$$

hold, where $\mu_1, \mu_2, \mu_3, \mu_4 = \text{const} \geqslant 0$, $\varphi_1 \in L^{m/(m-1)}(\Omega)$, $\varphi_2 \in L^{l/(l-1)}(\Omega)$ if $l < +\infty$, and

$$|B^{1/2}\mathfrak{L}(x, u, p)| \leqslant \mu_1(u)(|A^{1/2}p|^{m-1} + \varphi_1(x)),$$

$$|\mathfrak{L}_0(x, u, p)| \leqslant \mu_3(u)(|A^{1/2}p|^m + \varphi_2(x)), \tag{2.16a}$$

where $\mu_1(u)$ and $\mu_2(u)$ are nonnegative continuous functions of u and $\varphi_1 \in L^{m/(m-1)}(\Omega)$, $\varphi_2 \in L^{l/(l-1)}(\Omega)$, if $l = +\infty$. It is easy to see that conditions (2.16) and (2.16a) are equivalent to the statement that for almost all $x \in \Omega$ and any $u \in \mathbb{R}$, $p \in \mathbb{R}^n$ and $\xi \in \mathbb{R}^n$ the inequalities

$$|\mathfrak{L}^i(x, u, p)\xi_i| \leqslant (\mu_1|A^{1/2}p|^{m-1} + \mu_2|u|^{l(m-1)/m} + \varphi_1(x))|A^{1/2}\xi|,$$

$$|\mathfrak{L}_0(x, u, p)| \leqslant \mu_3|A^{1/2}p|^{m(l-1)/l} + \mu_4|u|^{l-1} + \varphi_2(x) \tag{2.16'}$$

hold for $l < +\infty$, and

$$|\mathfrak{L}^i(x, u, p)\xi_i| \leqslant \mu_1(u)(|A^{1/2}p|^{m-1} + \varphi_1(x))|A^{1/2}\xi|,$$

$$|\mathfrak{L}_0(x, u, p)| \leqslant \mu_3(u)(|A^{1/2}p|^m + \varphi_2(x)) \tag{2.16'a}$$

for $l = +\infty$.

In particular, (2.16) will hold if $r \geqslant n/2$ (then $l \geqslant m$) and for almost all $x \in \Omega$ and any $u \in \mathbb{R}$ and $p, \xi \in \mathbb{R}^n$ the inequalities

$$|\mathfrak{L}^i(x, u, p)\xi_i| \leqslant (\tilde{\mu}_1|A^{1/2}p|^{m-1} + a_1(x)|u|^{m-1} + \psi_1(x))|A^{1/2}\xi|,$$

$$|\mathfrak{L}_0(x, u, p)| \leqslant a_2(x)|A^{1/2}p|^{m-1} + a_3(x)|u|^{m-1} + \psi_2(x) \tag{2.17}$$

hold, where $\tilde{\mu}_1 = \text{const} \geqslant 0$, $a_1(x) \geqslant 0$, $a_2(x) \geqslant 0$, $a_3(x) \geqslant 0$, $a_1^{m/(m-1)} \in L^s(\Omega)$, $a_2^m \in L^s(\Omega)$, $a_3 \in L^s(\Omega)$, $1/s + m/l = 1$, $\psi_1 \in L^{m/(m-1)}(\Omega)$, $\psi_2 \in L^{l/(l-1)}(\Omega)$. In fact, using Young's inequality, we deduce from (2.17) the relation (2.16) with $\mu_1 = \tilde{\mu}_1$, $\mu_2 = \mu_3 = \mu_4 = 1$, $\varphi_1 = \psi_1 + a_1^s$, $\varphi_2 = \psi_2 + a_2^{ms(l-1)/l} + a_3^{s(l-1)/s}$.

4. *Algebraic criteria for the satisfaction of the coercivity condition.* It is easy to see that the condition of coercivity for the operator $\mathfrak{L}$ defined by (2.13) will hold if for almost all $x \in \Omega$ and any $u \in \mathbb{R}$ and $p \in \mathbb{R}^n$ the inequalities

$$\mathfrak{L}^i(x, u, p)p_i \geqslant \nu|A^{1/2}p|^m - \mu_5|u|^l - \varphi_3(x),$$

$$\mathfrak{L}_0(x, u, p)u \geqslant -\mu_6|A^{1/2}p|^m - \mu_7|u|^l - \varphi_4(x) \tag{2.18}$$

hold. Here $\nu = \text{const} \geqslant 0$, $\mu_5, \mu_6, \mu_7 = \text{const} \geqslant 0$, $\varphi_3, \varphi_4 \in L^1(\Omega)$, and

$$\nu - \mu_6 - (\mu_5 + \mu_7)c^l \|\lambda^{-1}\|_{r,\Omega}^{l/2} - \int_\Omega (\varphi_3 + \varphi_4)\,dx \geqslant 0, \tag{2.19}$$

where c is the constant from (2.7). In fact, by use of (2.18) and (2.7), we obtain

$$\langle \mathfrak{L}u, u\rangle = \int_\Omega \left[\mathfrak{L}^i(x, u, u_x)u_{x_i} + \mathfrak{L}_0(x, u, u_x)u \right] dx$$

$$\geqslant (\nu - \mu_6)\|u\|^m_{\dot{H}^1_m(A,\Omega)} - (\mu_5 + \mu_7)c^l\|\Lambda^{-1}\|^{l/2}_{r,\Omega}\|u\|^l_{\dot{H}^1_m(A,\Omega)} - \int_\Omega (\varphi_3 + \varphi_4)\, dx,$$

and from this and (2.19) with $\|u\|_{\dot{H}^1_m(A,\Omega)} = 1$ follows the inequality $\langle \mathfrak{L}u, u\rangle \geqslant 0$.

Note that under conditions (2.18), inequality (2.19) will hold provided that $\mu_6 < \nu$, $r < +\infty$ and mes Ω is sufficiently small. Thus under conditions (2.16), (2.18) and the assumption that the operator $\mathfrak{L}$ has semibounded variation, we obtain the solvability of the Dirichlet problem for (2.15) "in the small".

If, however, $\mu_5 = \mu_7 = 0$ in (2.18), then in place of (2.19) one need require only that the inequality $\mu_6 < \nu$ hold. Then

$$\langle \mathfrak{L}u, u\rangle \geqslant (\nu - \mu_6)\|u\|_{\dot{H}^1_m(A,\Omega)} - \int_\Omega (\varphi_3 + \varphi_4)\, dx,$$

from which it is clear that the condition $\|u\|_{\dot{H}^1_m(A,\Omega)} = \rho_0$ with sufficiently large ρ_0 will imply the inequality $\langle \mathfrak{L}u, u\rangle \geqslant 0$. In this case from (2.16), (2.18) and the semiboundedness of variation of $\mathfrak{L}$ we obtain the solvability of the Dirichlet problem "in the large".

We now consider the condition

$$\mathfrak{L}^i(x, u, p)p_i \geqslant \tilde{\nu}|A^{1/2}p|^m - a_4(x)|u|^m - \psi_3(x),$$
$$\mathfrak{L}_0(x, u, p)u \geqslant -\tilde{\mu}_6|A^{1/2}p|^m - b(x)|u|\,|A^{1/2}p|^{m-1} - a_5(x)|u|^m - \psi_4(x)|u|, \tag{2.20}$$

stricter than (2.18), where $\tilde{\nu} = \text{const} > 0$, $\tilde{\mu}_6 = \text{const} \geqslant 0$, $b(x) \geqslant 0$, a_4, a_5, $b^m \in L^s(\Omega)$, $1/s + m/l = 1$, $\psi_3 \in L^1(\Omega)$, $\psi_4 \in L^{l/(l-1)}(\Omega)$, and it will be assumed that $r \geqslant n/2$. Applying Young's inequality, it is easy to verify that (2.20) implies an inequality of the form (2.18). However, a sufficient condition for the coercivity of the operator under conditions (2.20) is best obtained not from (2.19), but rather by direct use of (2.20). Considering (2.20), we obtain

$$\langle \mathfrak{L}u, u\rangle \geqslant \int_\Omega \Big[\Big(\tilde{\nu} - \tilde{\mu}_6 - \frac{m-1}{m}\varepsilon_1\Big)|A^{1/2}u_x|^m$$
$$- \Big(a_4 + a_5 + \frac{1}{m}\varepsilon_1^{-(m-1)}b^m\Big)_+ |u|^m - \psi_3 - |\psi_4|\,|u| \Big]\, dx, \tag{2.21}$$

where $\varepsilon_1 > 0$ and f_+ denotes the positive part of the function f, i.e. $f_+(x) = f(x)$ at those points x where $f(x) > 0$, and $f(x) = 0$ elsewhere in Ω. Applying Young's inequality and using (2.7), we derive from (2.21) the inequality

$$\langle \mathfrak{L}u, u\rangle \geqslant \Big(\tilde{\nu} - \tilde{\mu}_6 - \frac{m-1}{m}\varepsilon_1\Big)\|u\|^m_{\dot{H}^1_m(A,\Omega)}$$
$$- c^m\Big\|\Big(a_4 + a_5 + \frac{1}{m}\varepsilon_1^{-(m+1)}b^m\Big)_+\Big\|_{s,\Omega}\|\Lambda^{-1}\|^{m/2}_{r,\Omega}\|u\|^m_{\dot{H}^1_m(A,\Omega)}$$
$$- \int_\Omega \psi_3\, dx - \varepsilon_2 c^l\|\Lambda^{-1}\|^{l/2}_{r,\Omega}\|u\|^l_{\dot{H}^1_m(A,\Omega)} - \varepsilon_2^{-1/(l-1)}\int_\Omega |\psi_4|^{l/(l-1)}\, dx, \tag{2.22}$$

where $\varepsilon_2 > 0$ and c is the constant in (2.7). We assume that

$$\delta = \inf_{\varepsilon_1>0}\left\{\tilde{\nu} - \tilde{\mu}_6 - \frac{m-1}{m}\varepsilon_1 - c^m \left\| \left(a_4 + a_5 + \frac{1}{m}\varepsilon_1^{-(m-1)}b^m\right)_+ \right\|_{s,\Omega} \|\lambda^{-1}\|_{r,\Omega}^{m/2}\right\} > 0. \tag{2.23}$$

We show that under conditions (2.20) and (2.23), the operator $\mathfrak{L}$ is coercive. In fact, denoting $\kappa = c^l\|\lambda^{-1}\|_{r,\Omega}^{l/2}$, $K_1 = \int_\Omega \psi_3\,dx$, $K_2 = \int_\Omega |\psi_4|^{l/(l-1)}\,dx$ and remembering that $\|u\|_{\mathring{H}_m^1(A,\Omega)} = \rho_0 > 0$, we may write, in place of (2.22), the inequality

$$\langle \mathfrak{L}u, u\rangle \geqslant \delta\rho_0^m - \varepsilon_2\kappa\rho_0^l - K_1 - K_2\varepsilon_2^{-1/(l-1)}, \tag{2.24}$$

valid for any $\varepsilon_2 > 0$. In particular, (2.24) holds for $\varepsilon_2 = \sigma\rho_0^{m-l}$, where $\sigma = \delta/2^\kappa$. For such values of ε_2, the equation $\delta\rho_0^m - \varepsilon_2\kappa\rho_0^l = \delta\rho_0^m/2$ holds. Therefore from (2.24) it follows that for all $u \in \mathring{H}_m^1(A, \Omega)$ with $\|u\|_{\mathring{H}_m^1(A,\Omega)} = \rho_0$ the inequality

$$\langle \mathfrak{L}u, u\rangle \geqslant \delta\rho_0^m/2 - K_1 - K_2\sigma^{-1/(l-1)}\rho_0^{(l-m)/(l-1)} \tag{2.25}$$

holds. Now since $(l - m)/(l - 1) < m$ for any $m > 1$, from (2.25) with sufficiently large ρ_0 follows the coercivity condition $\langle \mathfrak{L}u, u\rangle \geqslant 0$ for all u with $\|u\|_{\mathring{H}_m^1(A,\Omega)} = \rho_0$.

The following proposition is thereby proved.

PROPOSITION 2.1. *Under conditions* (2.20) *and* (2.23), *the operator* $\mathfrak{L}\colon \mathring{H}_m^1(A, \Omega) \to (\mathring{H}_m^1(A, \Omega))^*$ *is coercive.*

Hence under conditions (2.17) (and even (2.16)), (2.20) and (2.23) and under the assumption that $\mathfrak{L}$ has semibounded variation, the Dirichlet problem for (2.15) is solvable "in the large".

Conditions (2.17) and (2.20) will in particular be satisfied by the functions

$$\mathfrak{L}^i(x, u, p) \equiv a^{ij}(x)p_j + a^i(x)u + f^i(x), \qquad i = 1, \dots, n,$$
$$\mathfrak{L}_0(x, u, p) \equiv b^i(x)p_i + a_0(x)u + f_0(x), \tag{2.26}$$

corresponding to a (degenerating) linear elliptic equation

$$-\frac{\partial}{\partial x_i}(a^{ij}u_{x_j} + a^iu + f^i) + b^iu_{x_i} + a_0u + f_0 = 0, \tag{2.27}$$

if we assume that a^{ij}, a^i, b^i, a_0, f^i, and f_0 $(i, j = 1, \dots, n)$ are functions measurable in Ω, that the matrix $A = \|a^{ij}(x)\|$ is symmetric and satisfies (2.1) for $r > n/2$, and that

$$|B^{1/2}\mathbf{a}|^2,\ |B^{1/2}\mathbf{b}|^2,\ a_0 \in L^s(\Omega),\quad 1/s + 2/l = 1,\quad |B^{1/2}\mathbf{f}| \in L^2(\Omega),$$
$$f_0 \in L^{l/(l-1)}(\Omega), \tag{2.28}$$

where $B = A^{-1}$, $\mathbf{a} = (a^1, \dots, a^n)$, $\mathbf{b} = (b^1, \dots, b^n)$, $\mathbf{f} = (f^1, \dots, f^n)$, and the exponent l satisfies (2.2) with $m = 2$. In fact, with the aid of the Cauchy inequality and the inequality $|(\xi, \eta)| \leqslant |A^{1/2}\xi|\,|B^{1/2}\eta|$, $\xi, \eta \in \mathbb{R}^n$, we deduce from (2.26) and (2.28) the inequality (2.17) with $m = 2$, $\tilde{\mu}_1 = 1$, $a_1 = |B^{1/2}\mathbf{a}|$, $\psi_1 = |B^{1/2}\mathbf{f}|$, $a_2 = |B^{1/2}\mathbf{b}|$, $a_3 = |a_0|$, $\psi_2 = |f_0|$ and (2.20) with $m = 2$, $\tilde{\nu} = 1/2$, $a_4 = |B^{1/2}\mathbf{a}|^2$, $\psi_3 = |B^{1/2}\mathbf{f}|^2$, $\tilde{\mu}_6 = 0$, $b = |B^{1/2}\mathbf{b}|$, $a_5 = -a_0$, $\psi_4 = |f_0|$. In this case, (2.23) will have the form

$$\inf_{\varepsilon_1>0}\{1/2 - 1/2\,\varepsilon_1 - c^2\|(|B^{1/2}\mathbf{a}|^2 + 1/2\,\varepsilon_1|B^{1/2}\mathbf{b}|^2 - a_0)_+\|_{s,\Omega}\|\lambda^{-1}\|_{r,\Omega}\} > 0. \tag{2.29}$$

In particular, (2.29) holds if

$$1/4 - c^2 \|(|B^{1/2}\mathbf{a}|^2 + |B^{1/2}\mathbf{b}|^2 - a_0)_+\|_{s,\Omega} \|\lambda^{-1}\|_{r,\Omega} > 0. \tag{2.30}$$

Inequality (2.30) will hold in turn if

$$a_0 > |B^{1/2}\mathbf{a}|^2 + |B^{1/2}\mathbf{b}|^2 \quad \text{a.e. in } \Omega. \tag{2.31}$$

Since in the case of a linear operator $\mathfrak{L}$ its coercivity implies its monotonicity, one may conclude that under conditions (2.28) and (2.29) the Dirichlet problem for (2.27) has at least one (generalized) solution. Moreover, (under the conditions indicated) from an inequality of the form (2.25) (for $m = 2$) one easily deduces an estimate of the form

$$\|u\|_{\mathring{H}^1_2(A,\Omega)} \leqslant c\left(\|B^{1/2}\mathbf{f}\|^2_{2,\Omega} + \|f_0\|^2_{l/(l-1),\Omega}\right), \tag{2.32}$$

from which follows the uniqueness of the solution. Furthermore, if the condition on s in (2.28) is replaced by the condition $1/s + 2/e < 1$, it may be shown by our results that the operator $\mathfrak{L}$ defined by (2.26) and (2.28) satisfies a Fredholm alternative. Therefore from Theorem 2.1 one may derive results on the Fredholm solvability of linear elliptic equations admitting degeneracies characterized by conditions of the form (2.1). Results of this sort were obtained earlier in [9]–[11].

5. *Homeomorphism theorems for operators satisfying a definiteness condition.* We shall say that the operator $\hat{\mathfrak{L}}\colon \mathring{H}^1_m(A,\Omega) \to (\mathring{H}^1_m(A,\Omega))^*$ defined by (2.13) *satisfies the definiteness condition* (of its variation) if for any $u, v \in \mathring{H}^1_m(A,\Omega)$ the inequality(6)

$$\langle \hat{\mathfrak{L}}u - \hat{\mathfrak{L}}v, u - v\rangle \geqslant c_0 \|u - v\|^m_{\mathring{H}^1_m(A,\Omega)} \tag{2.33}$$

holds, where $c_0 = \text{const} > 0$.

It is clear that the condition of definiteness for an operator $\mathfrak{L}$ implies its strict monotonicity (and therefore uniqueness for solutions of (2.15)).

THEOREM 2.2. *Let the operator* $\hat{\mathfrak{L}}$ *defined by*

$$\langle \hat{\mathfrak{L}}u, \eta\rangle = \int_\Omega \left[\mathfrak{L}^i(x, u, u_x)\eta_{x_i} + \mathfrak{L}_0(x, u, u_x)\eta\right] dx,$$

$$u, \eta \in \mathring{H}^1_m(A,\Omega), \tag{2.34}$$

satisfy Condition I *and the condition of definiteness. Then the mapping* $\hat{\mathfrak{L}}\colon \mathring{H}^1_m(A,\Omega) \to (\mathring{H}^1_m(A,\Omega))^*$ *is a homeomorphism.*

PROOF. We first show that the equation

$$\hat{\mathfrak{L}}u = F \tag{2.35}$$

is uniquely solvable for any $F \in (\mathring{H}^1_m(A,\Omega))^*$. To do this we use the estimate

$$\langle \hat{\mathfrak{L}}u, u\rangle \geqslant c_1 \|u\|^m_{\mathring{H}^1_m(A,\Omega)} - K, \tag{2.36}$$

where $c_1 > 0$ and $K \geqslant 0$ are constants. Inequality (2.36) easily follows from (2.33) for the operator $\hat{\mathfrak{L}}$. In fact, setting $v = 0$ in (2.33), we obtain

$$\langle \hat{\mathfrak{L}}u, u\rangle \geqslant c_0 \|u\|^m_{\mathring{H}^1_m(A,\Omega)} - \langle \hat{\mathfrak{L}}0, u\rangle. \tag{2.37}$$

(6) Compare [6], Chapter II, §§1.3 and 2.1.

Since by (2.16)

$$|\mathfrak{L}^i(x, 0, 0)\xi_i| \leqslant \varphi_1(x)|A^{1/2}\xi|, \quad |\mathfrak{L}_0(x, 0, 0)| \leqslant \varphi_2(x), \tag{2.38}$$

where $\varphi_1 \in L^{m/(m-1)}(\Omega)$ and $\varphi_2 \in L^{l/(l-1)}(\Omega)$, it follows that

$$\begin{aligned}|\langle \hat{\mathfrak{L}}0, u\rangle| &\leqslant \|\varphi_1\|_{m/(m-1),\Omega}\|u\|_{\mathring{H}^1_m(A,\Omega)} + \|\varphi_2\|_{l/(l-1),\Omega}\|u\|_{l,\Omega} \\ &\leqslant \|u\|_{\mathring{H}^1_m(A,\Omega)}\left(\|\varphi_1\|_{m/(m-1),\Omega} + c\|\lambda^{-1}\|^{1/2}_{r,\Omega}\|\varphi_2\|_{l/(l+1),\Omega}\right) \leqslant \frac{c_0}{2}\|u\|_{\mathring{H}^1_m(A,\Omega)} \\ &\quad + \frac{m-1}{m}\left(\frac{2}{c_0 m}\right)^{1/(m-1)}\left(\|\varphi_1\|_{m/(m-1),\Omega} + c\|\lambda^{-1}\|^{1/2}_{r,\Omega}\|\varphi\|_{l/(l-1),\Omega}\right)^{m/(m-1)}.\end{aligned} \tag{2.39}$$

Now (2.37) and (2.39) imply (2.36) with $c_1 = c_0/2$ and $K = m - 1/m(2/c_0m)^{1/m} \times (\|\varphi_1\|_{m/(m-1),\Omega} + c\|\lambda^{-1}\|^{1/2}_{r,\Omega}\|\varphi_2\|)^{m/(m-1)}$. Furthermore, from (2.36) it follows that the operator $\tilde{\mathfrak{L}}\colon \mathring{H}^1_m(A, \Omega) \to (\mathring{H}^1_m(A, \Omega))^*$, defined by

$$\langle \tilde{\mathfrak{L}}u, \eta\rangle = \langle \hat{\mathfrak{L}}u, \eta\rangle - \langle F, \eta\rangle, \quad u \in \mathring{H}^1_m(A, \Omega), \quad \eta \in \mathring{H}^1_m(A, \Omega) \tag{2.40}$$

satisfies the coercivity condition, since for sufficiently large ρ_0 and all u with $\|u\|_{\mathring{H}^1_m(A,\Omega)} = \rho_0$ we have

$$\begin{aligned}\langle \tilde{\mathfrak{L}}u, u\rangle &\geqslant c_1\|u\|^m_{\mathring{H}^1_m(A,\Omega)} - K - \|F\|_{(\mathring{H}^1_m(A,\Omega))^*}\|u\|_{\mathring{H}^1_m(A,\Omega)} \\ &= c_1\rho_0^m - K - \|F\|_{(\mathring{H}^1_m(A,\Omega))^*}\rho_0 > 0.\end{aligned} \tag{2.41}$$

Since, clearly, the operator $\tilde{\mathfrak{L}}$, as well as $\hat{\mathfrak{L}}$, satisfies Condition I and (strict) monotonicity, it follows from Theorem 2.1 that the equation $\tilde{\mathfrak{L}}u = 0$, or equivalently $\hat{\mathfrak{L}}u = f$, has a solution in $\mathring{H}^1_m(A, \Omega)$. From the strict monotonicity it follows that this solution is unique. Thus $\hat{\mathfrak{L}}$ is bijective from $\mathring{H}^1_m(A, \Omega)$ onto $(\mathring{H}^1_m(A, \Omega))^*$. It remains to prove that $\hat{\mathfrak{L}}$ and $\hat{\mathfrak{L}}^{-1}$ are continuous. But the continuity of $\hat{\mathfrak{L}}$ has already been established in the proof of Theorem 2.1. The continuity of $\hat{\mathfrak{L}}^{-1}$ easily follows from an estimate of the type (2.33). In fact, it is clear that

$$\langle \hat{\mathfrak{L}}u - \hat{\mathfrak{L}}v, u - v\rangle \leqslant \|\hat{\mathfrak{L}}u - \hat{\mathfrak{L}}v\|_{(\mathring{H}^1_m(A,\Omega))^*}\|u - v\|_{\mathring{H}^1_m(A,\Omega)}$$

and, combining this estimate with (2.33) written for $\hat{\mathfrak{L}}$, we obtain

$$\|u - v\|^{m-1}_{\mathring{H}^1_m(A,\Omega)} \leqslant c_0^{-1}\|\hat{\mathfrak{L}}u - \hat{\mathfrak{L}}v\|_{(\mathring{H}^1_m(A,\Omega))^*},$$

which implies the continuity of $\hat{\mathfrak{L}}^{-1}$. Theorem 2.2 is proved.

We now give algebraic criteria for definiteness.

PROPOSITION 2.2. *Let $m \geqslant 2$, and let the functions $\mathfrak{L}^i(x, u, p)$, $i = 1, \ldots, n$, and $\mathfrak{L}_0(x, u, p)$ be differentiable with respect to u and p. Suppose, furthermore, that for almost all $x \in \Omega$ and any $u, \xi_0 \in \mathbb{R}$ and $p, \xi \in \mathbb{R}^n$*

$$\frac{\partial \mathfrak{L}^i(x, u, p)}{\partial p_j}\xi_i\xi_j + \frac{\partial \mathfrak{L}^i(x, u, p)}{\partial u}\xi_i\xi_0 + \frac{\partial \mathfrak{L}_0(x, u, p)}{\partial p_j}\xi_j\xi_0 + \frac{\partial \mathfrak{L}_0(x, u, p)}{\partial u}\xi_0^2 \geqslant \alpha_0|A^{1/2}p|^{m-2}|A^{1/2}\xi|^2. \tag{2.42}$$

Then the operator $\mathfrak{L}\colon \mathring{H}^1_m(A, \Omega) \to (\mathring{H}^1_m(A, \Omega))^*$ defined by (2.34) *satisfies the condition of definiteness.*

The proof is analogous to that of Theorem 2.2 in [6], Chapter II, §2. For simplicity we suppose that $\mathfrak{L}_0 \equiv 0$. Let $u, v \in \overset{\circ}{H}{}^1_m(A, \Omega)$. Then, using (2.42), we estimate

$$\begin{aligned}\langle \hat{\mathfrak{L}}u - \hat{\mathfrak{L}}v, u - v\rangle &= \int_\Omega (\mathfrak{L}^i(x, u, u_x) - \mathfrak{L}^i(x, v, v_x))(u_{x_i} - v_{x_i})\, dx \\ &= \int_\Omega\int_0^1 \left[\frac{\partial \mathfrak{L}^i(x, v + \tau(u - v), v_x + \tau(u_x - v_x))}{\partial p_j} (u_{x_j} - v_{x_j})(u_{x_i} - v_{x_i}) \right. \\ &\quad \left. + \frac{\partial \mathfrak{L}^i(x, v + \tau(u - v), v_x + \tau(u_x - v_x))}{\partial u} (u - v)(u_{x_i} - v_{x_i}) \right] d\tau dx \\ &\geqslant \alpha_0 \int_\Omega \int_0^1 |A^{1/2}v_x + \tau(A^{1/2}u_x - A^{1/2}v_x)|^{m-2}\, d\tau |A^{1/2}u_x - A^{1/2}v_x|^2\, dx.\end{aligned} \tag{2.43}$$

We now use the elementary inequality

$$\int_0^1 |\mathbf{a} + \tau(\mathbf{b} - \mathbf{a})|^{m-2}\, d\tau \geqslant \beta_0 |\mathbf{b} - \mathbf{a}|^{m-2},$$

where $\mathbf{a}$ and $\mathbf{b}$ are any fixed vectors in $\mathbb{R}^n$, and β_0 depends only on m, together with (2.43) to obtain the estimate

$$\langle \mathfrak{L}u - \mathfrak{L}v, u - v\rangle \geqslant \alpha_0\beta_0 \int_\Omega |A^{1/2}u_x - A^{1/2}v_x|^m\, dx = \alpha_0\beta_0 \|u - v\|^m_{\overset{\circ}{H}{}^1_m(A,\Omega)}.$$

Theorem 2.1 is proved.

6. *Algebraic criteria for semiboundedness of variation.*

PROPOSITION 2.3. *Assume the following conditions are fulfilled:*

a) *The operator* $\mathfrak{L}$ *may be written in the form* $\mathfrak{L} = \hat{\mathfrak{L}} + \mathfrak{L}'$, *where*

$$\hat{\mathfrak{L}}u = -\frac{d}{dx_i}(\hat{\mathfrak{L}}^i(x, u, u_x)), \quad \mathfrak{L}'u \equiv -\frac{d}{dx_i}\mathfrak{L}''(x, u) + \mathfrak{L}_0(x, u, u_x).$$

b) *The operators* $\hat{\mathfrak{L}} + \mathfrak{L}'$ *satisfy Condition* I, *i.e.* $\mathfrak{L}^i(x, u, p)$ *and* $\mathfrak{L}^i(x, u)$, $i = 1, \ldots, n$, *and also* $\mathfrak{L}_0(x, u, p)$ *satisfy the Carathéodory condition and for almost all* $x \in \Omega$ *and any* $u \in \mathbb{R}$, $p \in \mathbb{R}^n$ *and* $\xi \in \mathbb{R}^n$ *an inequality of the form* (2.16) *holds for the functions* $\mathfrak{L}^i(x, u, p) = \hat{\mathfrak{L}}^i(x, u, p) + \mathfrak{L}'(x, u)$ *and* $\mathfrak{L}_0(x, u, p)$.

c) *The operator* $\hat{\mathfrak{L}}$ *satisfies the condition of definiteness* (*see* (2.53)).

d) *The functions* $\mathfrak{L}^i(x, u)$, $i = 1, \ldots, n$, *and* $\mathfrak{L}_0(x, u, p)$ *are differentiable with respect to u and p, and for almost all* $x \in \Omega$ *and any* $u \in \mathbb{R}$, $p \in \mathbb{R}^n$ *and* $\xi \in \mathbb{R}^n$

$$\left| \frac{\partial \mathfrak{L}^i(x, u)}{\partial u} \xi_i \right| \leqslant \mu_2' |u|^{l'(m-1)/m-1} + \tilde{\varphi}_1(x)|A^{1/2}\xi|,$$

$$\left| \frac{\partial \mathfrak{L}_0(x, u, p)}{\partial u} \right| \leqslant \mu_3' |A^{1/2}p|^{m(l'-2)/l'} + \mu_4' |u|^{l'-2} + \tilde{\varphi}_2(x),$$

$$\begin{aligned}&\left| \frac{\partial \mathfrak{L}_0(x, u, p)}{\partial p_j} \xi_j \right| \\ &\leqslant \left(\mu_3'' |A^{1/2}p|^{m(l'-1)/l'-1} + \mu_4 |u|^{(m-1)l'/m-1} + \tilde{\tilde{\varphi}}_2(x) \right) |A^{1/2}\xi|, \; l' < l,\end{aligned} \tag{2.44}$$

where $\mu_2', \mu_3', \mu_4', \mu_3'', \mu_4'' > 0$ *are constants, and* $\tilde{\varphi}_1(x) \geqslant 0$, $\tilde{\varphi}_2(x) \geqslant 0$, $\tilde{\tilde{\varphi}}_2(x) \geqslant 0$, $\tilde{\varphi}_1 \in L^{ml'/(ml'-(m+l'))}(\Omega)$, $\tilde{\varphi}_2 \in L^{l'/(l'-2)}(\Omega)$, $\tilde{\tilde{\varphi}}_2 \in L^{ml'/(ml'-(m+l'))}$.

Then the operator $\mathfrak{L}\colon \mathring{H}^1_m(A,\Omega) \to (\mathring{H}^1_m(A,\Omega))^*$, *defined by*

$$\langle \mathfrak{L}u, \eta\rangle = \int_\Omega \left(\tilde{\mathfrak{L}}^i(x,u,u_x)\eta_{x_i} + \mathfrak{L}''^i(x,u)\eta_{x_i} + \mathfrak{L}_0(x,u,u_x)\eta\right) dx,$$

$$u \in \mathring{H}^1_m(A,\Omega), \quad \eta \in \mathring{H}^1_m(A,\Omega), \tag{2.45}$$

satisfies the condition of semibounded variation.

PROOF. We consider the difference $\langle \mathfrak{L}u - \mathfrak{L}v, u - v\rangle$. Using conditions a) and c), we estimate

$$\langle \mathfrak{L}u - \mathfrak{L}v, u-v\rangle = \langle \hat{\mathfrak{L}}u - \hat{\mathfrak{L}}v, u-v\rangle + \langle \mathfrak{L}'u - \mathfrak{L}'v, u-v\rangle$$

$$\geqslant c_0\|u-v\|^m_{\mathring{H}^1_m(A,\Omega)}$$

$$+ \int_\Omega \left[(\mathfrak{L}_i'(x,u) - \mathfrak{L}_i'(x,v))(u-v)_{x_i} + (\mathfrak{L}_0(x,u,\nabla u) - \mathfrak{L}_0(x,v,\nabla v))(u-v)\right] dx$$

$$= c_0\|u-v\|^m_{\mathring{H}^1_m(A,\Omega)} + \int_\Omega\int_0^1 \left[\frac{\partial \mathfrak{L}_i'(x, v+\tau(u-v))}{\partial u}(u-v)(u-v)_{x_i}\right.$$

$$+ \frac{\partial \mathfrak{L}_0(x, v+\tau(u-v), \nabla v + \tau(\nabla u - \nabla v))}{\partial u}(u-v)^2$$

$$\left. + \frac{\partial \mathfrak{L}_0(x, v+\tau(u-v), \nabla v + \tau(\nabla u - \nabla v))}{\partial p_j}(u-v)_{x_j}(u-v)\right] d\tau dx$$

$$\equiv c_0\|u-v\|^m_{\mathring{H}^1_m(A,\Omega)} + \int_\Omega\int_0^1 (J_1 + J_2 + J_3)\, d\tau dx. \tag{2.46}$$

Using condition d) and bearing in mind the bounds $\|u\|_{\mathring{H}^1_m(A,\Omega)} \leqslant \rho$, $\|v\|_{\mathring{H}^1_m(A,\Omega)} \leqslant \rho$, and (2.7), we estimate

$$|J_1| \equiv \left|\frac{\partial \mathfrak{L}''^i(x, v+\tau(u-v))}{\partial u}(u-v)(u-v)_{x_i}\right|$$

$$\leqslant \mu_2'\left(|u|+|v|\right)^{l'(m-1)/m-1}|u-v|\big)|A^{1/2}(u-v)_x|$$

$$+\varphi_1|u-v|\,|A^{1/2}(u-v)_x|,$$

$$\left|\int_\Omega\int_0^1 J_1\, d\tau dx\right| \leqslant \varepsilon\int_\Omega |A^{1/2}(u-v)_x|^m\, dx + c_1(\varepsilon,\rho)\left(\int_\Omega |u-v|^{l'}\, dx\right)^{m/l'(m-1)}; \tag{2.47}$$

$$|J_2| \equiv \left|\frac{\partial \mathfrak{L}_0(x, v+\tau(u-v), v_x + \tau(u_x - v_x))}{\partial u}(u-v)^2\right|$$

$$\leqslant \mu_3'\left(|A^{1/2}u_x| + |A^{1/2}v_x|\right)^{m(l'-2)/l'}|u-v|^2$$

$$+\mu_4'\left(|u|+|v|\right)^{l'-2}(u-v)\big)^2 + \tilde{\varphi}_2(u-v)^2,$$

$$\left|\int_\Omega\int_0^1 J_2\, d\tau dx\right| \leqslant c_2(\rho)\left(\int_\Omega |u-v|^{l'}\, dx\right)^{2/l}; \tag{2.48}$$

$$
\begin{aligned}
|J_3| \equiv &\left| \frac{\partial \mathfrak{L}_0(x, v + \tau(u - v), v_x + \tau(u_x - v_x))}{\partial p_j} (u - v)_{x_j} (u - v) \right| \\
&\leqslant \mu_3''(|A^{1/2}u_x| + |A^{1/2}v_x|)^{m(l-1)/l-1} |u - v|\, |A^{1/2}(u - v)_x| \\
&+ \mu_4''(|u| + |v|)^{(m-1)l'/m-1} |u - v| (|A^{1/2}u_x| + |A^{1/2}v_x|) \\
&+ \varphi_2 |u - v| (|A^{1/2}u_x| + |A^{1/2}v_x|),
\end{aligned}
$$

$$
\left| \int_\Omega \int_0^1 J_3 \, d\tau dx \right| \leqslant \varepsilon \int_\Omega |A^{1/2}(u - v)_x|^m \, dx + c_3(\varepsilon, \rho) \left(\int_\Omega |u - v|^{l'} \, dx \right)^{(1/l')(m/(m-1))}. \tag{2.49}
$$

Choosing $\varepsilon > 0$ sufficiently small, from (2.46)–(2.49) we derive

$$
\langle \mathfrak{L}u - \mathfrak{L}v, u - v \rangle \geqslant \frac{c_0}{2} \|u - v\|^m_{\mathring{H}^1_m(A,\Omega)} - \mu(\rho, \|u - v\|_{l',\Omega}), \tag{2.50}
$$

where $\mu(\rho, \tau) = c_1(c_0/2, \rho, m)\tau^{m/(m-1)} + c_2(\rho)\tau^2 + c_3(c_0/2, \rho, m)\tau^{m/(m-1)}$. Since the embedding $L^{l'}(\Omega) \hookrightarrow \mathring{H}^1_m(A, \Omega)$, $l' < l$, is compact, the norm $\|\cdot\|_{l,\Omega}$ is compact with respect to the norm $\|\cdot\|_{\mathring{H}^1_m(A,\Omega)}$. Moreover, the function $\gamma(\rho, \tau)$ is continuous, nonnegative, and such that $\lim_{\tau \to 0} \gamma(\rho, \sigma\tau)/\tau = 0$ for all $\rho, \tau > 0$. Therefore from (2.50) it follows that the operator $\mathfrak{L}$ has bounded variation. Proposition 2.3 is proved.

From Propositions 2.1 and 2.3 we obtain the following result.

THEOREM 2.3. *Let conditions* a) – d) *of Proposition* 2.3 *be fulfilled, and let the operator* $\mathfrak{L}$ *(acting from* $\mathring{H}^1_m(A, \Omega)$ *into* $(\mathring{H}^1_m(A, \Omega))^*$ *and defined by* (2.45)) *be coercive. Then the equation* $\mathfrak{L}u = 0$ *has a solution* $u \in \mathring{H}^1_m(A, \Omega)$. *In particular, the conditions of Theorem* 2.2 *are fulfilled if conditions* (2.20) *and* (2.23) *are satisfied along with those of Proposition* 2.2.

The proof follows in an obvious manner from Theorem 2.1 and Propositions 2.1 and 2.3.

7. *A specific class of operators*. We consider, for $m \geqslant 2$, the operator of the form

$$
\begin{aligned}
\mathfrak{L}u \equiv &-\frac{d}{dx_i}\left(|A^{1/2}u_x|^{m-2} \sum_{j=1}^n a^{ij}(x) u_{x_j} \right) \\
&+ \frac{d}{dx_i}(a^i(x)|u|^{m-1} + f^i(x)) + |A^{1/2}u_x|^{m-2} b^i(x) u_{x_i} + a_0(x)|u|^{m-2}u + f_0(x),
\end{aligned} \tag{2.51}
$$

where a^{ij}, a^i, b^i, a_0, f^i and f_0 $(i, j = 1, \ldots, n)$ are functions measurable in Ω, and $A = \|a^{ij}(x)\|$ is a symmetric matrix satisfying (2.1). We assume that the conditions

$$
\begin{gathered}
|B^{1/2}\mathbf{a}|^{m/(m-1)}, \ |B^{1/2}\mathbf{b}|^m, \ a_0 \in L^s(\Omega), \ 1/s + m/l = 1, \\
|B^{1/2}\mathbf{f}| \in L^{m/(m-1)}(\Omega), \ f_0 \in L^{l'/(l'-1)}(\Omega), \ l' < l,
\end{gathered} \tag{2.52}
$$

are fulfilled, where $B = A^{-1}$, $\mathbf{a} = (a^1, \ldots, a^n)$, $\mathbf{b} = (b^1, \ldots, b^n)$, $\mathbf{f} = (f^1, \ldots, f^n)$, and the exponent l satisfies (2.2).

We first check that all the conditions of Proposition 2.3 are satisfied for the operator (2.51). Denoting

$$\hat{\mathfrak{L}}u \equiv -\frac{d}{dx_i}\left(\left|A^{1/2}u_x\right|^{m-2}\sum_{j=1}^{n} a^{ij}(x)u_{x_j}\right),$$

$$\mathfrak{L}'u \equiv \frac{d}{dx_i}\mathfrak{L}''^{i}(x, u) + \mathfrak{L}_0(x, u, u_x) \equiv \frac{d}{dx_i}\left(a^i(x)|u|^{m-1} + f^i(x)\right)$$

$$+\left|A^{1/2}u_x\right|^{m-2}b^i(x)u_{x_i} + a_0(x)|u|^{m-2}u + f_0(x), \tag{2.53}$$

we see that condition a) of Proposition 2.3 holds. From the form of the functions

$$\hat{\mathfrak{L}}^i(x, p) \equiv \left|A^{1/2}p\right|^{m-2}\sum_{j=1}^{n} a^{ij}(x)p_j,$$

$$\mathfrak{L}''^{i}(x, u) \equiv a^i(x)|u|^{m-1} + f^i(x),$$

$$\mathfrak{L}_0(x, u, p) \equiv \left|A^{1/2}p\right|^{m-1}b^i(x)p_i + a_0(x)|u|^{m-2}u + f_0(x) \tag{2.54}$$

and condition (2.52), it is easy to see that b) also holds, the functions $\mathfrak{L}^i(x, u, p) \equiv \hat{\mathfrak{L}}^i(x, p) + \mathfrak{L}''^{i}(x, u)$, $i = 1, \dots, n$, and $\mathfrak{L}_0(x, u, p)$ satisfying (2.17) with $\tilde{\mu}_1 = 1$, $a_1 = |B^{1/2}\mathbf{a}|$, $\psi_1 = |B^{1/2}\mathbf{f}|$, $a_2 = |B^{1/2}\mathbf{b}|$, $a_3 = |a_0|$ and $\psi_2 = |f_0|$. Applying Young's inequality and taking (2.52) into account, we easily arrive at condition d). Finally, we verify that the operator $\hat{\mathfrak{L}}u \equiv -(d/dx_i)(|A^{1/2}u_x|^{m-2}\sum_{j=1}^{n} a^{ij}(x)u_{x_j})$ satisfies the condition of definiteness. For any $\xi \in \mathbb{R}^n$ we estimate

$$\frac{\partial\hat{\mathfrak{L}}^i(x, p)}{\partial p_j}\xi_i\xi_j = \frac{\partial}{\partial p_j}\left[\left(a^{kl}p_kp_l\right)^{m/2-1}\sum_{j-1}^{n} a^{ij}p_j\right]\xi_i\xi_j$$

$$= (m-2)(Ap, p)^{m/2-2}(Ap, \xi)^2 + (Ap, p)^{m/2-1}(A\xi, \xi) \geqslant |A^{1/2}p|^{m-2}|A^{1/2}\xi|^2. \tag{2.55}$$

And (2.55), by virtue of Proposition 2.3, implies the definiteness of the variation of $\hat{\mathfrak{L}}$, i.e. condition c). Thus all conditions of the proposition are satisfied.

On the other hand, by considering (2.52) and applying Young's inequality, one easily sees that for (2.51) we have (2.20) with $\tilde{\nu} = 1/2$, $a_4 = |B^{1/2}\mathbf{a}|^{m/(m-1)}$, $\psi_3 = |B^{1/2}\mathbf{f}|^{m/(m-1)}$, $\tilde{\mu}_6 = 0$, $b = 0$, $a_5 = -a_0 + |B^{1/2}\mathbf{b}|^m$ and $\psi_4 = |f_0|$. We assume that the condition

$$1/2 - c^m \left\|\left(-a_0 + |B^{1/2}\mathbf{a}|^{m/(m-1)} + |B^{1/2}\mathbf{b}|^m\right)_+\right\|_{s,\Omega}\left\|\lambda^{-1}\right\|_{r,\Omega}^{m/2} > 0. \tag{2.56}$$

holds. This inequality implies that (2.23) holds for the operator (2.51). In particular, (2.50) clearly holds if

$$a_0 > |B^{1/2}\mathbf{a}|^{m/(m-1)} + |B^{1/2}\mathbf{b}|^m \qquad \text{a.e. in } \Omega.$$

Thus the following result is obtained from Theorem 2.3.

THEOREM 2.4. *Assume conditions* (2.52) *and* (2.56) *hold for the operator* $\mathfrak{L}$: $\mathring{H}_m^1(A, \Omega) \to (\mathring{H}_m^1(A, \Omega))^*$, *defined by* (2.51). *Then* $\mathfrak{L}u = 0$ *has at least one solution* $u \in \mathring{H}_m^1(A, \Omega)$.

§3. Equations in the space $\overset{\circ}{H}{}^1_{\mathbf{m}}(A, \Omega)$

1. *The space* $\overset{\circ}{H}{}^1_{\mathbf{m}}(A, \Omega)$. In this section the results of §2 will be generalized somewhat. We note that the notation will differ from that of §2; it is taken to correspond with the notation used in the theory of linear degenerating equations. In particular, the matrix $A^{1/2}$ in §2 will be denoted here by A.

Let Ω be a bounded domain in $\mathbb{R}^n$, and let $A = \|a^{kj}(x)\|$ be a matrix whose elements $a^{kj}(x)$ are defined and measurable in Ω, with the matrix A invertible for almost all $x \in \Omega$. The elements a^{kj} and the elements b^{ik} of the matrix $B = A^{-1}$ are to satisfy the conditions

$$\begin{gathered} a^{kj} \in L^{m_k}(\Omega),\ m_k > 1,\ k,\ j = 1, \ldots, n, \\ b^{ik} \in L^{r_{ik}}(\Omega),\ r_{ik} \in (0, \infty],\ i,\ k = 1, \ldots, n. \end{gathered} \tag{3.1}$$

We note that if, for example, the matrix A is symmetric, $\lambda(x)$ and $\Lambda(x)$ its least and greatest eigenvalues (by modulus), and $\lambda^{-1}(x) \in L^r(\Omega)$, $r \in (0, +\infty]$ and $\Lambda(x) \in L^m(\Omega)$, $m > 1$, then $b^{ik} \in L^r(\Omega)$ and $a^{kj} \in L^m(\Omega)$, $i, k, j = 1, \ldots, n$ (compare with §2). In fact, for any vector $\xi = (\xi_1, \ldots, \xi_n)$ we have

$$\left|\sum_{i,j=1}^n b^{ij}\xi_i\xi_j\right| \leqslant |\lambda^{-1}||\xi|^2,$$

$$\left|\sum_{k,i,j=1}^n a^{kj}\xi_k\xi_j\right| \leqslant |\Lambda||\xi^2|.$$

(In this section summation is not assumed over repeated indices.) Choosing the vector ξ in a suitable manner, we establish the required assertion.

For an arbitrary matrix $T = \|t^{ij}\|$ and vector $\xi = (\xi_1, \ldots, \xi_n)$ we use the notation $T_i\xi = \sum_{j=1}^n t^{ij}\xi_j$.

LEMMA 3.1. *Let condition* (3.1) *be fulfilled, and let*

$$\begin{gathered} q_i = \min_k \left\{\frac{r_{ik}m_k}{r_{ik} + m_k}\right\} \geqslant 1,\quad 1/l = 1/n\left(\sum_{i=1}^n 1/q_i - 1\right) \quad \text{for } \sum_{i=1}^n 1/q_i > 1, \\ l \in (0, +\infty) \quad \text{for } \sum_{i=1}^n 1/q_i = 1,\quad l = +\infty \quad \text{for } \sum_{i=1}^n 1/q_i < 1. \end{gathered} \tag{3.2}$$

Then for any function $u \in \overset{\circ}{C}{}^1(\Omega)$

$$\|u\|_{l,\Omega} \leqslant c_1\|u\|_{\overset{\circ}{W}{}^1_{\mathbf{q}}(\Omega)} \leqslant c_2\sum_{k=1}^n \|A_k u_x\|_{m_k,\Omega}, \tag{3.3}$$

where $\|u\|_{\overset{\circ}{W}{}^1_{\mathbf{q}}(\Omega)} = \sum_1^n \|u_{x_i}\|_{q_i,\Omega}$, *and* c_1 *and* c_2 *are positive constants.*

PROOF. The proof of the left inequality in (3.3) may be found, for example, in [12]. For the proof of the right inequality in (3.3), we note that

$$u_{x_i} = \sum_{k=1}^n b^{ik}A_k u_x \tag{3.4}$$

and

$$\left(\int_\Omega |u_{x_i}|^{q_i}dx\right)^{1/q_i} \leqslant c\sum_{k=1}^n\left(\int_\Omega |b^{ik}|^{q_i}|A_k u_x|^{q_i}dx\right)^{1/q_i}, \quad i=1,\ldots,n. \tag{3.5}$$

Thus, applying to the integral on the right of (3.5) Hölder's inequality with index m_k/q_i for $m_k > q_i$ and estimating the functions b^{ik} by their maximum modulus with $m_k = q_i$ (in this case $r_{ik} = \infty$), we establish the right part of (3.3). The lemma is proved.

We denote by $\overset{\circ}{H}{}^1_{\mathbf{m}}(A, \Omega)$ the closure of the set $\overset{\circ}{C}{}^1(\Omega)$ with respect to the norm

$$\|u\|_{\overset{\circ}{H}{}^1_{\mathbf{m}}(A,\Omega)} = \sum_{k=1}^n \|A_k u_x\|_{m_k,\Omega}.$$

The satisfaction of the norm axiom is clear. It is also clear that $\overset{\circ}{H}{}^1_{\mathbf{m}}(A, \Omega)$ is a separable Banach space.

LEMMA 3.2. *Let conditions* (3.1) *and* (3.2) *be satisfied. Then* $\overset{\circ}{H}{}^1_{\mathbf{m}}(A, \Omega) \hookrightarrow \overset{\circ}{W}{}^1_{\mathbf{q}}(\Omega) \hookrightarrow L^l(\Omega)$, *and for any* $u \in \overset{\circ}{H}{}^1_{\mathbf{m}}(A, \Omega)$ *the inequality*

$$\|u\|_{l,\Omega} \leqslant c_1\|u\|_{\overset{\circ}{W}{}^1_{\mathbf{q}}(\Omega)} \leqslant c_2\|u\|_{\overset{\circ}{H}{}^1_{\mathbf{m}}(A,\Omega)} \tag{3.6}$$

is satisfied, where the constants c_1 *and* c_2 *are the same as in* (3.3).

The assertions of the lemma follow from (3.3), the definition of $\overset{\circ}{H}{}^1_{\mathbf{m}}(A, \Omega)$, and the properties of the spaces $L^l(\Omega)$ and $\overset{\circ}{W}{}^1_{\mathbf{q}}(\Omega)$.

LEMMA 3.3. *The space* $\overset{\circ}{H}{}^1_{\mathbf{m}}(A, \Omega)$ *is reflexive. Each linear functional* F *on* $\overset{\circ}{H}{}^1_{\mathbf{m}}(A, \Omega)$ *may be represented in the form*

$$\langle F, \eta\rangle = \int_\Omega (\mathbf{f}, \eta_x)\,dx, \quad \eta \in \overset{\circ}{H}{}^1_{\mathbf{m}}(A, \Omega), \tag{3.7}$$

where $B^*\mathbf{f} \in L^{m_1/(m_1-1)}(\Omega) \times \ldots \times L^{m_n/(m_n-1)}(\Omega)$, B^* *being the matrix adjoint to* B, $\mathbf{f} = (f^1, \ldots, f^n)$, *and* $\eta_x = (\eta_{x_1}, \ldots, \eta_{x_n})$. *Furthermore,*

$$\|F\|_{(\overset{\circ}{H}{}^1_{\mathbf{m}}(A,\Omega))^*} \leqslant K\|B^*\mathbf{f}\|_{L^{m_1/(m_1-1)}(\Omega)\times\ldots\times L^{m_n/(m_n-1)}(\Omega)}, \tag{3.8}$$

where K *is a positive constant.*

PROOF. The mapping

$$\pi : \overset{\circ}{H}{}^1_{\mathbf{m}}(A, \Omega) \to L^{m_1}(\Omega) \times \ldots \times L^{m_n}(\Omega)$$

defined by $\pi(\eta) = A\eta_x$ is a bicontinuous isomorphic mapping of $\overset{\circ}{H}{}^1_{\mathbf{m}}(A, \Omega)$ onto some closed subspace of the reflexive Banach space $L^{m_1}(\Omega) \times \ldots \times L^{m_n}(\Omega)$. Each linear functional in the latter may be represented in the form

$$\Phi(\mathbf{v}) = \int_\Omega (\mathbf{h}, \mathbf{v})\,dx, \quad \mathbf{v} = (v^1, \ldots, v^n) \in L^{m_1}(\Omega) \times \ldots \times L^{m_n}(\Omega),$$

$$\mathbf{h} = (h^1, \ldots, h^m) \in L^{m_1/(m_1-1)}(\Omega) \times \ldots \times L^{m_n/(m_n-1)}(\Omega).$$

The rest of the argument follows that of Lemma 2.3.

The following remark is analogous to Remark 2.1.

REMARK 3.1. If $f_0 \in L^{l/(l-1)}(\Omega)$, where l satisfies condition (3.2), then the integral

$$\int_\Omega f_0 \eta\, dx, \quad \eta \in \mathring{H}^1_{\mathbf{m}}(A, \Omega),$$

defines some linear functional on $\mathring{H}^1_{\mathbf{m}}(A, \Omega)$, the norm of this functional not surpassing $c_2 \| f_0 \|_{l/(l-1),\Omega}$, where c_2 is the constant in (3.6).

2. *Existence theorem.* Again we consider in Ω the quasilinear differential operator

$$\mathfrak{L}u \equiv \sum_{i=1}^{n} \frac{d}{dx_i} \mathfrak{L}^i(x, u, u_x) + \mathfrak{L}_0(x, u, u_x), \tag{3.9}$$

for which the following condition is now fulfilled.

CONDITION Ia. *The functions $\mathfrak{L}^i(x, u, p)$, $i = 1, \ldots, n$, and $\mathfrak{L}_0(x, u, p)$ satisfy the Carathéodory condition, and for any functions $u(x)$ and $p(x) = (p_1(x), \ldots, p_n(x))$ such that $u \in L^l(\Omega)$ and $Ap \in L^{m_1}(\Omega) \times \ldots \times L^{m_n}(\Omega)$ we have*

$$B^* \mathfrak{L}(x, u, p) \in L^{m_1/(m_1 - 1)}(\Omega) \times \ldots \times L^{m_n/(m_n - 1)}(\Omega),$$

$$\mathfrak{L}_0(x, u, p) \in L^{l(l-1)}(\Omega), \tag{3.10}$$

where $\mathfrak{L} = (\mathfrak{L}^1, \ldots, \mathfrak{L}^n)$.

It is now clear that for any $u, \eta \in \mathring{H}^1_{\mathbf{m}}(A, \Omega)$ the integral

$$\langle \mathfrak{L}u, \eta \rangle \equiv \int_\Omega \left[\mathfrak{L}(x, u, u_x)\eta_x + \mathfrak{L}_0(x, u, u_x)\, \eta \right] dx \tag{3.11}$$

converges; this defines the operator $\mathfrak{L}$ as a mapping from $\mathring{H}^1_{\mathbf{m}}(A, \Omega)$ into $(\mathring{H}^1_{\mathbf{m}}(A, \Omega))^*$.

We consider the equation

$$\mathfrak{L}u = F, \tag{3.12}$$

where $F \in (\mathring{H}^1_{\mathbf{m}}(A, \Omega))^*$. By virtue of Lemma 3.3, (3.12) may be reduced to the equation $\tilde{\mathfrak{L}}u = 0$, where

$$\tilde{\mathfrak{L}}u \equiv -\sum_{i=1}^{n} \frac{d}{dx_i} \tilde{\mathfrak{L}}^i(x, u, u_x), \quad \tilde{\mathfrak{L}}(x, u, p) = \mathfrak{L}(x, u, p) + \mathbf{f}, \quad \mathbf{f} = (f^1, \ldots, f^n),$$

$\int_\Omega (\mathbf{f}, \eta_x)\, dx = \langle F, \eta \rangle$, the operator $\tilde{\mathfrak{L}}$ satisfying Condition Ia.

THEOREM 3.1. *Let the operator $\mathfrak{L}: \mathring{H}^1_{\mathbf{m}}(A, \Omega) \to (\mathring{H}^1_{\mathbf{m}}(A, \Omega))^*$ defined by (3.11) satisfy Condition* Ia, *and let the operator $\mathfrak{L}$ be coercive and have semibounded variation. Then there exists at least one solution of the equation*

$$\mathfrak{L}u = 0.$$

PROOF. Consider the operators $\mathfrak{M}^i: \mathring{H}^1_{\mathbf{m}}(A, \Omega) \to L^{m_i/(m_i - 1)}(\Omega)$, $i = 1, \ldots, n$, and $\mathfrak{M}_0: \mathring{H}^1_{\mathbf{m}}(A, \Omega) \to L^{l/(l-1)}(\Omega)$, defined by

$$\mathfrak{M}^i u = B_i^* \mathfrak{L}(x, u, u_x), \qquad i = 1, \ldots, n,$$

$$\mathfrak{M}_0 u = \mathfrak{L}_0(x, u, u_x).$$

As in the proof of Theorem 2.1, by reworking suitably the reasoning in [8] (Chapter I, §2), one proves that:

1°) the operators $\mathfrak{M}^i$, $i = 1, \ldots, n$, and $\mathfrak{M}_0$ are continuous, and

2°) they are bounded.

The continuity and boundedness of these operators imply the continuity of $\mathfrak{L}$. Now the proof of Theorem 3.1 is finished by reference to Theorem 1.1.

3. *Sufficient conditions for Condition* Ia. Suppose that the functions $\mathfrak{L}^i(x, u, p)$, $i = 1, \ldots, m$, and $\mathfrak{L}_0(x, u, p)$ satisfy the Carathéodory condition, and for almost all $x \in \Omega$ and any $u \in \mathbb{R}$ and $p \in \mathbb{R}^n$

$$|B_i^*\mathfrak{L}(x, u, p)| \leqslant \mu_1\left(\sum_{k=1}^n |A_k p|^{m_k(m_i-1)/m_i} + |u|^{l(m_i-1)/m_i} + \varphi^i(x)\right),$$
$$i = 1, \ldots, n,$$
$$|\mathfrak{L}_0(x, u, p)| \leqslant \mu_2\left(\sum_{k=1}^n |A_k p|^{m_k(l-1)/l} + |u|^{l-1} + \psi(x)\right) \tag{3.13}$$

for $l < +\infty$, where μ_1 and μ_2 are positive constants,

$$\varphi^i(x) \in L^{m_i/(m_i-1)}(\Omega), \quad i = 1, \ldots, n, \quad \psi \in L^{l/(l-1)}(\Omega),$$
$$|B_i^*\mathfrak{L}(x, u, p)| \leqslant \tilde{\mu}_1(u)\left(\sum_{k=1}^n |A_k p|^{m_k(m_i-1)/m_i} + \tilde{\varphi}^i(x)\right), \qquad i = 1, \ldots, n,$$
$$|\mathfrak{L}_0(x, u, p)| \leqslant \tilde{\mu}_2(u)\left(\sum_{k=1}^n |A_k p|^{m_k} + \psi(x)\right) \tag{3.13a}$$

for $l = +\infty$, where $\tilde{\mu}_1(u)$ and $\tilde{\mu}_2(u)$ are continuous functions, $\tilde{\varphi}^i \in L^{m_i/(m_i-1)}(\Omega)$ and $\tilde{\psi} \in L^1(\Omega)$. Condition Ia clearly follows from this.

It is easy to show that (3.13) and (3.13a) are equivalent to the respective inequalities

$$|(\mathfrak{L}(x, u, p), \xi)| \leqslant \mu_1 \sum_{i=1}^n \left(\sum_{k=1}^n |A_k p|^{m_k(m_i-1)/m_i} + |u|^{l(m_i-1)/m_i} + \varphi^i(x)\right)|A_i\xi|,$$
$$|\mathfrak{L}_0(x, u, p)| \leqslant \mu_2\left(\sum_{k=1}^n |A_k p|^{m_k(l-1)/l} + |u|^{l-1} + \psi\right), \tag{3.13'}$$
$$|(\mathfrak{L}(x, u, p), \xi)| \leqslant \tilde{\mu}_1(u) \sum_{i=1}^n \left(\sum_{k=1}^n |A_k p|^{m_k(m_i-1)/m_i} + \tilde{\varphi}^i(x)\right)|A_i\xi|,$$
$$|\mathfrak{L}_0(x, u, p)| \leqslant \tilde{\mu}_2(u)\left(\sum_{k=1}^n |A_k p|^{m_k} + \tilde{\psi}(x)\right) \tag{3.13'a}$$

for almost all $x \in \Omega$ and any $u \in \mathbb{R}$ and $p, \xi \in \mathbb{R}^n$.

4. *Sufficient conditions for coercivity.*

PROPOSITION 3.1. *For any numbers $a_i \geqslant 0$ and $m_i > 1$, $i = 1, \ldots, m$, the inequalities*

$$\sum_{i=1}^n a_i^{m_i} \geqslant \frac{1}{n^{m-1}}\left(\sum_{i=1}^n a_i\right)^m \quad \textit{for} \quad r \equiv \sum_{i=1}^n a_i \geqslant n + \frac{n}{n-1}, \tag{3.14}$$

$$\sum_{i=1}^n a_i^{m_i} \geqslant \frac{1}{n^{M-1}}\left(\sum_{i=1}^n a_i\right)^M \quad \textit{for} \quad r \leqslant 1, \tag{3.15}$$

hold, where $m = \min_{i=1,\ldots,n}\{m_i\}$, $M = \max_{i=1,\ldots,n}\{m_i\}$.

PROOF. Let $r \geqslant n + n/(n-1)$. Then

$$\sum_{i=1}^{n} a_i^{m_i} = \sum_{\alpha_{\beta_i}<1} a_{\beta_i}^{m_{\beta_i}} + \sum_{a_{\gamma_i}\geqslant 1} a_{\gamma_i}^{m_{\gamma_i}} \geqslant \frac{1}{k^m}\Big(\sum_{a_{\gamma_i}\geqslant 1} a_{\gamma_i}\Big)^m \geqslant \frac{1}{k^{m-1}}\Big(\sum_{i=1}^{n}(a_i-(n-k)\Big)^m,$$

where k is the number of constants $a_i \geqslant 1$.

It is easy to see that for $r \geqslant n + n/(n-1)$ the function $f(k) = (1/k^{m-1}) \times (r-(n-k))^m$ decreases as k increases. Therefore

$$\frac{1}{k^{m-1}}\Big(\sum_{i=1}^{n}(a_i-(n-k))\Big)^m \geqslant \frac{1}{n^{m-1}}\Big(\sum_{i=1}^{n} a_i\Big)^m.$$

This proves (3.14).

Now let $r \leqslant 1$. Since in this case all $a_i \leqslant 1$, we have

$$\sum_{i=1}^{n} a_i^{m_i} \geqslant \sum_{i=1}^{n} a_i^{M} \geqslant \frac{1}{n^{M-1}}\Big(\sum_{i=1}^{n} a_i\Big)^M.$$

The proposition is proved.

Suppose for almost all $x \in \Omega$ and any $u \in \mathbb{R}$ and $p \in \mathbb{R}^n$

$$(\mathfrak{L}(x,u,p),p) + \mathfrak{L}_0(x,u,p)u \geqslant \nu \sum_{i=1}^{n} |A_i p|^{m_i} - \mu|u|^\alpha + \varphi_1(x), \tag{3.16}$$

where $\nu > 0$, $\mu \geqslant 0$, $0 \leqslant \alpha \leqslant l$, $\alpha < \infty$ and $\varphi \in L^1(\Omega)$. If $\alpha < m = \min_i\{m_i\}$, then $\mathfrak{L}$ is clearly coercive for any ν, μ and $\varphi_1 \in L^1(\Omega)$.

If $\alpha = m$, then in view of Proposition 3.1 it is easy to show that $\mathfrak{L}$ is coercive for any $\varphi_1 \in L^1(\Omega)$ if

$$\frac{1}{n^{m-1}}\nu \geqslant \mu c_\alpha^\alpha, \tag{3.17}$$

where the c_α are the constants from the inequality

$$\|u\|_{\alpha,\Omega} \leqslant c_\alpha \|u\|_{\mathring{H}^1_{\mathbf{m}}(A,\Omega)}, \tag{3.18}$$

the latter holding since

$$\mathring{H}^1_{\mathbf{m}}(A,\Omega) \hookrightarrow L^l(\Omega) \hookrightarrow L^\alpha(\Omega).$$

We note that the constants c_α are small for small domains Ω. If $\alpha \geqslant M$, then the condition of coercivity will hold for all $\varphi_1 \in L^1(\Omega)$ for which there exists a number ρ_{φ_1}, $0 < \rho_{\varphi_1} \leqslant 1$, such that

$$\frac{1}{n^{M-1}}\nu\rho_{\varphi_1}^M - \mu c_\alpha^\alpha \rho_{\varphi_1}^\alpha + \int_\Omega \varphi_1(x)\,dx \geqslant 0.$$

Clearly, the coercivity condition will be fulfilled for arbitrary ν, μ and α if $\int_\Omega \varphi_1(x)\,dx > 0$.

Now suppose that

$$(\mathfrak{L}(x,u,p),p) + \mathfrak{L}_0(x,u,p)u \geqslant \nu \sum_{i=1}^{n} |A_i p|^{m_i} - a(x)|u|^m - \varphi_2(x)|u| - \varphi_3(x), \tag{3.19}$$

where $\nu > 0$, $a \in L^s(\Omega)$, $1/s + m/l = 1$, $l > m$, $\varphi_2 \in L^{l/(l-1)}(\Omega)$, $\varphi_3 \in L^1(\Omega)$ and $m = \min_i\{m_i\}$.

Let $\rho_0 = \|u\|_{\mathring{H}^1_{\mathbf{m}}(A,\Omega)} \geqslant n + n/(n-1)$. Then bearing in mind Proposition 3.1 and (3.6) and applying the inequalities of Hölder and Young, we obtain

$$\langle \mathfrak{L}u, u\rangle \geqslant \left(\frac{\nu}{n^{m-1}} - c_2^m\|a_+\|_{s,\Omega}\right)\|u\|^m_{\mathring{H}^1_{\mathbf{m}}(A,\Omega)} - \varepsilon c_2^l\|u\|^l_{\mathring{H}^1_{\mathbf{m}}(A,\Omega)}$$

$$- \varepsilon^{-1/(l-1)}\int_\Omega |\varphi_2|^{l/(l-1)}\,dx - \int_\Omega \varphi_3\,dx, \tag{3.20}$$

where $\varepsilon > 0$, f_+, as in §2, denotes the positive part of f, and c_2 is the constant in (3.6).

Let

$$\delta = \frac{\nu}{n^{m-1}} - c_2^m\|a_+\|_{s,\Omega} > 0. \tag{3.21}$$

Introducing suitable notation, we write (3.20) in the form

$$\langle \mathfrak{L}u, u\rangle \geqslant \delta\rho_0^m - \varepsilon\kappa\rho_0^l - \varepsilon^{-1/(l-1)}K_1 - K_2, \tag{3.22}$$

which holds for every $\varepsilon > 0$.

We set $\varepsilon = \delta\rho_0^{m-l}/2\kappa$ and $\sigma = \delta/2\kappa$, and obtain, from (3.22),

$$\langle \mathfrak{L}u, u\rangle \geqslant \delta\rho_0^m/2 - \sigma^{-1(l-1)}K_1\rho_0^{(l-m)/(l-1)} - K_2.$$

As in §2 (see (2.25)), we obtain (noting that $(l-m)/(l-1) < m$ for any $m > 1$) the coercivity of $\mathfrak{L}$ under conditions (3.19) and (3.21).

5. *Theorems on the homeomorphism of operators satisfying the definiteness condition.* We shall say that the operator $\mathfrak{L}\colon \mathring{H}^1_{\mathbf{m}}(A,\Omega) \to (\mathring{H}^1_{\mathbf{m}}(A,\Omega))^*$ defined by (3.11) *satisfies the condition of definiteness* (of variation) if for any $u, v \in \mathring{H}^1_{\mathbf{m}}(A,\Omega)$

$$\langle \mathfrak{L}u - \mathfrak{L}v, u - v\rangle \geqslant c_0\sum_{i=1}^n \|A_i(u_x - v_x)\|^{m_i}_{m_i,\Omega}. \tag{3.23}$$

Bearing in mind Lemma 3.3, Proposition 3.1 and Theorem 3.1, one may prove the following theorem, analogous to Theorem 2.2.

THEOREM 3.2. *Let the operator* $\mathfrak{L}$, *defined by* (3.11), *satisfy Condition* Ia *and the condition of definiteness. Then the mapping* $\mathfrak{L}\colon \mathring{H}^1_{\mathbf{m}}(A,\Omega) \to (\mathring{H}^1_{\mathbf{m}}(A,\Omega))^*$ *is a homeomorphism.*

We give algebraic criteria for the definiteness condition for the operator $\mathfrak{L}$.

PROPOSITION 3.2. *Let* $m_i \geqslant 2$, $i = 1, \dots, m$, *and let the functions* $\mathfrak{L}^i(x,u,p)$, $i = 1, \dots, m$, *and* $\mathfrak{L}_0(x,u,p)$ *be differentiable with respect to* u *and* p, *and for almost all* $x \in \Omega$ *and any* $u, \xi_0 \in \mathbb{R}$ *and* $p, \xi \in \mathbb{R}^n$

$$\sum_{i,j=1}^n \frac{\partial\mathfrak{L}^i(x,u,p)}{\partial p_j}\xi_i\xi_j + \sum_{i=1}^n \frac{\partial\mathfrak{L}^i(x,u,p)}{\partial u}\xi_i\xi_0 + \sum_{j=1}^n \frac{\partial\mathfrak{L}_0(x,u,p)}{\partial p_j}\xi_j\xi_0$$

$$+ \frac{\partial\mathfrak{L}_0(x,u,p)}{\partial u}\xi_0^2 \geqslant \alpha_0\sum_{k=1}^n |A_kp|^{m_k-2}|A_k\xi|^2. \tag{3.24}$$

Then the operator $\mathfrak{L}$ *satisfies the condition of definiteness.*

The proof is analogous to that of Theorem 2.2 of [6] (Chapter II, §2) and Proposition 2.1 of the present paper.

6. *Algebraic criteria for the semiboundedness of the variation.*

PROPOSITION 3.3. *Let the following conditions hold*:

a) *The operator* $\mathfrak{L}$ *may be written in the form* $\mathfrak{L} = \hat{\mathfrak{L}} + \mathfrak{L}'$, *with* $\hat{\mathfrak{L}}$ *an operator of the form* (3.9), *and*

$$\mathfrak{L}'u = -\sum_{i=1}^{n} \frac{d}{dx_i} \mathfrak{L}'^i(x, u) + \mathfrak{L}'_0(x, u, u_x).$$

b) $\hat{\mathfrak{L}}$ *satisfies Condition* Ia.

c) $\hat{\mathfrak{L}}$ *also satisfies the condition of definiteness* (3.23).

d) *The functions* $\mathfrak{L}'^i(x, u)$, $i = 1, \ldots, m$, *and* $\mathfrak{L}'_0(x, u, p)$ *are differentiable with respect to u and p, and for almost all* $x \in \Omega$ *and any* $u \in \mathbb{R}$ *and* $p, \xi \in \mathbb{R}^n$

$$\left|\sum_{i=1}^{n} \frac{\partial \mathfrak{L}'^i(x, u)}{\partial u} \xi_i\right| \leqslant \mu'_1 \sum_{i=1}^{n} \left(|u|^{l'(m_i-1)/m_i - 1} + \tilde{\varphi}_1^i(x)\right)|A_i\xi|,$$

$$\left|\frac{\partial \mathfrak{L}'_0(x, u, p)}{\partial u}\right| \leqslant \mu'_2\left(\sum_{i=1}^{n} |A_i p|^{m_i(l'-2)/l'} + |u|^{l'-2} + \tilde{\varphi}_2(x)\right),$$

$$\left|\sum_{j=1}^{n} \frac{\partial \mathfrak{L}'_0(x, u, p)}{\partial p_j} \xi_j\right| \leqslant \mu'_3 \sum_{i=1}^{n}\left(\sum_{k=1}^{n} |A_k p|^{m_k(l'-1)/l' - m_k/m_i} + |u|^{l'(m_i-1)/m_i} + \tilde{\tilde{\varphi}}_2^i\right)|A_i\xi|, \tag{3.25}$$

where μ'_1, μ'_2 *and* μ'_3 *are constants, and* $\tilde{\varphi}_1^i$, $\tilde{\varphi}_2$, $\tilde{\tilde{\varphi}}_2^i \geqslant 0$, $i = 1, \ldots, n$, $\tilde{\varphi}_1^i \in L^{m_i l'/(m_i l' - (m_i + l'))}(\Omega)$, $\tilde{\tilde{\varphi}}_1^i \in L^{m_i l'/(m_i l' - (m_i + l'))}(\Omega)$, $i = 1, \ldots, n$, $\tilde{\varphi}_2 \in L^{l'(l'-2)}(\Omega)$, $l' < l$.

Then the operator $\mathfrak{L}$ *has semibounded variation.*

The proof is analogous to that of Proposition 2.3.

7. *Example of an operator satisfying the definiteness condition in the space* $\mathring{H}^1_{\mathbf{m}}(A, \Omega)$. We shall assume that $m_i \geqslant 2$, $i = 1, \ldots, n$. For simplicity we introduce an example of an operator without lower order terms. Namely, set

$$\hat{\mathfrak{L}}u \equiv -\sum_{i=1}^{n} \frac{d}{dx_i} \hat{\mathfrak{L}}^i(x, u_x), \tag{3.26}$$

where $\hat{\mathfrak{L}}(x, u_x) = (\mathfrak{L}^1(x, u_x), \ldots, \mathfrak{L}^n(x, u_x))$ is defined by the following conditions:

$$\begin{gathered} \hat{\mathfrak{L}}(x, u_x) = A^*\mathbf{y}(u), \\ \mathbf{y}(u) = (y_1(u), \ldots, y_n(u)), \\ y_k(u) = |A_k u_x|^{m_k-2} A_k u_x, \qquad k = 1, \ldots, n, \end{gathered} \tag{3.27}$$

i.e.

$$\hat{\mathfrak{L}}u \equiv -\sum_{i=1}^{n} \frac{d}{dx_i}\left(\sum_{j=1}^{n} a_{ji}|A_j u_x|^{m_j-2} A_j u_x\right).$$

We shall show that the operator $\hat{\mathfrak{L}}$ defined by the integral identity

$$\langle \hat{\mathfrak{L}}u, \eta\rangle \equiv \int_\Omega \left(\hat{\mathfrak{L}}(x, u_x), \eta_x\right) dx, \qquad \eta \in \mathring{H}^1_{\mathbf{m}}(A, \Omega), \tag{3.28}$$

satisfies the condition of definiteness. We first show that the integral (3.28) converges. In fact, considering (3.27), we have

$$\begin{aligned}\left|\int_\Omega(\hat{\mathfrak{L}}(x,u_x),\eta_x)\,dx\right| &\leqslant \int_\Omega\left|(A^{-1*}\hat{\mathfrak{L}}(x,u_x),A\eta_x)\right|dx\\ &\leqslant \sum_{i=1}^n\int_\Omega|A_iu_x|^{m_i-1}|A_i\eta_x|\,dx\\ &\leqslant \sum_{i=1}^n\left(\int_\Omega|A_i\eta_x|^{m_i}dx\right)^{1/m_i}\left(\int_\Omega|A_iu_x|^{m_i}dx\right)^{(m_i-1)/m_i}<\infty.\end{aligned}$$

Furthermore, considering (3.27), we obtain

$$\begin{aligned}\langle\mathfrak{L}_u-\mathfrak{L}_v,u-v\rangle &= \int_\Omega((\hat{\mathfrak{L}}(x,u_x)-\hat{\mathfrak{L}}(x,v_x)),u_x-v_x)\,dx\\ &= \int_\Omega(A^{-1*}(\hat{\mathfrak{L}}(x,u_x)-\hat{\mathfrak{L}}(x,v_x)),A(u_x-v_x))\,dx\\ &= \sum_{i=1}^n\int_\Omega(|A_iu_x|^{m_i-2}A_iu_x-|A_iv_x|^{m_i-2}A_iv_x)(A_iu_x-A_iv_x)\,dx\\ &\geqslant c_0\sum_{i=1}^n\|A_i(u_x-v_x)\|^{m_i}_{m_i,\Omega},\end{aligned}$$

where c_0 is a positive constant. Thus the definiteness of variation of $\hat{\mathfrak{L}}$ is proved.

BIBLIOGRAPHY

1. O. A. Ladyženskaja and N. N. Ural'ceva, *Linear and quasilinear elliptic equations*, 2nd ed., "Nauka", Moscow, 1973; English transl. of 1st ed., Academic Press, 1968.

2. Felix E. Browder, *Nonlinear elliptic boundary value problems*, Bull. Amer. Math. Soc. **69** (1963), 862–874.

3. George J. Minty, *On a "monotonicity" method for the solution of nonlinear equations in Banach spaces*, Proc. Nat. Acad. Sci. U.S.A. **50** (1963), 1038–1041.

4. R. I. Kačurovskiĭ, *Nonlinear monotone operators in Banach spaces*, Uspehi Mat. Nauk **23** (1968), no. 2(140), 121–168; English transl. in Russian Math. Surveys **23** (1968).

5. Jacques-Louis Lions, *Quelques méthodes de résolution des problèmes aux limites non linéaires*, Dunod, Paris, 1969.

6. Ju. A. Dubinskiĭ, *Nonlinear elliptic and parabolic equations*, Itogi Nauki: Sovremennye Problemy Mat., vol. 9, VINITI, Moscow, 1976, pp. 5–130; English transl. in J. Soviet Math. **12** (1979), no. 5.

7. O. V. Besov, V. P. Il'in and S. M. Nikol'skiĭ, *Integral representations of functions and embedding theorems*, "Nauka", Moscow, 1975; English transl., Vols. 1, 2, Wiley, 1979.

8. M. A. Krasnosel'skiĭ, *Topological methods in the theory of nonlinear integral equations*, GITTL, Moscow, 1956; English transl., Macmillan, New York, 1964.

9. M. K. V. Murthy and G. Stampacchia, *Boundary value problems for some degenerate-elliptic operators*, Ann. Mat. Pura Appl. (4) **80** (1968), 1–122.

10. S. N. Kružkov, *Boundary-value problems for degenerate second-order elliptic equations*, Mat. Sb. **77** (**119**) (1968), 299–334; English transl. in Math. USSR Sb. **6** (1978).

11. Neil S. Trudinger, *Linear elliptic operators with measurable coefficients*, Ann. Scuola Norm. Sup. Pisa (3) **27** (1973), 265–308.

12. V. P. Il'in, *The properties of some classes of differentiable functions of several variables defined in an n-dimensional region*, Trudy Mat. Inst. Steklov. **66** (1962), 227–363; English transl. in Amer. Math. Soc. Transl. (2) **81** (1969).

Translated by P. C. FIFE

Труды Матем. Инст. Стеклов.
147(1980)

Proceedings of the Steklov
Institute of Mathematics
1981, Issue 2

ON SECOND ORDER DIFFERENTIAL EQUATIONS WITH d-ELLIPTIC OPERATORS

UDC 517.9, 517.994

N. M. IVOČKINA

ABSTRACT. A nonlinear operator $F(u) = F(u_{xx})$, $x \in \mathbf{R}^h$, is said to be *d-elliptic* on a set $\mathfrak{M}_F$ if $F(u_{ij} + \lambda\xi^i\xi_j) - F(u_{ij})$ is a second-degree polynomial in ξ which is equal to zero only if $\xi = 0$, when $u(x) < \mathfrak{M}_F$, $\lambda \in \mathbf{R}$.

In this paper it is shown that d-elliptic operators are representable in divergence form, and integral comparison theorems for them are proved. Under the assumption that $\mathfrak{M}_F$ contains functions which are convex downwards, one-sided a priori estimates are obtained for the solution of the Dirichlet problem for equations with a d-elliptic operator and for the normal derivative of the solution.

Bibliography: 6 titles.

A significant portion of the theory of quasilinear second-order elliptic equations constructed in the monograph of O. A. Ladyženskaja and N. N. Ural'ceva [1] (see also [2]) is based on the method of integral inequalities. Up to now, this method has not been used in the study of essentially nonlinear equations. In the present paper, we introduce the concept of d-ellipticity for nonlinear operators of the form $F(u_{xx})$, $u(x) \in C^2(\Omega)$. On the basis of results in [3], we show that d-elliptic operators, and only those, are representable in divergence form. A study of the structure of first-order divergence operators of the form $F(x, \bar{v}, \bar{v}_x)$ can be found in [4] and [5]. A similar analysis for second-order operators of the form $F(x, u, u_x, u_{xx})$ has been given in [3]. An essentially similar circle of questions is considered in [6] from the point of view of contact geometry.

In §1 of this paper, we again prove the fundamental algebraic Lemma 1 of [3] at length and in more detail. In §2 we describe the structure of nonlinear differential operators of divergence type. For completeness, we reproduce two results (Lemmas 2.1 and 2.2) which are proved in [3]. In §3 we give the definition of d-ellipticity of a nonlinear operator and discuss the limits of its applicability. In §4 we show that the concept of a generalized solution makes sense for equations with d-elliptic operators, and we prove three comparison theorems for such solutions. A natural consequence of the comparison theorems is §5, which relates the proof of the uniqueness theorems and the construction of one-sided a priori estimates for generalized solutions of the Dirichlet problem and their normal derivatives. The work presented in this paper was greatly enhanced by the constant attention and generous counsel of O. A. Ladyženskaja, for which we are deeply grateful.

1980 *Mathematics Subject Classification.* Primary 35J65, 35J67; Secondary 35B45

§1. Auxiliary results of an algebraic character

1. We introduce some notation.

We let $\alpha^i = \binom{\alpha_{1i}}{\alpha_{2i}}$, where the α_{mi} are numbers from the set $\{1, \ldots, n\}$, and we put $A(k) = (\alpha^{i_1}, \ldots, \alpha^{i_k})$.

We will assume that the matrices $A(k)$ are invariant under permutations of the columns and elements in a column.

$A\begin{pmatrix} \cdot & i & \cdot \\ j & \cdot & l \end{pmatrix}$ is a matrix $A(k)$ containing i, j, and l in the configuration shown.

$\{A(\alpha^i \ldots)\}$ is the set of columns of A with the exception of those shown.

$\tilde{A}(k)$ is a matrix whose rows are identical.

$s^m(a_1 \ldots a_m)$ is the group of permutations of the elements $a_1, \ldots, a_m$.

$\{s^m(a_1 \ldots a_s)\}$ is the set of permutations of the elements $a_1, \ldots, a_s$ taken m at a time.

The following operations on the matrices $A(k)$ will be applied below:

$$A\begin{pmatrix} \alpha_{1i_l} & \cdot \\ \cdot & \underline{\alpha_{2i_l}} \end{pmatrix} = \frac{1}{2} \sum_{\{s^2(\alpha_{1i_l}\alpha_{2i_l})\}} A\begin{pmatrix} \beta_{1i_l} & \cdot \\ \cdot & \beta_{2i_l} \end{pmatrix}, \tag{1.1}$$

$$A\begin{pmatrix} \alpha_{1i_l} & \cdot \\ \alpha_{2i_l} & \alpha_{2i_{l+1}} \end{pmatrix}_{\underbrace{\qquad}} = \frac{1}{2} \sum_{\{c^3(\alpha_{1i_l}\alpha_{2i_l}\alpha_{2i_{l+1}})\}} A\begin{pmatrix} \beta_{1i_l} & \cdot \\ \beta_{2i_l} & \beta_{2i_{l+1}} \end{pmatrix}, \tag{1.2}$$

where $c^3(\tilde{a}_1\tilde{a}_2\tilde{a}_3)$ is a cyclic subgroup of order three of $s^4(a_1a_2a_3a_4)$,

$$A\begin{pmatrix} \alpha_{1i_1} \ldots \alpha_{1i_l} \\ [\alpha_{2i_1} \ldots \alpha_{2i_l}] \end{pmatrix} = \sum_{\{s^l(i_1 \ldots i_l)\}} \varepsilon(j) A\begin{pmatrix} \alpha_{1i_1} \ldots \alpha_{1i_l} \\ \alpha_{2j_1} \ldots \alpha_{2j_l} \end{pmatrix}. \tag{1.3}$$

2. We indicate some relations among the operations (1.1)–(1.3). From the invariance properties of A and the equality

$$\{c^3(1i_l, 2i_l, 2i_{l+1}) \cdot s^2(1i_l, 2i_l) \cdot s^2(1i_{l+1}, 2i_{l+1}) \\ \times s^2(i_l, i_{l+1})(\alpha_{1i_l}, \alpha_{2i_l}, \alpha_{1i_{l+1}}, \alpha_{2i_{l+1}})\} = \{s^4(\alpha_{1i_l}, \alpha_{2i_l}, \alpha_{1i_{l+1}}, \alpha_{2i_{l+1}})\}$$

it follows that

$$A\begin{pmatrix} \alpha_{1i_l} & \alpha_{1i_{l+1}} \\ \underline{\alpha_{2i_l} \;\; \alpha_{2i_{l+1}}} \end{pmatrix} = \frac{1}{8} \sum_{\{s^4(\alpha_{1i_l} \cdots \alpha_{2i_{l+1}})\}} A\begin{pmatrix} \beta_{1i_l} & \beta_{1i_{l+1}} \\ \beta_{2i_l} & \beta_{2i_{l+1}} \end{pmatrix}. \tag{1.4}$$

The operations (1.1) and (1.2) are related by

$$A\begin{pmatrix} \alpha_{1i_l} & \cdot \\ \alpha_{2i_{l+1}} & \underline{\alpha_{2i_l}} \end{pmatrix} = A\begin{pmatrix} \alpha_{1i_l} & \cdot \\ \alpha_{2i_l} & \alpha_{2i_{l+1}} \end{pmatrix}_{\underbrace{\qquad}} - \frac{1}{2} A\begin{pmatrix} \alpha_{1i_l} & \cdot \\ \alpha_{2i_l} & \alpha_{2i_{l+1}} \end{pmatrix}. \tag{1.5}$$

Using the iterative definition of alternation,

$$[a_1 \ldots a_l] = a_1[a_2 \ldots a_l] - \sum_{s=2}^{l} a_s[a_2 \ldots a_1 \ldots a_l]$$

and the invariance of A under permutations of the columns, we can rewrite (1.3) in the form

$$A\begin{pmatrix}\alpha_{1i_1}\ldots\alpha_{1i_l}\\ [\alpha_{2i_1}\ldots\alpha_{2i_l}]\end{pmatrix}=\frac{1}{l}\sum(-1)^{r-1}A\begin{pmatrix}\alpha_{1p_1}\ldots\alpha_{1p_r}&\alpha_{1p_{r+1}}\ldots\alpha_{1p_l}\\ \alpha_{2p_2}\ldots\alpha_{2p_1}&[\alpha_{2p_{r+1}}\ldots\alpha_{2p_l}]\end{pmatrix}. \quad (1.6)$$

We deduce a fundamental relationship among the operations (1.1)–(1.3) by induction.

LEMMA 1.1.

$$A\begin{pmatrix}\alpha_{1i_1}\ldots\alpha_{1i_l}\\ [\alpha_{2i_1}\ldots\alpha_{2i_l}]\end{pmatrix}\overset{\text{mod}\smile}{=}a^lA\begin{pmatrix}\alpha_{1i_1}\ldots\alpha_{1i_l}\\ \alpha_{2i_1}\ldots\alpha_{2i_l}\end{pmatrix}, \quad (1.7)$$

where $a^0 = 1$, and the a^l for $l \geqslant 1$ are determined inductively by

$$a^l=(l-1)!\sum_{r=1}^{l}\frac{a^{l-r}}{2^{r-1}r!}. \quad (1.8)$$

PROOF. For $l = 1$, the lemma is obvious. We assume that (1.7) holds for $m < l$. Then from (1.6) it follows that

$$\underline{A\begin{pmatrix}\alpha_{1i_1}\ldots\alpha_{1i_l}\\ [\alpha_{2i_1}\ldots\alpha_{2i_l}]\end{pmatrix}}$$

$$=\frac{1}{l}\sum_{\substack{\{s^r(i_1\ldots i_l)\}\\ p_{r+1}<p_{r+2}<\ldots<p_l}}(-1)^{r-1}a^{l-r}A\begin{pmatrix}\alpha_{1p_1}\ldots\alpha_{1p_{r-1}}\alpha_{1p_r}&\alpha_{1p_{r+1}}\ldots\alpha_{1p_l}\\ \underline{\alpha_{2p_2}\ldots\alpha_{2p_r}\alpha_{2p_1}}&\alpha_{2p_{r+1}}\ldots\alpha_{2p_l}\end{pmatrix}. \quad (1.9)$$

On the other hand, applying (1.5) $r - 1$ times gives

$$A\begin{pmatrix}\alpha_{1p_1}\ldots\alpha_{1p_{r-1}}\alpha_{1p_r}\ldots\alpha_{1p_l}\\ \underline{\alpha_{2p_2}\ldots\alpha_{2p_r}\ \alpha_{2p_1}}\ldots\alpha_{2p_l}\end{pmatrix}\overset{\text{mod}\smile}{=}\frac{(-1)^{r-1}}{2^{r-1}}A\begin{pmatrix}\alpha_{1p_1}\ldots\alpha_{1p_l}\\ \alpha_{2p_1}\ldots\alpha_{2p_l}\end{pmatrix}. \quad (1.10)$$

Taking (1.9) and (1.10) together gives

$$A\begin{pmatrix}\alpha_{1i_1}\ldots\alpha_{1i_l}\\ [\alpha_{2i_1}\ldots\alpha_{2i_l}]\end{pmatrix}\overset{\text{mod}\smile}{=}(l-1)!\sum_{r=1}^{l}\frac{a^{l-r}}{2^{r-1}r!}A\begin{pmatrix}\alpha_{1i_1}\ldots\alpha_{1i_l}\\ \alpha_{2i_1}\ldots\alpha_{2i_l}\end{pmatrix}.$$

This proves Lemma 1.1.

3. Let $\{t^{ij}\} \in R^{n(n-1)}$, $t^{ij} = t^{ji}$, $i, j = 1, \ldots, n$. For the matrices $A(k)$ introduced in subsection 1 above, $k \leqslant n$, we form the product $t^A = \Pi_{\{A(k)\}}\, t^{ij}$. With the linear combination $\Sigma\, h^A A$, $h^A \in R$, we associate the sum

$$S(h, t)=t^{\Sigma h^A A}=h^A t^A=h^{\Sigma t^A A}. \quad (1.11)$$

Here and in what follows, summation is to be carried out over repeated matrix indices. We put

$$\Delta^A=t^{\begin{pmatrix}\alpha_{1i_1}\ldots\alpha_{1i_k}\\ [\alpha_{2i_1}\ldots\alpha_{2i_k}]\end{pmatrix}} \quad (1.12)$$

and we consider the sums $S(c, \Delta) = c^A \Delta^A$, $c^A \in R$. Without loss of generality, we assume that the coefficients c^A and h^A are invariant with respect to permutations of rows and columns of A, that a transposition of elements of a row of A changes the sign of c^A, and that

$$h^{A\binom{i}{j}} = h^{A\binom{j}{i}}.$$

LEMMA 1.2. *The sum $S(h, t)$ is representable in the form*

$$S(h, t) = c^A \Delta^A, \tag{1.13}$$

if and only if for any triple $\{\alpha_{1i_l} \alpha_{2i_l} \alpha_{2i_{l+1}}\}$, $l = 1, \ldots, k$, $k \leqslant n$,

$$h^{A\underbrace{\begin{pmatrix}\alpha_{1i_l} & \cdot \\ \alpha_{2i_l} & \alpha_{2i_{l+1}}\end{pmatrix}}} = 0. \tag{1.14}$$

PROOF. The sums $S^k(h, t) = h^{A(k)} t^{A(k)}$ are linearly independent; therefore it suffices to prove Lemma 1.2 for S^k with fixed $k \leqslant n$. Suppose that (1.13) holds. Then

$$h^{A(k)} t^{A(k)} = c^{A(k)} \Delta^{A(k)}, \quad k = 1, \ldots, n.$$

Transforming Δ^A by (1.3) and (1.11) and reducing similar terms by taking into account the symmetries of t^A, h^A, and c^A, we get

$$h^A = c^{\underline{\begin{pmatrix}\alpha_{1i_1} \cdots \alpha_{1i_k} \\ \alpha_{2i_1} \cdots \alpha_{2i_k}\end{pmatrix}}}.$$

We calculate

$$h^{A\underbrace{\begin{pmatrix}\alpha_{1i_l} & \cdot \\ \alpha_{2i_l} & \alpha_{2i_{l+1}}\end{pmatrix}}} = \frac{1}{8} \sum_{\{s'(\alpha_{1i_l} \cdots \alpha_{2i_{l+1}})\}} c^{\begin{pmatrix}\alpha_{1i_1} \cdots \alpha_{1i_{l-1}} & \beta_{1i_l}\beta_{1i_{l+1}} & \alpha_{1i_{l+2}} \cdots \alpha_{1i_k} \\ \underline{\alpha_{2i_1} \cdots \alpha_{2i_{l-1}}} & \beta_{2i_l}\beta_{2i_{l+1}} & \underline{\alpha_{2i_{l+2}} \cdots \alpha_{2i_k}}\end{pmatrix}}. \tag{1.15}$$

We have used (1.4) here. Since a transposition of elements of a row of A changes the sign of c^A, (1.14) follows from (1.15).

We assume that (1.14) holds. For an arbitrary set of numbers h^A we form $S^k(c, \Delta)$, putting

$$c^{\begin{pmatrix}\alpha_{1i_1} \cdots \alpha_{1i_k} \\ \alpha_{2i_1} \cdots \alpha_{2i_k}\end{pmatrix}} = \frac{1}{a^k} h^{\begin{pmatrix}\alpha_{1i_1} \cdots \alpha_{1i_k} \\ [\alpha_{2i_1} \cdots \alpha_{2i_k}]\end{pmatrix}},$$

where a^k is the coefficient in Lemma 1.1. Transforming $S^k(c, \Delta)$ just as above, we get

$$c^A \Delta^A = \frac{1}{a^k} h^{\underline{\begin{pmatrix}\alpha_{1i_1} \cdots \alpha_{1i_k} \\ [\alpha_{2i_1} \cdots \alpha_{2i_k}]\end{pmatrix}}} t^{\begin{pmatrix}\alpha_{1i_1} \cdots \alpha_{1i_k} \\ \alpha_{2i_1} \cdots \alpha_{2i_k}\end{pmatrix}}.$$

From Lemma 1.1 it follows that

$$\frac{1}{a^k} h\begin{pmatrix} \alpha_{1i_1} \dots \alpha_{1i_k} \\ [\alpha_{2i_1} \dots \alpha_{2i_k}] \end{pmatrix} = h\begin{pmatrix} \alpha_{1i_1} \dots \alpha_{1i_k} \\ \alpha_{2i_1} \dots \alpha_{2i_k} \end{pmatrix},$$

and hence that (1.13) holds. This proves Lemma 1.2.

REMARK. By definition,

$$\Delta\begin{pmatrix} 1 \dots n \\ 1 \dots n \end{pmatrix} = \begin{vmatrix} t^{11} & \dots & t^{1n} \\ \cdot & \cdot & \cdot \\ t^{n1} & \dots & t^{nn} \end{vmatrix}.$$

The set of all elements of Δ^A coincides, except possibly in sign, with the set of all minors of the determinant

$$\Delta\begin{pmatrix} 1 \dots n \\ 1 \dots n \end{pmatrix}.$$

§2. The structure of second-order nonlinear divergence operators

1. We consider a second-order nonlinear differential operator with "constant coefficients," i.e. an operator of the form

$$F(u) \equiv \mathcal{F}(u_{xx}), \quad \mathcal{F}(0) = 0, \tag{2.1}$$

where $x = (x^1, \dots, x^n) \in \Omega$, a bounded region in R^n, and $u(x) \in C^2(\Omega)$. It is well known that for certain special choices of $\mathcal{F}(u_{xx})$ the functional $I(u) = \int_\Omega \mathcal{F}(u_{xx})\,dx$ does not depend on the values of $u(x)$ at interior points of Ω. It is natural to expect that this property holds for those and only those functions $\mathcal{F}(u_{xx})$ which are representable in divergence form identically in u. More precisely, there exists a vector $G(u_x, u_{xx})$ such that

$$\mathcal{F}(u_{xx}) = (G^i(u_{xx}, u_x))_i; \tag{2.2}$$

here and in what follows, subscripts indicate differentiation, $(\cdot)_i = d(\cdot)/dx^i$.

THEOREM 2.1. *The following four statements are equivalent*:

I)
$$\frac{\partial^2 \mathcal{F}}{\partial u_{ij} \partial u_{kl}} + \frac{\partial^2 \mathcal{F}}{\partial u_{ik} \partial u_{lj}} + \frac{\partial^2 \mathcal{F}}{\partial u_{il} \partial u_{jk}} = 0, \quad i, j, k, l = 1, \dots, n; \tag{2.3}$$

II) *$I(u)$ does not depend on the values of $u(x)$ at interior points of Ω.*

III)
$$\mathcal{F}(u_{xx}) = c^A \Delta_A; \tag{2.4}$$

here

$$c^A \in R, \quad \Delta_A = u\begin{pmatrix} \alpha_{1i_1} \dots \alpha_{1i_k} \\ [\alpha_{2i_1} \dots \alpha_{2i_k}] \end{pmatrix},$$

where $u_A = \sum_{\{A\}} u_{ij}$.

IV) *The equality* (2.2) *holds for* $\mathcal{F}(u_{xx})$.

2. A partial proof of this theorem is given in [2]. We give that part here in a form which is convenient for our purposes.

LEMMA 2.1. *Statements* I *and* II *are equivalent.*

PROOF. Suppose that $u(x) \in (C^4(\Omega))$. Then $I(u)$ is independent of the values of $u(x)$ at interior points of Ω if and only if the Euler equation for $\mathfrak{F}(u_{xx})$ is satisfied identically in $u(x)$:

$$\left(\frac{\partial \mathcal{F}}{\partial u_{ij}}\right)_{ij} = \frac{\partial^3 \mathcal{F}}{\partial u_{ij} \partial u_{kl} \partial u_{st}} u_{kli} u_{stj} + \frac{\partial^2 \mathcal{F}}{\partial u_{ij} \partial u_{kl}} u_{ijkl} = 0. \tag{2.5}$$

At a fixed point $x_0 \in \Omega$ the values of $u_x, \ldots, u_{xxxx}$ are independent; therefore (2.5) splits into the system

$$\frac{\partial^3 \mathcal{F}}{\partial u_{ij} \partial u_{kl} \partial u_{st}} u_{kli} u_{stj} = 0,$$

$$\frac{\partial^2 \mathcal{F}}{\partial u_{ij} \partial u_{kl}} u_{ijkl} = 0,$$

which in turn is equivalent to (2.3).

Since the fact that $u(x)$ belongs to $C^4(\Omega)$ is essential only in the intermediate steps of the proof of Lemma 2.1, the lemma remains true for $u(x) \in C^2(\Omega)$.

LEMMA 2.2. *Statements* I *and* III *of Theorem* 2.1 *are equivalent.*

PROOF. It is clear that (2.3) is possible only for power functions $\mathfrak{F}(u_{xx})$, i.e. for functions of the form

$$\mathcal{F}(u_{xx}) = h^A u_A, \tag{2.6}$$

where A denotes all possible matrices specified in §1.1.

Taking into account the independence of the values of u_{xx} at a fixed point $x_0 \in \Omega$, we see that the system of differential identities (2.3) calculated for the functions (2.6) is equivalent to the system of algebraic equations at each point $x \in \Omega$

$$h^{\underbrace{A\begin{pmatrix} \alpha_{1i_l} & \cdot \\ \alpha_{2i_l} & \alpha_{2i_{l+1}} \end{pmatrix}}} = 0,$$

which, as was shown in Lemma 1.2, is a necessary and sufficient condition for the existence of a representation (2.4).

3. PROOF OF THEOREM 2.1. The equivalence of the first three statements follows from Lemmas 2.1 and 2.2. We turn to the fourth. Suppose that $\mathfrak{F}(u_{xx})$ has the form (2.2). Then the functional $I(u)$ does not depend on the values of $u(x)$ at interior points of Ω, and (2.3) is fulfilled.

We assume that (2.3) holds. From Lemma 2.2 it follows that $\mathfrak{F}(u_{xx})$ can be written in the form (2.4). Taking into account the fact that (see [3], pp. 39 and 44)

$$\frac{\partial \Delta_{A(k)}}{\partial u_{ij}} u_{ij} = k\Delta_{A(k)}, \quad \left(\frac{\partial \Delta_{A(k)}}{\partial u_{ij}}\right)_i = 0, \tag{2.7}$$

we see that the vector $G(u_x, u_{xx})$ solves the problem

$$G^i = \frac{c^{A(k)}}{k} \frac{\partial \Delta_{A(k)}}{\partial u_{ij}} u_j.$$

Thus statement IV is equivalent to statement I and hence to all of the others. This proves Theorem 2.1.

REMARK. The function $\mathcal{F}(u_{xx})$ of Theorem 2.1 is representable in the form (see (1.6))

$$\mathcal{F}(u_{ij}) = \frac{c^{A(k)}}{k} \left(\frac{\partial \Delta_{A(k)}}{\partial u_{ij}} u \right)_{ij},$$

which is symmetric in the indices of differentiation.

§3. Definition of *d*-ellipticity of second-order nonlinear differential operators; limits of its validity

1. From the operator $F(u) = \mathcal{F}(u_{xx})$ we construct the polynomial

$$P_2(u, \xi) = \frac{\partial \mathcal{F}(u_{xx})}{\partial u_{ij}} \xi^i \xi^j, \quad \xi = (\xi^1 \ldots \xi^n).$$

We formulate the well-known definitions of ellipticity in the following form.

DEFINITION 3.1. An operator F is *elliptic on a set* $\mathfrak{M}_F \subset C^2(\Omega)$ if it follows from

$$P_2(u, \xi) = 0 \tag{3.1}$$

that $\xi = 0$ on $\mathfrak{M}_F$. We call $\mathfrak{M}_F$ the *ellipticity set* of F.

DEFINITION 3.2. An operator F is *globally elliptic* if for any $u(x) \in C^2(\Omega)$ it follows from (3.1) that $\xi = 0$.

In the theory of linear equations, not only the behavior of $P_2(\xi)$ but also the form of the operator has decisive significance. For example, the method of integral inequalities can be applied because a linear operator is a divergence operator. Definition 3.1 does not contain information concerning this property. As will be shown below, a globally elliptic operator is a divergence operator only in the linear case.

2. We give the definition of *d*-ellipticity of a nonlinear operator. For this, we introduce the function

$$P(u, \lambda, \xi) = F\left(u + \frac{\lambda}{2} (x^i \xi^i)^2\right) - F(u).$$

DEFINITION 3.3. A second-order differential operator is said to be *d-elliptic on the set* $\mathfrak{M}_F \subset C^2(\Omega)$ if $P(u, \lambda, \xi)$ is a second-degree polynomial in ξ for $\lambda \neq 0$, and $P = 0$ implies that $\xi = 0$ at elements of $\mathfrak{M}_F$ for $\lambda \neq 0$.

LEMMA 3.1. *The function $P(u, \lambda, \xi)$ is a second-order polynomial in ξ for $\lambda \neq 0$ if and only if $F(u)$ is representable in divergence form.*

PROOF. The equality

$$P(u, \lambda, \xi) = \lambda \frac{\partial \mathcal{F}(u_{xx})}{\partial u_{ij}} \xi^i \xi^j - \int_0^\lambda (\tau - \lambda) \frac{\partial^2 \mathcal{F}}{\partial w_{ij} \partial w_{kl}} d\tau \xi^i \xi^j \xi^k \xi^l, \tag{3.2}$$

where $w = u + \frac{1}{2}\tau(x^i\xi^i)^2$, is a consequence of the second-degree Taylor expansion in ξ for $F(u + \frac{1}{2}\lambda(x^i\xi^i)^2)$. Let $\mathfrak{F}$ be of divergence form; then (2.3) is satisfied, and we get from (3.2) that

$$P(u, \lambda, \xi) = \lambda \frac{\partial \mathcal{F}}{\partial u_{ij}} \xi^i \xi^j \doteq \lambda P_2(u, \xi). \tag{3.3}$$

We assume now that $P(u, \lambda, \xi)$ is a second-degree polynomial in ξ. Then

$$\int_0^\lambda (\tau - \lambda) \frac{\partial^2 \mathcal{F}}{\partial w_{ij} \partial w_{kl}} d\tau \xi^i \xi^j \xi^k \xi^l = Q_2(u, \lambda, \xi) = 0.$$

We differentiate $Q_2(u, \lambda, \xi)$ twice with respect to λ and evaluate the result at $\lambda = 0$:

$$\left(\frac{\partial^2 \mathcal{F}}{\partial u_{ij} \partial u_{kl}} + \frac{\partial^2 \mathcal{F}}{\partial u_{ik} \partial u_{lj}} + \frac{\partial^2 \mathcal{F}}{\partial u_{il} \partial u_{jk}}\right) \xi^i \xi^j \xi^k \xi^l = Q_2''(u, 0, \xi).$$

The identity (2.3) follows, and, as was proved in Theorem 2.1, this is equivalent to the representation (2.2).

THEOREM 3.1. *Definitions* 3.1 *and* 3.3 *are equivalent for operators of the form* (2.2).

The proof of this theorem is (3.3).

THEOREM 3.2. *A globally elliptic operator of the form* (2.2) *is linear.*

PROOF. The theorem concerns itself with operators of the form (2.4) for which $\mathfrak{M}_F = C^2(\Omega)$. We calculate $P_2(u, \xi)$ on the set $U^\lambda = \{u^\lambda(x) = \lambda^{ij} x^i x^j / 2, \lambda^{ij} \in R, \lambda^{ij} = \lambda^{ji}\}$:

$$P_2(u^\lambda, \xi) = c^A \frac{\partial \Delta_A}{\partial \lambda^{ij}} \xi^i \xi^j.$$

Let $\xi(i) = (0 \ldots 1 \ldots 0)$; then

$$P_2(u^\lambda, \xi(i)) = c^{A\binom{i}{i}} \frac{\partial \Delta_{A\binom{i}{i}}}{\partial \lambda^{ii}} \neq 0.$$

We choose any coefficient

$$c^{A\binom{k;\, i}{i}} \neq 0, \quad k > 1,$$

and from the matrix A_0 form two sets of numbers λ^{st}:

$$\lambda^{st} = \begin{cases} M, & \binom{s}{t} \in \left\{A_0\binom{i}{i}\right\}, \\ 0, & \binom{s}{t} \notin \left\{A_0\binom{i}{i}\right\}, \end{cases}$$

$$\tilde{\lambda}^{st} = \begin{cases} -M, & \binom{s}{t} = \binom{r}{q}, \\ M, & \binom{s}{t} \in \left\{A_0\binom{i\ r}{i\ q}\right\}, \\ 0, & \binom{s}{t} \notin \left\{A_0\binom{i\ r}{i\ q}\right\}. \end{cases}$$

We calculate

$$P_2(u^{\lambda}, \xi(i)) = c^{A_0\binom{i}{i}} M^{k-1} + \hat{Q}_{k-2}(M),$$
$$P_2(u^{\lambda}, \xi(i)) = -c^{A_0\binom{i}{i}} M^{k-1} + \hat{\hat{Q}}_{k-2}(M).$$

From these equalities it follows that $P_2(u, \xi)$ can have constant sign on $C^2(\Omega)$ only if

$$c^{A\left(k;\,\substack{i\\i}\right)} = 0 \quad \text{for } k > 1.$$

Putting $\xi = \xi(i, j) = (0 \ldots 1 \ldots 1 \ldots 0)$ and arguing as we did for $\xi(i)$, we conclude that the operator (2.4), and consequently (2.2), can be elliptic on $C^2(\Omega)$ if all $c^{A(k)} = 0$ for $k > 1$, i.e. the operator under consideration is linear. This proves the theorem.

3. Let W^- be the set of all functions in $C^2(\Omega)$ which are convex downwards. We describe what restriction the requirement that $W^- \subset \mathfrak{M}_F$ imposes on the structure of a d-elliptic operator F. Equations with such an operator are considered in §5. We introduce some notation (see §1):

$$k^i = \begin{cases} \max k, \\ c^{\bar{A}\left(k;\,\substack{i\\i}\right)} \neq 0, \\ 0, \qquad \text{if } c^{\bar{A}\left(k;\,\substack{i\\i}\right)} = 0 \quad \text{for all } k, \end{cases}$$
$$k_0 = \min_{i \in \{1 \ldots n\}} k^i.$$

THEOREM 3.3. *Let F be a d-elliptic operator on $\mathfrak{M}_F$. Then $W^- \subset \mathfrak{M}_F$ if and only if*

$$c^{\bar{A}(k)} \geqslant 0 \ (\leqslant 0), \ k = 1, \ldots, n, \tag{3.4}$$
$$k_0 \neq 0. \tag{3.4'}$$

PROOF. We calculate $P_2(u, \xi)$ for $u^a(x) = \frac{1}{2} a \Sigma_1^n (x^i)^2$, $a \in (0, \infty)$, and $\xi(i) = (0 \ldots 1 \ldots 0)$:

$$P_2(u^a, \xi(i)) = \sum_{k=1}^{n} \sum_{\bar{A}\left(k;\,\substack{i\\i}\right)} c^{\bar{A}\left(k;\,\substack{i\\i}\right)} a^{k-1}. \tag{3.5}$$

This polynomial will have constant sign for all a only if

$$c^{\bar{A}\left(k;\,\substack{i\\i}\right)} \geqslant 0 \ (\leqslant 0), \quad k^i \neq 0.$$

Since the choice of i in (3.5) is arbitrary and $u^a \in W^- \subset \mathfrak{M}_F$, (3.4) and (3.4′) must hold for d-elliptic operators.

We derive two inequalities for the operators in Theorem 3.3 which will be useful in §5. We assume for definiteness that the coefficients $c^{\bar{A}(k)}$ are nonnegative. Let

$$c_0 = \min_{i \in \{1 \ldots n\}} c^{\bar{A}(k^i)}, \ a \geqslant 1;$$

then

$$\mathcal{F}(u^a_{xx}) = \sum_{k=1}^{n} \sum_{\bar{A}(k)} c^{\bar{A}(k)} a^k \geqslant c_0 a^k. \tag{3.6}$$

and

$$\frac{\partial \mathcal{F}}{\partial u_{ij}}\bigg|_{u=u^a(x)} \xi^i\xi^j = \sum_{k=1}^{n} \sum_{\bar{A}(k)} c^{\bar{A}(k;\,{}^i_i)} a^{k-1} (\xi^i)^2 \geqslant c_0 a^{k_0-1} \sum (\xi^i)^2. \tag{3.7}$$

4. The structure of the operators (2.2) is especially convenient for the application of the method of auxiliary functions. In fact, we have the following result.

LEMMA 3.2. *The value of the operator* (2.1) *at* $u = h(v)$, *where* $u, v \in C^2(\Omega)$ *and* h', $h'' \neq 0$, *is a second-order polynomial in* v_x *if and only if* F *is of divergence type. In addition,*

$$F(h(v)) = \mathcal{F}(h' v_{xx}) + h'' \frac{\partial \mathcal{F}(h' v_{xx})}{\partial h' v_{ij}} v_i v_j. \tag{3.8}$$

PROOF. As in Lemma 1.2, we represent $F(h(v))$ in the form

$$F(h(v)) = \mathcal{F}(h' v_{xx}) + h'' \frac{\partial \mathcal{F}(h' v_{xx})}{\partial h' v_{ij}} v_i v_j - \int_0^{h''} (\tau - h'') \frac{\partial^2 \mathcal{F}}{\partial w_{ij} \partial w_{kl}} d\tau\, v_i v_j v_k v_l. \tag{3.9}$$

Since the differential character of the indices in (2.3) is not essential, (3.8) follows from (3.9) for divergence operators F.

Now we suppose that $F(h(v))$ is a second-order polynomial in v_x. Then $\tilde{Q}_2(h', h'', v_{xx}, v_x) = F(h(v)) - \mathcal{F}(h' v_{xx})$ is also a second-order polynomial in v_x. We consider $\tilde{Q}_2$ on $\{v^\lambda\}$, where $v^\lambda = \lambda^{st}x^s x^t/2 + \lambda^s x^s + \lambda_0$. Let $x_0 \in \Omega$ be arbitrary. The numbers λ^{st}, λ^s, and λ^0 can be chosen so that v_{ij}, v_i, and v, and consequently also $u_{ij} = h' v_{ij} + h'' v_i v_j$, have any previously chosen values at x_0. We differentiate $\tilde{Q}_2(h', h'', v^\lambda_{xx}, v^\lambda_x)|_{x_0}$ twice with respect to $\mu = h''(v^\lambda(x_0))$. We get

$$\left(\frac{\partial^2 \mathcal{F}}{\partial u_{ij} \partial u_{kl}} + \frac{\partial^2 \mathcal{F}}{\partial u_{ik} \partial u_{lj}} + \frac{\partial^2 \mathcal{F}}{\partial u_{il} \partial u_{jk}}\right) v^\lambda_i v^\lambda_j v^\lambda_k v^\lambda_l \Big|_{x_0} = 0. \tag{3.10}$$

Let $\lambda^s = -\lambda^{st} x_0^t$ for $s \neq i, j, k, l$. Then $v_s(x_0) = 0$, and (3.10) is equivalent to (2.3); therefore F is of divergence type. This proves Lemma 3.2.

§4. Some consequences of the structural properties of *d*-elliptic operators

1. We consider the differential equation with a *d*-elliptic operator:

$$F(u) \equiv \operatorname{div} G = f(x, u, u_x), \tag{4.1}$$

$$G^i = \frac{c^{A(k)}}{k} \frac{\partial \Delta_{A(k)}}{\partial u_{ij}} u_j \equiv \bar{A}^{ij} u_j. \tag{4.1'}$$

The ellipticity condition on F can be written in the form

$$\frac{\partial \mathcal{F}}{\partial u_{ij}} \xi^i \xi^j > 0 \quad \text{for } |\xi| \neq 0,\ u(x) \in \mathfrak{M}_r. \tag{4.2}$$

The structure of F allows us to seek generalized solutions of (4.1), i.e. solutions which satisfy the integral identity

$$\int_{\Omega} [\bar{A}^{ij} u_i \eta_j + f(x, u, u_x)\eta]\, dx = 0, \tag{4.3}$$

where $\eta(x)$ is an arbitrary sufficiently smooth function with support in Ω. The set of functions among whose elements the generalized solution must be sought depends first of all on the properties of $\mathfrak{M}_F$. A discussion of these properties lies outside of the scope of this paper. We mention only one result which is useful in this connection.

LEMMA 4.1. *Suppose that both $u(x)$ and $w^\tau = \tau u(x)$, $\tau \in (0, 1]$, belong to $\mathfrak{M}_F$. Then*

$$\bar{A}^{ij}\xi^i\xi^j > 0 \quad for \ |\xi| \neq 0 \tag{4.4}$$

on $\mathfrak{M}_F$.

PROOF. We write the ellipticity condition for W^τ:

$$\frac{\partial \mathcal{F}}{\partial w^\tau_{ij}} \xi^i\xi^j = \tau^{k-1} c^{A(k)} \frac{\partial \Delta_{A(k)}}{\partial u_{ij}} \xi^i\xi^j > 0. \tag{4.5}$$

Integrating (4.5) with respect to τ from 0 to 1, we get (4.4).

In what follows it will be assumed that the generalized solution $u(x)$ of (4.1), (4.2) is such that all of the integrals encountered below exist, and $u(x) \in C^1(\Omega)$.

A maximum principle holds for solutions of (4.3), (4.2) with $f(x, u, u_x) > 0$.

THEOREM 4.1. *Suppose that $u(x) \in \mathfrak{M}_F$ is a generalized solution of* (4.1), (4.2) *and that $f(x, u, u_x) \geqslant 0$. Under the hypotheses of Lemma* 4.1, *$u(x) < m_0$, where $\Omega u(x) < m_0$, $m_0 = \max_{\partial\Omega} u(x)$, everywhere in Ω.*

PROOF. In (4.3) we put $\eta(x) = \max\{u - m; 0\}$, $m > m_0$,

$$\int_{\Omega^m} [\bar{A}^{ij} u_i u_j + f(x, u, u_x)(u - m)]\, dx = 0, \tag{4.6}$$

and $\Omega^m = \{x \in \Omega;\ u(x) > m\}$. But (4.6) is possible only for $\eta(x) \equiv 0$, which proves the theorem.

2. Since $F(u)$ is of divergence type, we can speak of a monotonicity set of F.

DEFINITION 4.1. We call $\mathfrak{N}_F$ a *monotonicity set* of the operator F if for any u, $v \in \mathfrak{N}_F$,

$$\langle F(u) - F(v),\ u - v\rangle|_{\Omega'} = \int_{\Omega'} [G^i(u) - g^i(v)](u - v)_i\, dx \geqslant 0, \tag{4.7}$$

where $\Omega' \subseteq \Omega$ is arbitrary.

LEMMA 4.2. *Let $\tilde{\mathfrak{M}}_F$ be a d-ellipticity set (not necessarily strict) of F. Then F is monotone on every convex subset of $\tilde{\mathfrak{M}}_F$.*

PROOF. We choose arbitrary functions u and v from any convex subset of $\tilde{\mathfrak{M}}_F$. We represent the difference $F(u) - F(v)$ in the form

$$F(u) - F(v) = \int_0^1 \frac{\partial \mathcal{F}}{\partial w^\tau_{ij}}\, d\tau\, (u - v)_{ij},$$

where $w^\tau = \tau u + (1 - \tau)v$, $\tau \in [0, 1]$. From (2.7) it follows that $(\partial\mathcal{F}/\partial w^\tau_{ij})_i = 0$. Therefore

$$F(u) - F(v) = (A^{ij}(u - v)_j)_i,$$

$$A^{ij} = \int_0^1 \frac{\partial\mathcal{F}}{\partial w^\tau_{ij}}\, d\tau.$$

Then from the condition $w^\tau \in \tilde{\mathfrak{M}}_F$, we have that

$$[G^i(u) - G^i(v)](u - v)_i = A^{ij}(u - v)_j(u - v)_i \geqslant 0,$$

which is necessary and sufficient for (4.7).

3. The monotonicity of F allows us to prove comparison theorems for solutions of (4.3). We state three results of this type. The first two will be used in §5, and the third comports with the "classical" comparison theorem for quasilinear elliptic operators in [2].

THEOREM 4.2. *Suppose that $u(x) \in \mathfrak{N}_F$, where $\mathfrak{N}_F$ is a monotonicity set of F and $u(x)$ is a solution of* (4.3), *and $v(x) \in \mathfrak{N}_F \cap C^2(\Omega)$. Assume that*

$$f(x, u, p) = f_1(x, u) + f_2(x, u, p),$$

where f_2 is a continuous function of its arguments, f_1 is monotone, and f_2 is a strictly monotone increasing function of u. If

$$\begin{aligned} F(v) &\geqslant f(x, v, v_x), \\ u - v|_{\partial\Omega} &\geqslant 0, \end{aligned} \tag{4.8}$$

then $(u - v)|_\Omega \geqslant 0$.

REMARK. Theorem 4.2 remains true for the case $f_2(x, u, p) \equiv 0$ if we replace $\mathfrak{N}_F$ by $\tilde{\mathfrak{N}}_F$, a strict monotonicity set of F, in the hypotheses.

THEOREM 4.3. *Suppose that $u(x) \in \mathfrak{N}_F$ is a solution of* (4.3), *that $v(x) \in \mathfrak{N}_F \cap C^2(\Omega)$, and that $(u - v)|_{\partial\Omega} \geqslant 0$. Assume that*

$$|f(x, u, p)| \leqslant g(p), \tag{4.9}$$

where $g(p)$ is continuous, and

$$\mathcal{F}(v_{xx}) > g(v_x). \tag{4.10}$$

Then $u \geqslant v$ everywhere in Ω.

THEOREM 4.4. *Suppose that $u(x) \in \mathfrak{N}_F$ is a solution of* (4.3), *that $v(x) \in \mathfrak{N}_F \cap C^2(\Omega)$, and that $(u - v)|_{\partial\Omega} \geqslant 0$. Suppose that*

$$F(v) > f(x, v - c, v_x)$$

for any $c \geqslant 0$, and that $f(x, u, p)$ is continuous on its domain. Then $u \geqslant v$ everywhere in Ω.

Since the proofs of Theorems 4.2–4.4 are similar, we prove only the first.

PROOF OF THEOREM 4.2. We calculate $\langle F(u) - F(v), \eta^m\rangle$ for $\eta^m = \max\{v - u - m, 0\}$, $m \geqslant m_0 > 0$:

$$\langle F(u)-F(v),\ \eta^m\rangle|_\Omega = -\langle F(u)-F(v),\ u-v\rangle|_{\Omega^m}$$
$$= \int_{\Omega^m} [\mathcal{F}(v_{xx}) - f(x, u, u_x)]\,\eta^m dx, \quad (4.11)$$

where $\Omega^m = \{x \subset \Omega;\ v - u > m\}$. We represent the integral on the right-hand side of (4.11) as a sum of three terms: $(I_1 + I_2 + I_3)\eta^m$; $I_1 = \mathcal{F}(v_{xx}) - f(x, v, v_x)$, $I_2 = f(x, v, v_x) - f(x, u, v_x)$, $I_3 = f(x, u, v_x) - f(x, u, u_x)$. By hypothesis, $I_1 \geqslant 0$. From the continuity and monotonicity of $f(x, u, p)$ it follows that, for fixed $u(x)$, $v(x)$, and m_0, and for all $x \in \Omega$ and all $m \geqslant m_0$, we can find $\varepsilon > 0$ such that $I_2 > \varepsilon$. We turn to I_3. If $\eta^m \not\equiv 0$, the region Ω^m contains a maximum point, x_0, of the difference $(v - u)$, and $v_x(x_0) = u_x(x_0)$. Let $m_1 = \sup_\Omega(v - u)$. We choose m_0 close enough to m_1 so that $|I_3| \leqslant \varepsilon/2$. Then for $m_0 \leqslant m < m_1$ we have

$$\langle F(u)-F(v),\ u-v\rangle|_{\Omega^m} \leqslant -\frac{\varepsilon}{2}\int_{\Omega^m} \eta^m dx,$$

which, by virtue of the monotonicity of F, is impossible if $\eta^m \not\equiv 0$. Consequently $(v - u) \leqslant m_0$ everywhere in Ω. Since m_0 can be chosen arbitrarily close to 0, $(u - v)|_\Omega \leqslant 0$. This proves the theorem.

§5. Uniqueness theorems and a priori estimates for a solution of the Dirichlet problem and its normal derivative for an equation with d-elliptic operator

1. We consider the Dirichlet problem for an equation with d-elliptic operator

$$F(u) = f(x, u, u_x), \quad (5.1)$$

$$u|_{\partial\Omega} = \varphi(x). \quad (5.2)$$

in a bounded region $\Omega \subset R^n$.

As was shown in §4.1, the problem (5.1), (5.2) admits a generalized formulation, and, in what follows, when we speak of its solutions, we will always mean generalized solutions.

THEOREM 5.1. *Suppose that*

$$f(x, u, u_x) = f_1(x, u) + f_2(x, u, u_x),$$

where $f_1(x, u)$ is a monotone and $f_2(x, u, p)$ a strictly monotone increasing function of u, and $f_2(x, u, p)$ is continuous on the set $\{x \in \Omega, |u| + |p| < \infty\}$. Then the problem (5.1), (5.2) *has not more than one solution $u(x) \in \mathfrak{N}_F \cap C^1(\Omega)$ which is continuous up to the boundary, where $\mathfrak{N}_F$ is a monotonicity set of F.*

PROOF. Let $u_1(x)$, $u_2(x) \in \mathfrak{N}_F \cap C^1(\Omega)$ be solutions of (5.1), (5.2) which are continuous up to the boundary. The hypotheses of Theorem 4.2 are satisfied for these two functions, and they are invariant under replacement of u_1 by u_2. Therefore the inequalities $(u_1 - u_2)|_\Omega \leqslant 0$ and $(u_2 - u_1)|_\Omega \geqslant 0$ hold simultaneously, and consequently $(u_1 - u_2)|_\Omega = 0$. This proves the theorem.

A consequence of the remark after Theorem 4.2 is still another uniqueness theorem.

THEOREM 5.1′. *Let $f(x, u, p) \equiv \tilde{f}(x, u)$, and suppose that $\tilde{f}(x, u)$ is a nondecreasing function of u. Then the problem* (5.1), (5.2) *has no more than one solution $u(x) \in \tilde{\mathfrak{N}}_F \cap C^1(\Omega)$ which is continuous up to the boundary, where $\tilde{\mathfrak{N}}_F$ is a strict monotonicity set of F.*

2. In applications of comparison theorems, it is necessary to have solutions of the inequalities in these theorems, called *barrier functions*. Following the method of constructing barriers for equations with quasilinear elliptic operators, we prove an auxiliary result.

LEMMA 5.1. *The function $h(t, h_0, h_1)$ given parametrically by*

$$h(t, h_0, h_1) = \int_0^\rho \frac{(\rho + h_1)}{\omega(\rho)}\, d\rho + h_0, \quad h_1 \geqslant 0, \tag{5.3}$$

$$\sqrt{t} = r = \int_0^\rho \frac{d\rho}{\omega(\rho)} + R = \left[k_0 \int_0^\rho \frac{(\rho + h_1)^{k_0 - 1}\, d\rho}{\Psi(\rho + h_1)} + R^{k_0}\right]^{1/k_0}, \tag{5.4}$$

$$\omega(\rho) = \Psi(\rho + h_1)\left\{\frac{1}{(\rho + h_1)^{k_0}}\left[k_0 \int_0^\rho \frac{(\rho + h_1)^{k_0 - 1}}{\Psi(\rho + h_1)}\, d\rho + R^{k_0}\right]\right\}^{(k_0 - 1)/k_0}, \tag{5.5}$$

$\rho \in [0, \infty)$, is the general solution of the equation

$$(2h')^{k_0 - 1}(2h''t + h') = \frac{1}{2}\Psi(2h'\sqrt{t}), \tag{5.6}$$

where $h' \geqslant \frac{1}{2}$, $h'' \geqslant 0$ for $t \geqslant R^2$ if $\Psi(\rho) \geqslant 1$ is a nondecreasing function, and $h_1 \Psi^{-1/k_0}(h_1) \leqslant R \leqslant h_1$.

PROOF. We put $r = \sqrt{t}$. We calculate

$$h'_t = h'_r/2\sqrt{t} = (\rho + h_1)/2r, \quad h''_t = 1/4t\,(h''_r - h'_r/\sqrt{t}) = (1/4r^2)(\omega(\rho) - (\rho + h_1)/r),$$

and for the functions (5.3)–(5.5) equation (5.6) becomes an identity:

$$(\rho + h_1)^{k_0 - 1}\omega(\rho) = \Psi(\rho + h_1)\, r^{k_0 - 1}.$$

The inequality $h''_t \geqslant 0$ is equivalent to $\Psi(\rho + h_1) r^{k_0} \geqslant (\rho + h_1)^{k_0}$, which is satisfied if $R \geqslant h_1/\Psi(h_1)^{1/k_0}$. Indeed, for nondecreasing functions Ψ we have

$$\Psi(\rho + h_1)\, r^{k_0} = \Psi(\rho + h_1)\left[\frac{1}{k_0}\int_0^\rho \frac{(\rho + h_1)^{k_0 - 1}}{\Psi(\rho + h_1)}\, d\rho + R^{k_0}\right]$$
$$\geqslant (\rho + h_1)^{k_0} + \Psi(h_1) R^{k_0} - h_1^{k_0} \geqslant (\rho + h_1)^{k_0}.$$

The condition $h'_t \geqslant \frac{1}{2}$ is equivalent to the inequality $(\rho + h_1)^{k_0} \geqslant r^{k_0}$, which is true for $R \leqslant h_1$, $\Psi(h_1) \geqslant 1$, which in turn follows from the estimate

$$r^{k_0} = \frac{1}{k_0}\int_0^\rho \frac{(\rho + h_1)^{k_0 - 1}\, d\rho}{\Psi(\rho + h_1)} + R^{k_0} \leqslant (\rho + h_1)^{k_0} + R^{k_0} - h_1^{k_0} \leqslant (\rho + h_1)^{k_0}.$$

This proves Lemma 5.1.

We address the problem of finding a solution $v(x)$ of the inequality

$$F(v) > \psi(|v_x|). \tag{5.7}$$

with d-elliptic operator. To solve this problem effectively, it is necessary to find a concrete set in $\mathfrak{M}_F$ from whose elements it is possible to choose the barrier functions $v(x)$. In what follows, this set will be W^-; more precisely, we require that

$$\mathfrak{M}_F \cap \mathfrak{N}_F \cap W^- = W^-, \tag{5.8}$$

where $\mathfrak{M}_F$ and $\mathfrak{N}_F$ are d-ellipticity and monotonicity sets, respectively, and W^- is a set of functions which are convex from below. This requirement makes lower barriers available in general, barriers which realize the inequality $v \leqslant u$. We recall that (5.8) imposes additional restrictions on the structure of F (see Theorem 3.3). In particular, the numerical characteristics $c_0 > 0$ and $k_0 = [k_0] \geqslant 1$ occur.

LEMMA 5.2. *Let k_0 and c_0 be the characteristics of the d-elliptic operator F for which* (5.8) *holds, and suppose that $\psi(|p|) \geqslant c_0$ is a nondecreasing function. Choose $R_1 \geqslant 0$ and $R_2 > 0$ so that*

$$R_2^{k_0} - R_1^{k_0} < \frac{k_0 c_0}{1+\varepsilon} \int_{R_1}^{\infty} \frac{\rho^{k_0-1} d\rho}{\psi(\rho)}. \tag{5.9}$$

Then the function $v(x) = h(t, h_0, R_1)$, $t = r^2$, $R_1 \leqslant r \leqslant R_2$, obtained from (5.3)–(5.5) *with $\Psi = (1+\varepsilon)\psi(\rho)/c_0$, and $\varepsilon > 0$ an arbitrary number, is a solution in W^- of* (5.7) *in the spherical shell $B_{R_2} \setminus B_{R_1}$.*

PROOF. We calculate the value of F at the function v (see (3.8)):

$$F(v) = \mathcal{F}(h' t_{xx}) + h'' \frac{\partial \mathcal{F}(h' t_{xx})}{\partial h' t_{ij}}.$$

Since $v(x) \in W^-$ and $h'_t \geqslant \frac{1}{2}$ and $h'' \geqslant 0$, (3.6) and (3.7) hold with $a = 2h'_t$, and

$$F(v) \geqslant 2c_0 (2h')^{k_0-1} (2h'' t + h').$$

On the other hand, $\psi(|v_x|) = \psi(2h'\sqrt{t})$, and (5.7) will be satisfied if

$$(2h')^{k_0-1}(2h''t + h') > \frac{1}{2c_0} \psi(2h'\sqrt{t}).$$

From Lemma 5.1 it follows that $h(t, h_0, R_1)$, constructed from $\Psi = (1+\varepsilon)\psi(\rho)/c_0$, satisfies this last inequality for all values of h_0.

3. We derive a priori estimates for a generalized solution of (5.1), (5.2) and its normal derivative. Since we have only lower barriers, the estimates are one-sided.

THEOREM 5.2. *Suppose that $u(x) \in \mathfrak{N}_F \cap C^1(\Omega)$ is a solution of* (5.1), (5.2), (5.8) *and that c_0 and k_0 are characteristics of F. Assume that*

$$f(x, u, p) \leqslant \psi(|p|), \tag{5.10}$$

for $u > -M_0$, $-M_0 \leqslant \min_{\partial\Omega} \varphi(x)$, where

$$\psi(|p_2|) - \psi(|p_1|) \geqslant 0, \quad \textit{if } |p_2| \geqslant |p_1|, \tag{5.11}$$

$$\psi(0) = c_0. \tag{5.11'}$$

If Ω can be imbedded in the spherical shell determined by (5.9) *for any $\varepsilon > 0$, then*

$$u \geqslant -\int_0^{\rho_0} \frac{(\rho + R_1)}{\omega(\rho)} d\rho - M_0, \tag{5.12}$$

where ρ_0 is a root of the equation

$$R_2^{k_0} - R_1^{k_0} = \frac{k_0 c_0}{1+\varepsilon} \int_0^{\rho_0} \frac{\rho^{k_0-1}}{\psi(\rho + R_1)}, \tag{5.13}$$

and $\omega(\rho)$ is given by (5.5) *with $h_1 = R = R_1$ and $\psi = (1 + \varepsilon)\psi/c_0$.*

PROOF. The condition (5.9) is necessary and sufficient for the solvability of (5.13). Let $v(x) = h(t, h_0, R_1)$ be the function in Lemma 5.2 with

$$h_0 = -\int_0^{\rho_0} \frac{\rho + R_1}{\omega(\rho)} d\rho - M_0.$$

Then u and v satisfy the hypotheses of Theorem 4.3 on the set $\Omega^{M_0} = \{x \in \Omega;\ u > -M_0\}$ with $g(p) = \psi(|p|)/c_0$, and $v \leqslant u$ in Ω^{M_0}; hence (5.12) follows.

REMARK. If $\psi(|p|) > 0$, then (5.11′) is a normalization of (5.1) with respect to similarity transformations in R^n.

THEOREM 5.3. *Let $u(x) \in \mathfrak{N}_F \cap C^1(\Omega)$ be a solution of* (5.1), (5.2), (5.8), (5.10), (5.11) *in $\Omega \subset R^n$ which is continuous up to the boundary. Assume that the following conditions hold*:

1) *For $x_0 \in \partial\Omega$ there exists a sphere B_R such that $\Omega \subset B_R$ and $x_0 \in \partial B_R$.*

2) *The hypersurface determined by the function $\varphi(x)$ lies "above" the plane $S(x) = u(x_0) + (s, x - x_0)$, $s = (s^1 \ldots s^n)$.*

3) $$R < \frac{c_0 k_0}{1+\varepsilon} \int_0^{\infty} \frac{\rho^{k_0-1} d\rho}{\psi(\rho + |s|)} = R_0.$$

Then

$$\frac{\partial u}{\partial n} < \rho_0 + |s|, \tag{5.14}$$

where ρ_0 is a root of the equation

$$R = \frac{c_0 k_0}{1+\varepsilon} \int_0^{\rho_0} \frac{\rho^{k_0-1} d\rho}{\psi(\rho + |s|)},$$

and n is the outward-directed normal to $\partial\Omega$.

PROOF. The function $\tilde{u}(x) = u(x) - S(x)$ is a solution of (5.1) with $f(x, \tilde{u} + s, \tilde{u}_x + s) \leqslant \psi(|\tilde{u}_x| + |s|)$, where $\tilde{u}|_{\partial\Omega} \geqslant 0$ and $\tilde{u}(x_0) = 0$. Let $v(x) = h(r^2, h_0, 0)$ be the function in Lemma 5.2 constructed from $\psi(|p| + |s|)$, where $h_0 = -\int_0^{\rho_0}(\rho/\omega(\rho))\, d\rho$. Then $v|_{\partial\Omega} \leqslant 0$, and from Theorem 4.3 it follows that $(u - v)|_{\partial\Omega} \leqslant 0$. Since $\tilde{u}(x_0) = v(x_0)$, (5.14) holds.

To construct an estimate for u_n in the case $R > R_0$, we need a solution (ρ, r) of the system

$$M = \int_0^{\rho} \frac{t+r}{\omega(t, r)} dt, \tag{5.15}$$

$$R^{k_0} - r^{k_0} + \frac{k_0 c_0}{1+\varepsilon} \int_r^{\rho+r} \frac{t^{k_0-1} dt}{\psi(t+|s|)}. \tag{5.16}$$

We deduce a necessary and sufficient condition for the solvability of this system for $R > R_0$.

LEMMA 5.3. *The system* (5.15), (5.16) *has a solution* (ρ, r), $0 < \rho < \infty$, $r_0 < r < R$, $R > R_0$, *if and only if*

$$M \in [0, M_R), \quad M_R = \int_0^{\infty} \frac{(t+r_0)\, dt}{\omega(t, r_0)},$$

where r_0 is a root of the equation

$$R^{k_0} = r_0^{k_0} + \frac{c_0 k_0}{1+\varepsilon} \int_0^{\infty} \frac{t^{k_0-1} dt}{\psi(t+|s|)},$$

and ω is given by (5.5) *with* $\Psi(\rho) = ((1+\varepsilon)/c_0)\psi(\rho + |s|) > 1$.

PROOF. We choose any $R \in [R_0, \infty)$. Then (5.16) determines a function $r = r(\rho)$, and $0 < -r' \leq 1$, $r \in (r_0, R)$. The inequality $M'(\rho, r(\rho)) > -r'r/\psi(r) > 0$ and the fact that the values of $M(\rho, r(\rho))$ fill out the interval $[0, M_R)$ for $0 < \rho < \infty$ prove Lemma 5.3.

THEOREM 5.4. *Suppose that the hypotheses of Theorem* 5.3, *except for* 3), *are fulfilled, and that*

3′) $R > R_0$, $\min_\Omega u(x) > -(M + |u(x_0)|) > -(M_R + |u(x_0)|)$.

Then

$$\frac{\partial u}{\partial n} \leq \rho_1 + r_1 + |s|, \tag{5.17}$$

where (ρ_1, r_1) is a solution of (5.15), (5.16).

PROOF. Let $\tilde{u}(x) = u(x) - S(x)$, let $v(x)$ be the function in Lemma 5.2, constructed from

$$\psi(|p| + |s|), \; h_0 = -\int_0^{\rho_1} (t + r_1)\, dt/\omega(t, r_1),$$

let (ρ_1, r_1) be a solution of the system (5.15), (5.16), and let $\Omega' = \Omega \cap B_R \setminus B_{r_1}$. Since

$$\tilde{u}|_{\partial\Omega} \geq 0, \quad \tilde{u}|_\Omega \geq -M, \quad v|_{\partial\Omega \cap B_R \setminus B_{r_1}} \leq 0, \quad v|_{\Omega \cap \partial B_{r_1}} = -M,$$

it follows that the hypotheses of Theorem 4.3 are satisfied for $\tilde{u}$, v in Ω', and $(v - \tilde{u})|_{\Omega'} \leq 0$. Then (5.17) follows from the fact that $\tilde{u}(x_0) = v(x_0)$. This proves the theorem.

COROLLARY 5.1. *The assumption* 3′) *of Theorem* 5.4 *is necessary for the existence of the a priori estimate of u_n.*

PROOF. The function $v(x)$ of Theorem 5.4 with $|s| = 0$ is a solution of the Dirichlet problem for the equation of the type under consideration

$$F(v) \equiv \sum_{\bar{I}(k_0)} \Delta_{\bar{I}(k_0)} = \frac{1+\varepsilon}{c_0} \psi(|v_x|)$$

in the spherical shell $B_R \setminus B_{r_1}$. It is clear that $v_n|_{\partial B_R} = \rho_1 + r_1 \uparrow \infty$, if

$$\min_{B_R \setminus B_{r_1}} v(x) = -M = \int_0^{\rho_1} \frac{(t+r_1)\,dt}{\omega(t, r_1)} \to M_R = \int_0^{\infty} \frac{(t+r_0)\,dt}{\omega(t, r_0)}.$$

COROLLARY 5.2. *Suppose that B_{R_2} is formed for $x_0 \in \partial\Omega$ just as in Theorems 5.3 and 5.4. Suppose that all the hypotheses of Theorem 5.3 except 3) are satisfied, and that*

3″) $$R_2 \geqslant R_0, \quad R_1 < r_0, \quad \Omega \subset B_{R_2} \setminus B_{R_1}.$$

Then

$$\left.\frac{\partial u}{\partial n}\right|_{x_0} = \rho_1(R_1) + R_1 + |s|, \tag{5.18}$$

where $\rho_1(R_1)$ is determined by (5.13).

PROOF. The hypotheses of Theorem 5.2 with $M_0 = 0$ are satisfied for $\tilde{u}(x) = u(x) - S(x)$, and therefore

$$\min \tilde{u} \geqslant -\left[\int_0^{\rho_1(R_1)} \frac{t+R_1}{\omega(t, R_1)}\,dt + |u(x_0)|\right] > -M_{R_1} + |u(x_0)|).$$

But then (5.18) follows from Theorem 5.4.

BIBLIOGRAPHY

1. O. A. Ladyženskaja and N. N. Ural'ceva, *Linear and quasilinear elliptic equations*, 2nd ed., "Nauka", Moscow, 1974; English transl. of 1st ed., Academic Press, 1968.

2. J. Serrin, *The problem of Dirichlet for quasilinear elliptic differential equations with many independent variables*, Philos. Trans. Royal Soc. London Ser. A **264** (1969), 413–496.

3. N. M. Ivočkina, *On the possibility of integral formulas in R^n*, Zap. Naučn. Sem. Leningrad. Otdel. Mat. Inst. Steklov. (LOMI) **52** (1975), 35–51; English transl. in J. Soviet Math. **9** (1978), no. 5.

4. Hanno Rund, *Integral formulae associated with the Euler-Lagrange operators of multiple integral problems in the calculus of variations*, Aequationes Math. **11** (1974), 212–229.

5. John M. Ball, *Convexity conditions and existence theorems in nonlinear elasticity*, Arch. Rational Mech. Anal. **63** (1976/7), 337–403.

6. V. V. Lyčagin, *Nonlinear differential equations and contact geometry*, Dokl. Akad. Nauk SSSR **238** (1978), 273–276; English transl. in Soviet Math. Dokl. **19** (1978).

Translated by H. T. JONES

Труды Матем. Инст. Стеклов.
147(1980)

Proceedings of the Steklov
Institute of Mathematics
1981, Issue 2

ASYMPTOTICS OF SOLUTIONS OF A CLASS OF DIFFERENTIAL EQUATIONS WITH A SMALL PARAMETER MULTIPLYING THE DERIVATIVE IN THE CASE OF CHANGE OF STABILITY OF A FOCUS

UDC 517.91

S. KARIMOV

ABSTRACT. The asymptotics of solutions of a system of two ordinary differential equations of first order, containing a small parameter multiplying the derivative, are constructed in the case of change of stability of a focus.

Bibliography: 4 titles.

We consider the system of differential equations

$$\varepsilon \dot{x}_1 = (a_1 y + b_1)(x_1 - y) + (a_2 y + b_2) x_2 + \tilde{F}_1(y, x_1, x_2), \tag{1}$$

$$\varepsilon \dot{x}_2 = (a_3 y + b_3)(x_1 - y) + (a_4 y + b_4) x_2 + \tilde{F}_2(y, x_1, x_2),$$

$$\dot{y} = 1, \tag{2}$$

where ε is a small positive parameter, a_k and b_k ($k = 1, 2, 3, 4$) are given real numbers, and $\tilde{F}_k(y, x_1, x_2) = f_k(y, t) g_k(y, x_1, x_2)$ ($k = 1, 2$). The functions $f_k(y, t)$ and $g_k(y, x_1, x_2)$ ($k = 1, 2$) are defined and continuous together with their partial derivatives of first order in a region

$$\Delta_0 = \{|y + \rho| \leqslant t_0,\ |x_1 - y| \leqslant \delta_1,\ |x_2| \leqslant \delta_2\}.$$

In addition, $a_1 + a_4 > 0$ and

$$f_k(y, 0) \equiv g_k(y, y, 0) \equiv 0 \quad (k = 1, 2);$$

δ_1 and δ_2 are positive numbers, t_0 is a nonnegative number, which is defined below, and $\rho = (b_1 + b_4)/(a_1 + a_4)$.

We introduce the following notation:

$$a = \frac{a_1 + a_4}{2}, \quad b = \frac{b_1 + b_4}{2},$$

$$\alpha_{11} = (a_1 - a_4)^2 + 4a_2 a_3,$$

$$\alpha_{12} = 2(a_1 - a_4)(b_1 - b_4) + 4(a_3 b_2 + a_2 b_3),$$

$$\alpha_{22} = (b_1 - b_4)^2 + 4 b_2 b_3,$$

$$\Delta = \alpha_{12}^2 - 4\alpha_{11}\alpha_{22},$$

$$\beta_0 = \frac{\alpha_{12}}{2\alpha_{11}} \ (\alpha_{11} \neq 0), \quad d_1 = c(\beta_0 - \rho),$$

$$c = \frac{1}{2}\sqrt{-\alpha_{11}}, \quad p = a + ic, \quad q = id_1.$$

1980 *Mathematics Subject Classification.* Primary 34E15.

By $\sqrt{a}$ $(a \geqslant 0)$ we shall mean the numerical value of the square root of the number a.

Let $\Delta = 0$, $(a_1 - a)/c_1 = a_2/c_2 = c/d_1$, and $\alpha_{11} < 0$. Then it can be shown that the focus $x_1 = y$, $x_2 = 0$ of the system (1) is stable if $y < -\rho$ and unstable if $y > -\rho$, i.e., as y passes through $-\rho$, the focus becomes unstable. We introduce new unknown functions $x_1 - y = \xi_1$ and $x_2 = \xi_2$, and take solutions of (2) in the form $y = t - \rho$. Then in place of (1), (2) we get the system

$$\varepsilon\dot{\xi}_1 = (a_1 t + c_1)\xi_1 + (a_2 t + c_2)\xi_2 + F_1(t, \xi_1, \xi_2) - \varepsilon,$$
$$\varepsilon\dot{\xi}_2 = (a_3 t + c_3)\xi_1 + (a_4 t + c_4)\xi_2 + F_2(t, \xi_1, \xi_2), \tag{3}$$

where

$$F_k(t, \xi_1, \xi_2) \equiv \tilde{F}_k(t - \rho, t - \rho + \xi_1, \xi_2) \ (k = 1, 2),$$
$$c_k = b_k - \rho a_k \quad (k = \overline{1, 4}).$$

Let $\xi(t) = \binom{\xi_1(t)}{\xi_2(t)}$ be a solution of (3) which satisfies an initial condition such that

$$\|\xi(-t_0)\| = O(\varepsilon). \tag{4}$$

In [1] a concrete example of the system (1) was considered. For a solution of this example with initial condition (4) an asymptotic estimate was obtained with an arbitrary degree of precision with respect to the small parameter ε. In [2] the system (1) was studied in the case that $a_1 = a_4$, $b_1 = b_4$, $a_3 = -a_2$ and $b_3 = -b_2$. For a solution of this system with initial condition (4) an asymptotic estimate was found with error $O(\varepsilon^{2-3\lambda})$ $(1/6 < \lambda < 1/3)$ for $-t_0 + \sqrt{\varepsilon} \leqslant t \leqslant t_0 - \varepsilon^{\tilde{\lambda}}$ $(1/2 \leqslant \tilde{\lambda} < 1)$.

In the present paper we consider solutions of (3) with initial condition (4), and obtain for these solutions an asymptotic estimate with an arbitrary degree of precision on the interval $[-t_0, t_0]$. The construction of this approximate solution is carried out in the complex plane. The principal difficulty in the study of such systems is that their solutions do not have the usual power asymptotics [3], [4].

We replace the differential equations (3) by the integral equations

$$v = f_1(t) + \frac{1}{\varepsilon}\int_{-t_0}^{t} E_1(t, \tau)\, Q_1(\tau, v(\tau), w(\tau))\, d\tau,$$

$$w = g_1(t) + \frac{1}{\varepsilon}\int_{-t_0}^{t} E_2(t, \tau)\, Q_2(\tau, v(\tau), w(\tau))\, d\tau, \tag{5}$$

where

$$t_0 = \frac{|d_1|}{\sqrt{a^2 - c^2}}, \quad v = -\xi_2 + \bar{\gamma}_1\xi_1, \quad w = \xi_2 - \gamma_1\xi_1,$$

$$Q_1 = -F_2 + \bar{\gamma}_1 F_1, \quad Q_2 = F_2 - \gamma_1 F_1, \quad \gamma_1 = \frac{-c_1 + id_1}{c_2},$$

$$E_1(t, \tau) = e^{m_1(t) - m_1(\tau)}, \quad E_2(t, \tau) = e^{m_2(t) - m_2(\tau)},$$

$$m_1(t) = \frac{(pt + q)^2}{2\varepsilon p}, \quad m_2(t) = \frac{(\bar{p}t + \bar{q})^2}{2\varepsilon \bar{p}}, \quad p = a + ic, \quad q = id_1,$$

$$f_1(t) \stackrel{\varepsilon}{=} v(-t_0)\, E_1(t, -t_0) - \int_{-t_0}^{t} E_1(t, \tau)\, d\tau, \tag{a_1}$$

$$g_1(t)=w(-t_0)E_2(t,\ -t_0)+\int_{-t_0}^{t} E_2(t,\ \tau)\,d\tau. \qquad (b_1)$$

We consider the functions

$$f_n(t)=\frac{1}{\varepsilon}\int_{-t_0}^{t} P_{n-1}(\tau)E_1(t,\ \tau)\,d\tau, \qquad (a_n)$$

$$g_n(t)=\frac{1}{\varepsilon}\int_{-t_0}^{t} S_{n-1}(\tau)E_2(t,\ \tau)\,d\tau\ (n=2,\ 3,\ \ldots), \qquad (b_n)$$

where

$$P_n(t)=Q_1(t,\ V_n(t),\ W_n(t))-Q_1(t,\ V_{n-1}(t),\ W_{n-1}(t)),$$
$$S_n(t)=Q_2(t,\ V_n(t),\ W_n(t))-Q_2(t,\ V_{n-1}(t),\ W_{n-1}(t)),$$
$$V_n(t)=f_0(t)+f_1(t)+\ \ldots\ +f_n(t),\ \ W_n(t)=g_0(t)+g_1(t)+\ \ldots\ +g_n(t),$$
$$f_0(t)\equiv 0,\ \ g_0(t)\equiv 0.$$

Let $v = V_n(t) + v^{(n)}$ and $w = W_n(t) + w^{(n)}$, where $v^{(n)}$ and $w^{(n)}$ $(n = 1, 2, \ldots)$ are new unknown functions. We then obtain from (5) the system

$$v^{(n)}(t)=\frac{1}{\varepsilon}\int_{-t_0}^{t} B_n(v^{(n)}(\tau),\ w^{(n)}(\tau))E_1(t,\ \tau)\,d\tau,$$

$$w^{(n)}(t)=\frac{1}{\varepsilon}\int_{-t_0}^{t} K_n(v^{(n)}(\tau),\ w^{(n)}(\tau))E_2(t,\ \tau)\,d\tau, \qquad (d_n)$$

where

$$B_n=Q_1(t,\ V_n(t)+v^{(n)},\ \ W_n(t)+w^{(n)})-Q_1(t,\ V_{n-1}(t),\ W_{n-1}(t)),$$
$$K_n=Q_2(t,\ V_n(t)+v^{(n)},\ \ W_n(t)+w^{(n)})-Q_2(t,\ V_{n-1}(t),\ W_{n-1}(t)).$$

We solve (d_n) by the method of successive approximations. Let $v_0^{(n)}(t) \equiv 0$ and $w_0^{(n)}(t) \equiv 0$; the $(k + 1)$st approximation is defined in terms of the kth approximation by the following formulas:

$$v_{k+1}^{(n)}(t)=\frac{1}{\varepsilon}\int_{-t_0}^{t} B_n(v_k^{(n)}(\tau),\ \ w_k^{(n)}(\tau))E_1(t,\ \tau)\,d\tau,$$

$$w_{k+1}^{(n)}(t)=\frac{1}{\varepsilon}\int_{-t_0}^{t} K_n(v_k^{(n)}(\tau),\ w_k^{(n)}(\tau))E_2(t,\ \tau)\,d\tau,$$

$$(n=1,\ 2,\ \ldots;\ \ k=0,\ 1,\ 2,\ \ldots). \qquad (b_k^{(n)})$$

We shall consider the functions $f_n(t)$, $g_n(t)$, $v_k^{(n)}(t)$ and $w_k^{(n)}(t)$ $(n = 1, 2, \ldots; k = 0, 1, 2, \ldots)$ for all values of t in an octagon H_0, which is defined below in the plane of the complex variable t. Let $t = t_1 + it_2$ and $\tau = \tau_1 + i\tau_2$ where t_1, t_2, τ_1 and τ_2 are real variables.

We introduce the new variables

$$\tilde{\varphi} = h(t_1, t_2) \equiv t_1 + \left(\frac{1}{\alpha} - \frac{c}{a}\right)t_2 + t_0,$$

$$\tilde{\psi} = S(t_1, t_2) \equiv -t_1 + \left(\frac{1}{a} + \frac{c}{d}\right)t_2 + t_0,$$

$$\varphi = h(\tau_1, \tau_2), \quad \psi = S(\tau_1, \tau_2).$$

Here and later

$$\frac{a}{\sqrt{a^2 + c^2}} = \alpha, \quad \frac{c}{\sqrt{a^2 + c^2}} = \beta,$$

$$\beta_1 = 1 - \beta, \quad \beta_2 = 1 + \beta, \quad a_0 = 2t_0, \quad \gamma = \frac{1}{1 + \beta},$$

$$\delta_0 = \sqrt{\varepsilon}, \quad \delta = \varepsilon^{\lambda}\left(\frac{1}{2} \leqslant \lambda < 1\right).$$

The vertices of the octagon H_0 in the new variables have the following coordinates:

$$\mathscr{A}(0, a_0), \; B_2(\delta_0, 0), \; B_3(0, \delta_0), \; B_4(\delta_0, \delta_0),$$

$$C(a_0, 0), \; \mathscr{D}_2(a_0\beta_1 + \beta\delta_0, \beta_2(a_0 - \delta_0)),$$

$$\mathscr{D}_3(\beta_1(a_0 - \delta_0), \; a_0\beta_2 - \beta\delta_0), \; \mathscr{D}(\beta_1(a_0 - \delta_0) + \beta\delta_0, \; \beta_2(a_0 - \delta_0) - \beta\delta_0).$$

We also consider the points

$$\mathscr{A}_1(0, b_0), \; \mathscr{A}_2(\beta_1(a_0 - b_0), \; a_0\beta_2 - \beta b_0), \; B_1(\delta_0, b_0),$$

$$C_1(a_0 - b_0, b_0), \; C_2(a_0 - \gamma b_0, 0), \; C_3(a_0 - \beta\gamma b_0, b_0),$$

$$\mathscr{D}_1(\beta_1(a_0 - b_0) + \beta\delta_0, \; \beta_2(a_0 - \delta_0) - \beta\delta_0), \; E(a_0 - \delta, \delta),$$

$$E_1(b_0 - \beta\gamma b_0 - \gamma\delta, \; b_0), \; E_2(a_0 - \gamma b_0 - \beta\delta\gamma, \; \delta_0),$$

$$E_3(a_0 - \beta\gamma\delta, \delta), \; E_4(a_0 - \gamma\delta, 0),$$

where $0 < b_0 < a_0$. In what follows we agree to denote a quadrilateral with vertices at the points $\mathscr{A}, B, C, \mathscr{D}$ by $(\mathscr{A}BC\mathscr{D})$.

We introduce the following quadrilaterals:

$$K = (\mathscr{A}\mathscr{A}_1C_1\mathscr{A}_2), \quad H_2 = (\mathscr{A}_2\mathscr{D}_1\mathscr{D}_4\mathscr{D}_3), \quad P_2 = (\mathscr{D}_1C_1C_3\mathscr{D}_2),$$

$$H_1 = (\mathscr{A}_1B_3B_4B_1), \quad P_1 = (B_1B_2C_2C_1), \quad M_1 = (E_2C_2E_4E),$$

$$K_2 = (EE_4CE_3), \quad M_2 = (E_1EE_3C_3), \quad K_1 = (C_1E_2EE_1).$$

Then $H_0 = N_1 \cup N_2$, where $N_1 = K \cup K_1 \cup H_2 \cup P_2 \cup M_2$ and $N_2 = K_2 \cup H_1 \cup P_1 \cup M_1$. We further rewrite the equality (a_1) in the coordinates $(\tilde{\varphi}, \tilde{\psi})$. We then obtain

$$|f_1(\tilde{\varphi}, \tilde{\psi})| \leqslant O(\varepsilon)\, l^{-\alpha\tilde{\varphi}\tilde{\psi}/2\varepsilon} + \left|\int_{l_1^*} l^{\alpha(\varphi\psi - \tilde{\varphi}\tilde{\psi})/2\varepsilon}\, dl\right|, \qquad (\mathscr{A}_1)$$

where the path of integration l_1^* is defined as follows: if $(\tilde{\varphi}, \tilde{\psi}) \in K \cup H_1 \cup H_2 \cup P_2$, l_1^* is a broken-line path and consists of straight line segments which connect in sequence the points $(0, a_0)$, $(-b_0, a_0)$, $(-b_0, \tilde{\psi})$ and $(\tilde{\varphi}, \tilde{\psi})$; if $(\tilde{\varphi}, \tilde{\psi}) \in K_1 \cup K_2 \cup P_1 \cup M_1 \cup M_2$, l_1^* consists of straight line segments which connect in sequence the points $(0, a_0)$, $(-b_0, a_0)$, $(-b_0, \delta_0)$, (δ_0, δ_0), $(\delta_0, -b_0)$, $(\tilde{\varphi}, -b_0)$ and $(\tilde{\varphi}, \tilde{\psi})$.

We have that

$$|f_1(\tilde{\varphi}, \tilde{\psi})| \leqslant C_1 \omega_1(\tilde{\varphi}, \tilde{\psi}), \tag{α_1}$$

where C_1 is a positive constant and

$$\omega_1(\tilde{\varphi}, \tilde{\psi}) = \begin{cases} \varepsilon & \text{if } (\tilde{\varphi}, \tilde{\psi}) \in N_1, \\ \sqrt{\varepsilon} & \text{if } (\tilde{\varphi}, \tilde{\psi}) \in N_2. \end{cases}$$

Similarly, if we rewrite (b_1) in the coordinates $(\tilde{\varphi}, \tilde{\psi})$, we obtain

$$|g_1(\tilde{\varphi}, \psi)| \leqslant O(\varepsilon) e^{a\tilde{\varphi}^*\tilde{\psi}^*/2\varepsilon} + \left| \int_{l_2^*} e^{a(\varphi^*\psi^* - \tilde{\varphi}^*\tilde{\psi}^*)/2\varepsilon} dl \right|, \tag{B_1}$$

where

$$\tilde{\varphi}^* = (a_0 - \tilde{\psi}) - \beta(a_0 - \tilde{\varphi} - \psi),$$
$$\tilde{\psi}^* = (a_0 - \tilde{\varphi}) + \beta(a_0 - \tilde{\varphi} - \psi),$$
$$\varphi^* = (a_0 - \psi) - \beta(a_0 - \varphi - \psi),$$
$$\psi^* = (a_0 - \varphi) + \beta(a_0 - \varphi - \psi);$$

the path of integration l_2^* is defined as follows: if $(\tilde{\varphi}, \tilde{\psi}) \in K \cup H_1 \cup H_2 \cup P_1$, then l_2^* consists of straight line segments which connect in sequence the points

$$(0,\ a_0),\ (-b_0\beta,\ a_0 + b_0\beta_2),\ (\beta_1(a_0 - \tilde{\psi}) - \beta b_0,\ \beta_2(a_0 + b_0) - \beta\tilde{\psi}),\ (\tilde{\varphi}^*,\ \tilde{\psi}^*);$$

if $(\tilde{\varphi}, \tilde{\psi}) \in K_1 \cup K_2 \cup P_2 \cup M_1 \cup M_2$, l_2^* consists of straight-line segments which connect in sequence the points

$$(0,\ a_0),\ (-\beta b_0,\ a_0 + \beta_2 b_0),\ (\beta_1(a_0 - \delta_0) - \beta\delta_0,\ \beta_2(a_0 + b_0) - \beta\delta_0),$$
$$(\beta_1(a_0 - \delta_0) + \beta\delta_0,\ \ \beta_2(a_0 - \delta_0) - \beta\delta_0),\ \ (\beta_1(a_0 + \delta_0) + \beta\delta_0,\ \beta_2(a_0 - \delta_0) + \beta b_0),$$
$$(\beta_1(a_0 + \delta_0) + \beta\tilde{\varphi},\ \ \beta_2(a_0 - \tilde{\varphi}) + \beta b_0,\ \ (\tilde{\varphi}^*,\ \tilde{\psi}^*).$$

We have further that

$$|g_1(\tilde{\varphi}, \tilde{\psi})| \leqslant c_1 \omega_1(\tilde{\varphi}^*, \tilde{\psi}^*). \tag{β_1}$$

We now rewrite (a_n) and (b_n) $(n = 2, 3, \ldots)$ in the coordinates $\tilde{\varphi}, \tilde{\psi}$. On going to the modulus we obtain

$$|f_n(\tilde{\varphi}, \tilde{\psi})| \leqslant O\left(\frac{1}{\varepsilon}\right) \left| \int_{l_1} e^{a(\varphi\psi - \tilde{\varphi}\tilde{\psi})/2\varepsilon} |P_{n-1}(\varphi, \psi)|\, dl \right|, \tag{$\mathscr{A}_n$}$$

$$|g_n(\tilde{\varphi}, \tilde{\psi})| \leqslant O\left(\frac{1}{\varepsilon}\right) \left| \int_{l_2} e^{a(\varphi^*\psi^* - \tilde{\varphi}^*\tilde{\psi}^*)/2\varepsilon} |S_{n-1}(\varphi^*, \psi^*)|\, dl \right|, \tag{B_n}$$

where the paths of integration l_1 and l_2 are defined as follows: if $(\tilde{\varphi}, \tilde{\psi}) \in K \cup H_2 \cup P_2$, l_1 consists of straight-line segments which connect in sequence the points $(0, a_0)$, $(0, b_0)$, $(\tilde{\varphi}, b_0)$ and $(\tilde{\varphi}, \tilde{\psi})$; if $(\tilde{\varphi}, \tilde{\psi}) \in H_1$, l_1 consists of straight-line segments which connect in sequence the points $(0, a_0)$, $(0, \tilde{\psi})$ and $(\tilde{\varphi}, \tilde{\psi})$; if $(\tilde{\varphi}, \tilde{\psi}) \in K_1 \cup N_2 \cup M_2$, l_1 consists of straight-line segments which connect in sequence the points $(0, a_0)$, $(0, \delta_0)$, (δ_0, δ_0), $(\delta_0, 0)$, $(\tilde{\varphi}, 0)$ and $(\tilde{\varphi}, \tilde{\psi})$.

If $(\tilde{\varphi}, \tilde{\psi}) \in K \cup H_1 \cup P_1$, l_2 consists of straight-line segments which connect in sequence the points

$$(0,\ a_0),\ (\beta_1(a_0-b_0),\ b_2a_0-\beta b_0),\ (\beta_1(a_0-b_0)+\beta\tilde{\varphi},\ \beta_2(a_0-\tilde{\varphi})-\beta b_0),\ (\tilde{\varphi}^*,\ \tilde{\psi}^*);$$

if $(\tilde{\varphi}, \tilde{\psi}) \in H_2$, l_2 consists of straight-line segments which connect in sequence the points

$$(0,\ a_0),\ (\beta_1(a_0-\tilde{\psi}),\ \beta_2 a_0-\beta\tilde{\psi}),\ (\tilde{\varphi}^*,\ \tilde{\psi}^*);$$

if $(\tilde{\varphi}, \tilde{\psi}) \in K_1 \cup K_2 \cup P_2 \cup M_1 \cup M_2$, l_2 consists of straight-line segments which connect in sequence the points

$$(0,\ a_0),\ (\beta_1(a_0-\delta_0),\ \beta_2 a_0-\beta\delta_0),\ (\beta_1(a_0-\delta_0)+\beta\delta_0,\ \beta_2(a_0-\delta_0)-\beta\delta_0),$$
$$(\beta_1 a_0+\beta\delta_0,\ \beta_2(a_0-\delta_0)),\ (\beta_1 a_0+\beta\tilde{\varphi},\ \beta_2(a_0-\tilde{\varphi})),\ (\tilde{\varphi}^*,\ \tilde{\psi}^*).$$

From $(\mathscr{A}_n)$ and (B_n) for $n = 2$, we obtain on the basis of the estimates (α_1) and (β_1)

$$|f_2(\tilde{\varphi},\tilde{\psi})| \leqslant O\left(\frac{1}{\varepsilon}\right)\left|\int_{l_1} l^{a(\varphi\psi-\tilde{\varphi}\tilde{\psi})/2\varepsilon}(\omega_1(\varphi,\ \psi)+\omega_1(\varphi^*,\ \psi^*))\,\omega_1(\varphi,\ \psi)\,\omega_1(\varphi^*,\ \psi^*)\,dl\right|,$$

$$|g_2(\tilde{\varphi},\tilde{\psi})| \leqslant O\left(\frac{1}{\varepsilon}\right)\left|\int_{l_2} l^{a(\varphi^*\psi^*-\tilde{\varphi}^*\tilde{\psi}^*)/2\varepsilon}(\omega_1(\varphi^*,\ \psi^*)+\omega_1(\varphi,\ \psi))\,\omega_1(\varphi,\ \psi)\,\omega_1(\varphi^*,\ \psi^*)\,dl\right|.$$

Thus

$$|f_2(\tilde{\varphi},\ \tilde{\psi})| \leqslant c_2\omega_2(\tilde{\varphi},\ \tilde{\psi}), \qquad (\alpha_2)$$
$$|g_2(\tilde{\varphi},\ \tilde{\psi})| \leqslant c_2\omega_2(\tilde{\varphi}^*,\ \tilde{\psi}^*), \qquad (\beta_2)$$

where c_2 is a positive constant.

Similarly from $(\mathscr{A}_n)$ and (B_n) for $n = 3$ we have

$$|f_3(\tilde{\varphi},\ \tilde{\psi})| \leqslant c_3\omega_3(\tilde{\varphi},\ \tilde{\psi}), \qquad (\alpha_3)$$
$$|g_3(\tilde{\varphi},\ \tilde{\psi})| \leqslant c_3\omega_3(\tilde{\varphi}^*,\ \tilde{\psi}^*), \qquad (\beta_3)$$

where c_3 is a positive constant and

$$\omega_3(\tilde{\varphi},\ \tilde{\psi}) = \begin{cases} \varepsilon^3 & \text{if } (\tilde{\varphi},\ \tilde{\psi}) \in K, \\ \varepsilon^2\sqrt{\varepsilon} & \text{if } (\tilde{\varphi},\ \tilde{\psi}) \in K_1 U H_2 U P_2, \\ \varepsilon^2 & \text{if } (\tilde{\varphi},\ \tilde{\psi}) \in M_2, \\ \varepsilon\sqrt{\varepsilon} & \text{if } (\tilde{\varphi},\ \tilde{\psi}) \in N_2. \end{cases}$$

Further, we have the estimates

$$|f_n(\tilde{\varphi},\ \tilde{\psi})| \leqslant c_n\omega_n(\tilde{\varphi},\ \tilde{\psi}), \qquad (\alpha_n)$$
$$|g_n(\tilde{\varphi},\ \tilde{\psi})| \leqslant c_n\omega_n(\tilde{\varphi}^*,\ \tilde{\psi}^*), \qquad (\beta_n)$$

where c_n is a positive constant, and $\omega_n(\tilde{\varphi}, \tilde{\psi})$ is defined recurrently as follows:

$$\omega_n(\tilde{\varphi},\ \tilde{\psi}) = \begin{cases} \varepsilon\omega_{n-1}(\tilde{\varphi},\ \tilde{\psi}) & \text{if } (\tilde{\varphi},\ \tilde{\psi}) \in K, \\ \sqrt{\varepsilon}\,\omega_{n-1}(\tilde{\varphi},\ \tilde{\psi}) & \text{if } (\tilde{\varphi},\ \tilde{\psi}) \in H_0 \setminus K \end{cases}$$
$$(n = 4,\ 5,\ \ldots).$$

We now find from $(b_k^{(n)})$ that

$$|v_{k+1}^{(n)}(\tilde{\varphi}, \tilde{\psi})| \leqslant \alpha_{k+1}^{(n)}(\varepsilon)\,\omega_{n+1}(\tilde{\varphi}, \tilde{\psi}),$$
$$|w_{k+1}^{(n)}(\tilde{\varphi}, \tilde{\psi})| \leqslant \alpha_{k+1}^{(n)}(\varepsilon)\,\omega_{n+1}(\tilde{\varphi}^*, \tilde{\psi}^*),$$

where $\alpha_{k+1}^{(n)}(\varepsilon)$ is defined recurrently as follows:

$$\alpha_{k+1}^{(n)}(\varepsilon) = c_{n+1}\left[1 + \sqrt{\varepsilon}\,\alpha_k^{(n)}(\varepsilon)(1 + \alpha_k^{(n)}(\varepsilon))^2\right], \quad \alpha_1^{(n)}(\varepsilon) = c_{n+1}.$$

From this it follows that for sufficiently small ε and for arbitrary fixed n we have $\alpha_k^{(n)}(\varepsilon) < q_{n+1}$ $(k = 0, 1, 2, \ldots)$, where q_{n+1} is a number independent of ε.

Thus on the octagon H_0 we have the estimates

$$|v_k^{(n)}(\tilde{\varphi}, \tilde{\psi})| \leqslant q_{n+1}\omega_{n+1}(\tilde{\varphi}, \tilde{\psi}),$$
$$|w_k^{(n)}(\tilde{\varphi}, \tilde{\psi})| \leqslant q_{n+1}\omega_{n+1}(\tilde{\varphi}^*, \tilde{\psi}^*)$$
$$(n = 1, 2, \ldots;\ k = 0, 1, 2, \ldots).$$

In addition we can obtain the estimates

$$|v_{k+1}^{(n)}(\tilde{\varphi}, \tilde{\psi}) - v_k^{(n)}(\tilde{\varphi}, \tilde{\psi})| \leqslant \beta_{k+1}^{(n)}(\varepsilon)\,\omega_{n+1}(\tilde{\varphi}, \tilde{\psi}),$$
$$|w_{k+1}^{(n)}(\tilde{\varphi}, \tilde{\psi}) - w_k^{(n)}(\tilde{\varphi}, \tilde{\psi})| \leqslant \beta_{k+1}^{(n)}(\varepsilon)\,\omega_{n+1}(\tilde{\varphi}^*, \tilde{\psi}^*),$$

where $\beta_{k+1}^{(n)}(\varepsilon)$ is defined recurrently as follows:

$$\beta_{k+1}^{(n)}(\varepsilon) = c_{n+1}\sqrt{\varepsilon}\,\beta_k^{(n)}(\varepsilon),$$

with $\beta_1^{(n)}(\varepsilon) = c_{n+1}$ $(n = 1, 2, \ldots)$. For sufficiently small values of ε and for arbitrary fixed n we have $c_{n+1}\sqrt{\varepsilon} < 1$, so that the geometrical progression $\{\beta_k^{(n)}(\varepsilon)\}$ converges.

Consequently the sequence of functions $\{v_k^{(n)}(\tilde{\varphi}, \tilde{\psi})\}$ converges uniformly on H_0 to a function $v^{(n)}(\tilde{\varphi}, \tilde{\psi}) \equiv v^{(n)}(t)$, and the sequence $\{w_k^{(n)}(\tilde{\varphi}, \tilde{\psi})\}$ converges uniformly on H_0 to a function $w^{(n)}(\tilde{\varphi}, \tilde{\psi}) \equiv w^{(n)}(t)$; and these functions give the solution of (d_n) with initial conditions

$$v^{(n)}(-t_0) = 0, \quad w^{(n)}(-t_0) = 0 \quad (n = 1, 2, \ldots)$$

and the estimates

$$|v^{(n)}(t)| \leqslant q_{n+1}\omega_{n+1}(\tilde{\varphi}, \tilde{\psi}),$$
$$|w^{(n)}(t)| \leqslant q_{n+1}\omega_{n+1}(\tilde{\varphi}^*, \tilde{\psi}^*). \qquad (\mathfrak{D}_n)$$

It follows from $(\mathfrak{D}_n)$ that on $[-t_0, t_0]$ we have

$$|v^{(n)}(t)| = |w^{(n)}(t)| \leqslant q_{n+1}\tilde{\omega}_{n+1}(t) \quad (n = 1, 2, \ldots),$$

where

$$\tilde{\omega}_{n+1}(t) = \begin{cases} \varepsilon^{n+1} & \text{if } -t_0 \leqslant t \leqslant t_0 - b_0, \\ \varepsilon^{(n+3)/2} & \text{if } t_0 - b_0 \leqslant t \leqslant t_0 - \delta, \\ \varepsilon^{(n+1)/2} & \text{if } t_0 - \delta \leqslant t \leqslant t_0. \end{cases}$$

Let

$$G_n(t) = \frac{1}{\bar{\gamma}_1 - \gamma_1}\begin{pmatrix} V_n(t) + W_n(t) \\ \gamma_1 V_n(t) + \bar{\gamma}_1 W_n(t) \end{pmatrix}.$$

Then on $[-t_0, t_0]$ we have

$$\|\xi(t) - G_n(t)\| \leqslant \tilde{q}_{n+1}\tilde{\omega}_{n+1}(t), \tag{6}$$

where $\tilde{q}_{n+1}$ is a positive constant.

We have thus proved the following theorem.

THEOREM. *Let the functions $f_k(y, z)$ and $g_k(y, x_1, x_2)$ $(k = 1, 2)$ be defined and continuous along with their first partial derivatives in the region Δ_0. Let*

$$t=(x_1-y)^2+x_2^2, \quad f_k(y, 0)\equiv 0, \quad g_k(y, y, 0)\equiv 0 \quad (k=1, 2),$$

$$\Delta=0, \quad \frac{a_1-a}{c_1}=\frac{a_2}{c_2}=\frac{c}{d_1}, \quad \alpha_{11}<0 \ u \ a_1+a_4>0.$$

Then the system (3) *with initial condition* (4) *has a solution, which satisfies* (6).

A similar result holds for the systems (1), (2).

BIBLIOGRAPHY

1. S. Karimov, *Asymptotics of the solutions of a system of differential equations with a small parameter multiplying the derivatives in the case of instability of the focus*, Studies on Integrodifferential Equations in Kirgizia, Vyp. 11, "Ilim", Frunze, 1977, pp. 405–412. (Russian)

2. ______, *Asymptotic behavior of the solutions of a system of differential equations with a small parameter multiplying the derivatives in the case of instability of the focus*, Studies on Integrodifferential Equations in Kirgizia, Vyp. 11, "Ilim", Frunze, 1977, pp. 77–95. (Russian)

3. A. B. Vasil'eva and V. F. Butuzov, *Asymptotic expansions of the solutions of singularly perturbed equations*, "Nauka", Moscow, 1973. (Russian)

4. M. Imanaliev, *Asymptotic methods in the theory of singularly perturbed integrodifferential systems*, "Ilim", Frunze, 1972. (Russian)

Translated by W. S. LOUD

Труды Матем. Инст. Стеклов.
147(1980)

Proceedings of the Steklov
Institute of Mathematics
1981, Issue 2

UDC 519.46

THE PLANCHEREL FORMULA FOR HYPERBOLOIDS

V. F. MOLČANOV

ABSTRACT. The object of study is the quasiregular representation of the group $SO_0(p, q)$ on the hyperboloid X: $-x_1^2 - \cdots - x_p^2 + x_{p+1}^2 + \cdots + x_{p+q}^2 = 1$, $p > 1$, $q > 2$. The main results are the inversion formula that recovers a compactly supported infinitely differentiable function on X from its Fourier components, and the corresponding Plancherel formula.

Bibliography: 13 titles.

§0. Introduction

Let G denote the group $SO_0(p, q)$, in other words the connected group of linear mappings of real n-space that preserve the bilinear form $[x, y] = -x_1y_1 - \cdots - x_py_p + x_{p+1}y_{p+1} + \cdots + x_ny_n$, where $n = p + q$. In this article we study the quasiregular representation of G on a hyperboloid. We can limit ourselves to the hyperboloid X defined by $[x, x] = 1$. Moreover, we leave aside the well-known case $q = 1$ of a hyperboloid with compact stabilizer subgroup; we assume that $p \geqslant 1$ and $q \geqslant 2$. The main results are the inversion formula that recovers an infinitely differentiable function with compact support from its Fourier components, and the corresponding Plancherel formula (see Theorems 0.1 and 0.2 below).

The inversion formula and the Plancherel formula for a hyperboloid with noncompact stabilizer subgroup were first obtained by Gel'fand and Graev [1] (see also [2]). They studied the case $p = 1$, $q = 3$, i.e. the imaginary Lobachevsky space. In their investigation they used the horosphere method. In the general case (of arbitrary p and q) the horosphere method does not seem to work: a discrete series is not obtained from the horospheres of the second kind; for $n < 4$ they are too few, and for $n > 4$ they are too many. We shall apply the method of (zonal) spherical functions. These functions were considered in [3]. The key formula is (4.1), where we expand the delta function on the hyperboloid with respect to spherical functions. The deduction of this formula is founded on certain integral transformations. In this way we overcome the considerable technical difficulties involved in the justification of term by term integration, inverting the order of integration, passing to the limit, and so on.

My articles [4], for $p = 1$, and [5], for odd p, contain brief announcements of the results of the present article, which is a direct continuation of [3]. We shall use without explanation the concepts and notation of that paper. By different methods, and in different forms, the Plancherel formula for hyperboloids has been obtained by Shintani [6] for $p = 1$, by Limić, Niederle and Rączka [7], and by Strichartz [8].

Let us formulate the results of the present paper. First let $p > 1$.

1980 *Mathematics Subject Classification.* Primary 43A80, 43A90; Secondary 43A30.

THEOREM 0.1. *Let the function f belong to $\mathfrak{D}(X)$. Let $\tilde{F}_r f$ and $F_\chi f$ be its Fourier components* (*see* [3], §3) *of the first discrete and continuous series, respectively. Then f is recovered from its Fourier components in the following way:*

$$f(x) = \sum_{r=-[\nu]}^{\infty} \omega_r \overline{\left(T'_{\chi_r^*}(g^{-1})\theta_{r^*}, F_r^\vee f\right)} + \int_{-\infty}^{\infty} \sum_{\varepsilon=0,1} \omega(\rho, \varepsilon)\overline{(T'_\chi(g^{-1})\theta_\chi, F_\chi f)}\, d\rho, \qquad (0.1)$$

where $\chi = (\sigma, \varepsilon)$, $\sigma = -n' + i\rho$, $n' = (n-2)/2$, $x = x^0 g$, $x^0 = (0, \ldots, 0, 1)$, $\nu = (n-3)/2$, *the bar denotes complex conjugation*, $F_r^\vee f$ *is a representative of the equivalence class* $\tilde{F}_r f$,*

$$\omega_r = \frac{1}{16}\pi^{1-n}(2r+n-2)\Gamma\left(\frac{\varepsilon+n-2+r}{2}\right), \quad \varepsilon \equiv p + r, \qquad (0.2)$$

$$\omega(\rho, \varepsilon) = \frac{1}{8}\pi^{3-n}\frac{\operatorname{ch}\rho\pi - (-1)^\varepsilon \cos(p\pi/2)\cos(q\pi/2)}{\operatorname{ch}\rho\pi + (-1)^\varepsilon \cos(p\pi/2)\cos(q\pi/2)} \times \frac{\rho \operatorname{sh}\rho\pi}{\operatorname{ch} 2\rho\pi - (-1)^n}\left|\Gamma\left(\frac{\sigma-\varepsilon+2}{2}\right)\right|^{-2}, \qquad (0.3)$$

the congruence sign denotes congruence mod 2, *and the remaining notation is as in* [3].

THEOREM 0.2. *For every $f \in \mathfrak{D}(X)$*

$$\int_X |f(x)|^2\, dx = \sum_{r=-[\nu]}^{\infty} \omega_r j_r^{-1}\|\tilde{F}_r f\|_r^2 + \int_{-\infty}^{\infty} \sum_{\varepsilon=0,1} \omega(\rho, \varepsilon)\|F_\chi f\|^2\, d\rho, \qquad (0.4)$$

where $\chi = (-n' + i\rho, \varepsilon)$, *and the norms correspond to the inner products in the representation spaces of the first discrete and continuous series* (*see* [3], (1.3), (1.1); *for* j_r *see* [3], §2).

Since $\mathfrak{D}(X)$ is dense in $L^2(X, dx)$, the correspondence $f \mapsto \{\tilde{F}_r f, F_\chi f\}$, $\chi = (-n' + i\rho, \varepsilon)$, extends from $\mathfrak{D}(X)$ to $L^2(X, dx)$ by Theorem 0.2; by the same token we obtain the expansion of the quasiregular representation of the group G on X. The continuous and the first discrete series appear in it with multiplicity 1.

Let us remark that in the case $q = 2$, when the representations of the first discrete series decompose into two irreducible ones (see [9]), the formulas (0.1) and (0.4) contain the Fourier components corresponding to the complete representations of this series.

For $p = 1$ the same assertions are true subject to the following changes: suppress the word "first" before the word "discrete"; instead of T'_χ write T'_σ; the variable ε appearing in the pair $\chi = (\sigma, \varepsilon)$ serves as an index for two Fourier components of the continuous series F_χ and two generalized functions θ_χ that correspond to one representation T_σ. Thus, for $p = 1$, the continuous and discrete series appear with multiplicities two and one, respectively.

The formula that expands the delta function with respect to spherical functions is proved according to the following plan. We derive two formulas, which we call the *principal integral relation* (§2) and the *principal sum expansion* (§3). The first formula expands a certain generalized function $H_{-\nu-1,\nu}$ into an integral over spherical functions of the continuous series and a finite sum of spherical functions of the (first) discrete series. The terms of this sum are the residues belonging to the

---

* *Editor's note.* In the Russian literature, sh and ch denote the hyperbolic sine and cosine.

poles of the integrand that cross the path of integration as a certain complex parameter μ approaches the point $\nu = (n - 3)/2$. The second formula expands another generalized function S_ν into an infinite sum with respect to the remaining spherical functions of the (first) discrete series. The functions $H_{-\nu-1,\nu}$ and S_ν are chosen in such a way that a certain linear combination of them gives the delta function.

The author expresses his deep gratitude to F. A. Berezin for stimulating discussions and his interest in this work.

§1. The function $E_{\sigma\mu}$

The results of this section will be applied in §2. Let us consider the generalized function $E_{\sigma\mu}$ on the hyperboloid X. It depends on two complex parameters, σ and μ. In terms of the generalized functions $\varphi_{\sigma\mu}^{(i)}$ and $\psi_{\sigma\mu}^{(i)}$ introduced in [3], §5, it is defined as follows:

$$E_{\sigma\mu} = \sin\left(\frac{p}{2} + \sigma + \mu\right)\pi \cdot \varphi_{\sigma\mu}^{(1)} + \sin\frac{p\pi}{2} \cdot \hat{\varphi}_{\sigma\mu}^{(2)} + \sin\left(\frac{p}{2} + \mu\right)\pi \cdot \varphi_{\sigma\mu}^{(3)} + 2\sin\frac{p\pi}{2} \cdot \Gamma^{-1}(\sigma + 2\mu + 1)\,\psi_{\sigma\mu}^{(1)} \tag{1.1}$$

(for the operator ^ see [3], §3). Comparing this expression with formula (5.7) of [3], which defines the generalized function $I_{\sigma\mu}$, we see that for uneven p the functions $E_{\sigma\mu}$ and $I_{\sigma\mu}$ coincide and for even p they differ in sign in the region $x_n < 0$; in other words, $E_{\sigma\mu} = \operatorname{sign}^{p-1} x_n \cdot I_{\sigma\mu}$. Let the function $\mathring{E}_{\sigma\mu}(c)$ of the real variable c correspond to $E_{\sigma\mu}$ in the sense of [3], §4. From the computations of [3], §6, it is easy to obtain the following integral representation for $\mathring{E}_{\sigma\mu}(c)$:

$$\mathring{E}_{\sigma\mu}(c) = d_{\sigma\mu} \int_{-\infty}^{\infty} g_\mu(c, w)\,[a(c, w)]_+^\sigma\, dw,$$

where

$$d_{\sigma\mu} = \frac{\Gamma(1/2 - \mu)}{2^\mu \sqrt{\pi}\,\Gamma(\sigma + 1)\,\Gamma(-\sigma - 2\mu)}, \tag{1.2}$$

$$g_\mu(c, w) = \begin{cases} \sin\left(\frac{p}{2} - \mu\right)\pi \cdot (w^2 - 1)_+^{\mu - 1/2} + \sin\frac{p-1}{2}\pi \cdot (w^2 - 1)_-^{\mu - 1/2}, & c > 1, \\ \sin\frac{p\pi}{2} \cdot (w^2 + 1)^{\mu - 1/2}, & -1 < c < 1, \\ \sin\left(\frac{p}{2} + \mu\right)\pi\,(w^2 - 1)_+^{\mu - 1/2} + \sin\frac{p+1}{2}\pi \cdot (w^2 - 1)_-^{\mu - 1/2}, & c < -1, \end{cases} \tag{1.3}$$

$$a(c, w) = c + w\,|c^2 - 1|^{1/2}. \tag{1.4}$$

This representation holds if

$$\operatorname{Re}\sigma > -1, \quad \operatorname{Re}\mu > -1/2, \quad \operatorname{Re}(\sigma + 2\mu) < 0. \tag{1.5}$$

LEMMA 1.1. *Under the conditions* (1.5), *for every function* $f \in \mathfrak{D}(X)$

$$(E_{\sigma\mu}, f) = d_{\sigma\mu} \int_{-\infty}^{\infty} dw \int_{-\infty}^{\infty} g_\mu(c, w)\,[a(c, w)]_+^\sigma\, \overline{\mathring{f}(c)}\, dc, \tag{1.6}$$

where $\mathring{f}(c)$ *is the mean of* f *over* G_0 *(see* [3], §4).

PROOF. The iterated integral

$$\int_{-\infty}^{\infty} \overset{\circ}{f} dc \int_{-\infty}^{\infty} g_\mu a_+^\sigma dw$$

converges absolutely under the conditions (1.5). This we see from the following facts: The integral of the absolute value of the integrand converges under these conditions; it is, moreover, a linear combination of the functions $\varphi_{\mathrm{Re}\,\sigma,\mathrm{Re}\,\mu}(c)$ and $\psi_{\mathrm{Re}\,\sigma,\mathrm{Re}\,\mu}(c)$; and these functions are integrable for $\mathrm{Re}\,\mu < 1$. We may thus invert the order of integration. The lemma is proved.

Let us now give the inner integral in (1.6) special consideration.

$$G_{\sigma\mu}(w) = \int_{-\infty}^{\infty} g_\mu(c, w)\,[a(c, w)]_+^\sigma\, \overline{\overset{\circ}{f}(c)}\, dc. \tag{1.7}$$

As c runs through any one of the intervals $\mathbf{R}^{(1)} = (1, \infty)$, $\mathbf{R}^{(2)} = (-1, 1)$ or $\mathbf{R}^{(3)} = (-\infty, -1)$, the value of $g_\mu(c, w)$ is constant. Thus $G_{\sigma\mu}(w)$ is a linear combination of the integrals

$$G_\sigma^{(i)}(w) = \int_{\mathbf{R}^{(i)}} [a(c, w)]_+^\sigma\, \overline{\overset{\circ}{f}(c)}\, dc, \quad i = 1, 2, 3.$$

The latter integrals converge absolutely for $\mathrm{Re}\,\sigma > -1$ and define continuous functions of w.

LEMMA 1.2. *If* $-1 < \mathrm{Re}\,\sigma < 0$, *then*

$$|G_\sigma^{(i)}(w)| \leqslant C\,|\rho|^{-1/2}\,|w^2 - 1|^{1/2\,\mathrm{Re}\,\sigma}, \quad \rho = \mathrm{Im}\,\sigma,$$

where the constant C *is independent of* ρ *and* w.

We give the proof for $i = 1$, $0 < w < 1$. The function $\varphi(t) = \overline{\overset{\circ}{f}(\mathrm{ch}\, t)}\,\mathrm{sh}\, t$, given for $t > 0$, is infinitely differentiable and has compact support; moreover, $\varphi(t) = o(t)$ as $t \to 0$. Introducing the substitution $a = \sqrt{1 - w^2}\, e^v$, we obtain

$$G_\sigma^{(1)}(w) = (1 - w^2)^{\sigma/2} \int_{v_0}^{\infty} e^{i\rho v} \varphi(t)\, h(v)\, v^{-1/2} dv,$$

where

$$t = \ln\left\{(e^v + \sqrt{e^{2v} - 1}) \sqrt{\frac{1-w}{1+w}}\right\}, \quad h(v) = e^{(1+\mathrm{Re}\,\sigma)v}, \quad v_0 = \ln \frac{1}{\sqrt{1-w^2}}.$$

Setting $\varphi(t)$ equal to zero where it is not yet defined, we extend the integral over $v > 0$ and apply Titchmarsh's theorem (see [10], Theorem 126) for the exponent $\alpha = \frac{1}{2}$. The total variation of the function φh depends only on $\mathrm{Re}\,\sigma$, from which the assertion of the lemma follows.

An immediate consequence of Lemma 1.2 is the following lemma.

LEMMA 1.3. *If* $-1 < \mathrm{Re}\,\sigma < 0$, *then* $G_{\sigma\mu}(w)$ *is continuous in* w *everywhere except at* $w = \pm 1$, *and*

$$|G_{\sigma\mu}(w)| \leqslant C\,|\rho|^{-1/2}\,|w^2 - 1|^{\mathrm{Re}(\sigma + 2\mu - 1)/2}, \quad \rho = \mathrm{Im}\,\sigma,$$

where the constant C *is independent of* ρ *and* w.

We now estimate the behavior of $E_{\sigma\mu}$ as $\rho \to \infty$. For this we consider the following regions of two complex variables:

$$\mathfrak{B}_0 : \operatorname{Re}\mu < 1, \quad \operatorname{Re}\sigma < 0, \quad \operatorname{Re}(\sigma + 2\mu + 1) > 0,$$
$$\mathfrak{B}_j = \{(\sigma, \mu) : (\sigma + j, \mu - j) \in \mathfrak{B}_0\}, \quad j = 0, 1, \ldots,$$
$$\mathfrak{B} = \bigcup_{j=0}^{[\nu]} \mathfrak{B}_j.$$

THEOREM 1.4. *If $(\sigma, \mu) \in \mathfrak{B}$, then for every $f \in \mathfrak{D}(X)$*

$$(E_{\sigma\mu}, f) = O(e^{\pi|\rho|} |\rho|^{2\mu-2} \ln^2 |\rho|), \quad \rho \to \infty.$$

PROOF. For every $j = 0, 1, \ldots, [\nu]$ consider the functional on $\mathfrak{D}(X)$ defined by

$$f \mapsto (\mathring{E}_{\sigma\mu}, \mathring{f}^{(j)}), \text{ where } \mathring{f}^{(j)}(c) = \left(\frac{d}{dc}\right)^j \mathring{f}(c).$$

For Re $\mu < 1$ this functional is defined by the absolutely convergent integral

$$\int_{-\infty}^{\infty} \mathring{E}_{\sigma\mu}(c) \overline{\mathring{f}^{(j)}(c)}\, dc;$$

for the remaining σ and μ it is taken in the sense of analytical continuation. We have

$$(E_{\sigma\mu}, f) = (-1)^j (E_{\sigma+j,\mu-j}, \mathring{f}^{(j)}), \quad j = 0, 1, \ldots, [\nu]. \tag{1.8}$$

This we see integrating by parts and using the equality $d\mathring{E}_{\sigma\mu}(c)/dc = \mathring{E}_{\sigma-1,\mu+1}(c)$ (see [3], (5.8)). The equation (1.8) reduces the proof of Theorem 1.4 to the following lemma.

LEMMA 1.5. *If $(\sigma, \mu) \in \mathfrak{B}_0$, then for every function $f \in \mathfrak{D}(X)$*

$$(\mathring{E}_{\sigma\mu}, \mathring{f}^{(j)}) = O(e^{\pi|\rho|} |\rho|^{2\mu-2} \ln^2 |\rho|), \quad \rho \to \infty.$$

Let us prove this for $j = [\nu]$, which is the hardest case. Write $f_\nu = \mathring{f}^{([\nu])}$. The function $f_\nu(c)$ has compact support and is infinitely differentiable everywhere except at the points $c = \pm 1$. The asymptotic expansions at these points are obtained by $[\nu]$-fold differentiation of the expansions (4.1) and (4.2) of [3]. Let us write $(E_{\sigma\mu}, f_\nu)$ as the sum of the three integrals

$$E^{(i)} = \int_{R^{(i)}} \mathring{E}_{\sigma\mu}(c) f_\nu(c)\, dc, \quad i = 1, 2, 3,$$

and estimate each of them separately.

Let $i = 2$. We apply the following integral representation of the function $\mathring{E}_{\sigma\mu}(c)$ for $-1 < c < 1$:

$$\mathring{E}_{\sigma\mu}(c) = -\sin\frac{p\pi}{2} \Gamma(\mu + 1/2)\, 2^{-1/2} \pi^{-3/2} \int_{-\infty}^{\infty} e^{(\sigma+\mu+1/2)u}$$
$$\times \{\sin(\sigma + \mu)\pi \cdot (\operatorname{ch} u - c)^{-\mu-1/2} - \sin\mu\pi \cdot (\operatorname{ch} u + c)^{-\mu-1/2}\}\, du;$$

this is obtained from the book [11], 8.714(2) with the aid of [12], 3.4, (17), (14), and (15). We multiply both its members by $f_\nu(c)$, integrate with respect to $c \in (-1, 1)$ and invert the order of integration under the conditions $\operatorname{Re} \sigma < 0$, $\operatorname{Re} \mu < \frac{1}{2}$ and $\operatorname{Re}(\sigma + 2\mu + 1) > 0$ (the legitimacy of this is proved as in Lemma 1.1). We obtain

$$E^{(2)} = -\sin \frac{p\pi}{2} \Gamma(\mu + {}^1/_2) 2^{-1/2} \pi^{-3/2} \int_{-\infty}^{\infty} e^{(\sigma+\mu+1/2)u} \times \{\sin(\sigma+\mu)\pi \cdot D_\mu^+(u) - \sin \mu\pi \cdot D_\mu^-(u)\} du, \tag{1.9}$$

where

$$D_\mu^\pm(u) = \int_{-1}^{1} (\operatorname{ch} u - c)^{-\mu-1/2} f_\nu(\pm c)\, dc.$$

The function $D_\mu^+(u)$ has the following properties (analogous assertions hold for $D_\mu^-(u)$): if $u \neq 0$ it is entire with respect to μ, for every μ it is infinitely differentiable with respect to $u \neq 0$,

$$\frac{d}{du} D_\mu^\pm(u) = (-\mu - {}^1/_2) D_{\mu+1}^\pm(u) \operatorname{sh} u,$$

$$D_\mu^+(u) \sim \text{const} \cdot e^{(-\mu-1/2)u} \quad \text{as} \quad u \to \infty,$$

the leading term of the asymptotics for $u \to +0$ has the form (for $p \equiv 1$ and $\operatorname{Re} \mu > -\frac{1}{2}$)

$$A + (B + C \ln u + D \ln^2 u) u^{1-2\mu},$$

the remaining terms look similar except for multiplication by positive integral powers of u, and the quotient of $D_\mu^+(u)$ and the leading term has absolutely integrable derivative. The properties listed here are proved by completely elementary means, so we omit their deduction. On the basis of these properties we may, firstly, extend (1.9) to the region $\operatorname{Re} \sigma < 0$, $\operatorname{Re}(\sigma + 2\mu + 1) > 0$ and, secondly, apply to $E^{(2)}$ Theorem 126 of [10]. We obtain the estimate of Lemma 1.5 for σ and μ belonging to the region just indicated.

The cases $i = 1, 3$ are treated similarly. Here we apply for $c > 1$ and $c < -1$, respectively, the integral representations

$$\mathring{E}_{\sigma\mu}(c) = \frac{1}{\sqrt{2\pi}\, \Gamma({}^1/_2 - \mu)} \left\{ \sin\left(\frac{p}{2} + \sigma + \mu\right)\pi \cdot \int_{-\infty}^{\infty} e^{-(\sigma+\mu+1/2)u} (\operatorname{ch} u - c)_-^{-\mu-1/2} du \right.$$
$$\left. + 2 \sin \frac{p\pi}{2} \sin(-\sigma - 2\mu)\pi \int_0^\infty e^{-(\sigma+\mu+1/2)u} (\operatorname{ch} u - c)_+^{-\mu-1/2} du \right\},$$

$$\mathring{E}_{\sigma\mu}(c) = \frac{1}{\sqrt{2\pi}\, \Gamma({}^1/_2 - \mu)} \sin\left(\frac{p}{2} + \mu\right)\pi \cdot \int_{-\infty}^{\infty} e^{-(\sigma+\mu+1/2)u} (\operatorname{ch} u + c)_-^{-\mu-1/2} du,$$

which are deduced from the formulas 3.7 (4) and (8) of [12]. The estimate of Lemma 1.5 is obtained under the conditions $\operatorname{Re} \mu < 1$ and $\operatorname{Re}(\sigma + 2\mu + 1) > 0$. We combine the estimates of the integrals $E^{(i)}$. The proof of Lemma 1.5 and, consequently, Theorem 1.4 is complete.

To conclude, let us indicate for special values of the parameters how $E_{\sigma\mu}$ is written in terms of spherical functions. We have

$$\left.\frac{E_{r\mu}}{\sin \mu\pi}\right|_{\mu=\nu} = (-1)^{r+1}\frac{1}{j_r\gamma_r}\Phi_r, \quad r=-1, -2, \ldots, -[\nu], \tag{1.10}$$

$$E_{\sigma\nu} = \sum_{s=0,\,1} \alpha(\rho, \varepsilon)\Phi_{\rho, s}, \quad \sigma=-n'+i\rho, \tag{1.11}$$

where

$$\alpha(\rho, \varepsilon) = \begin{cases} \frac{1}{2}a_\chi^{-1}, & p=1 \text{ or } q=1, \\ -\frac{1}{2}a_\chi^{-1}\operatorname{ctg}^2\frac{\sigma-\varepsilon}{2}\pi, & p=q=0; \end{cases} \tag{1.12}$$

for a_χ see [3], (6.5); for j_r and γ_r see [3], §§2 and 7.

Formula (1.10) is obtained by evaluating the left member with the help of (1.1) and formulas 3.3(4) and 3.4(13) of [12] and comparing with (7.4) and (7.5) of [3]. We deduce (1.11) from (6.13) of [3] using the following relation, which is easily verified, for $\sigma = -n' + i\rho$:

$$E_{\sigma\nu} = \begin{cases} I_{\sigma\nu}, & \text{if } p=1 \text{ or } q=1, \\ \frac{\operatorname{ch}^2\pi\rho+1}{\operatorname{sh}^2\pi\rho}I_{\sigma\nu} + \frac{2(-1)^{n'}\operatorname{ch}\pi\rho}{\operatorname{sh}^2\pi\rho}\hat{I}_{\sigma\nu}, & \text{if } p=q=0. \end{cases}$$

§2. The principal integral relation

It is the aim of this section to prove Theorem 2.6. First we study the generalized function $H_{\lambda\mu}$ that appears in the theorem. It is defined on the hyperboloid X and depends on two complex parameters λ and μ. We consider the following function of a real variable:

$$h_{\lambda\mu}(t) = \mathrm{B}^{-1}(\lambda+{}^1/_2, -\lambda-\mu)\,|t-1|^{2\lambda}t_+^{-\lambda-\mu},$$

where B denotes Euler's beta function. We use an integral to define the function $\mathring{H}_{\lambda\mu}(c)$ of the real variable c:

$$\mathring{H}_{\lambda\mu}(c) = \int_{-\infty}^{\infty} g_\mu(c, w)\,h_{\lambda\mu}(a(c, w))\,dw, \tag{2.1}$$

where $g_\mu(c, w)$ and $a(c, w)$ are given by (1.3) and (1.4), respectively. The integral converges absolutely under the conditions $\operatorname{Re}\mu > -\frac{1}{2}$, $\operatorname{Re}\lambda > -\frac{1}{2}$ and $\operatorname{Re}(\lambda+\mu) < 0$. The function $h_{\lambda\mu}$ is chosen in such a way that the integral (2.1) can be evaluated explicitly. The appearance of this function is inspired by a formula in Chapter V, §2.3 of the book [2]. We have

$$\mathring{H}_{\lambda\mu}(c) = \begin{cases} 2^{\lambda+\mu}\sin\left(\frac{p}{2}+\lambda\right)\pi\cdot(c-1)^\lambda(c+1)^{-\mu}, & c>1, \\ 2^{\lambda+\mu}\sin\frac{p\pi}{2}(1-c)^\lambda(c+1)^{-\mu}, & -1<c<1, \\ 2^{\lambda+\mu}\sin\left(\frac{p}{2}+\mu\right)\pi\cdot(1-c)^\lambda(-c-1)^{-\mu}, & c<-1. \end{cases} \tag{2.2}$$

Let us show how the integral (2.1) is evaluated if, e.g., $c > 1$. Instead of w we introduce the new variable $u = 2^{-1} \ln a(c, w)$. Then

$$\mathring{H}_{\lambda\mu}(c) = \mathrm{B}^{-1}(\lambda + 1/2, -\lambda - \mu)\, 2^{2\lambda+\mu+1/2}(c^2 - 1)^{-\mu} \int_{-\infty}^{\infty} |\operatorname{sh} u|^{2\lambda}$$
$$\times \left\{ \sin\left(\frac{p}{2} - \mu\right)\pi \cdot (\operatorname{ch} 2u - c)_+^{\mu - 1/2} + \sin \frac{p-1}{2}\pi \cdot (\operatorname{ch} 2u - c)_-^{\mu-1/2} \right\} e^u du.$$

The factor e^u in the integral may be replaced by ch u because the rest of the integrand is an even function. Thus, writing sh $u = s$, we can write the integral as a linear combination of the two integrals

$$\int_0^\infty s^{2\lambda}(2s^2 - c + 1)_\pm^{\mu - 1/2}\, ds,$$

each of which is easy to compute.

If

$$\operatorname{Re}\lambda > -1, \quad \operatorname{Re}\mu < 1, \tag{2.3}$$

then $\mathring{H}_{\lambda\mu}(c)$ is locally integrable, and the formula

$$(H_{\lambda\mu}, f) = \int_{-\infty}^{\infty} H_{\lambda\mu}(c)\overline{f(c)}\, dc \tag{2.4}$$

defines on X a generalized function $H_{\lambda\mu}$ that depends analytically on λ and μ in the region (2.3). We now extend $H_{\lambda\mu}$ beyond this region. Consider the following open subsets of X: $\dot{X} = \{|x_n| \neq 1\}$, $X^+ = \{x_n > 0\}$ and $X^- = \{x_n < 0\}$. Together they cover X. It is clear that on the space $\mathfrak{D}(\dot{X})$ the generalized function $H_{\lambda\mu}$ is analytic for all λ and μ. In the two regions $X^\pm$ we take $x_1, \ldots, x_{n-1}$ as coordinates. We write $x' = (x_1, \ldots, x_{n-1})$ and $P = x_1^2 + \cdots + x_p^2 - x_{p+1}^2 - \cdots - x_{n-1}^2$. Then the restrictions $H_{\lambda\mu}^\pm$ of the generalized function $H_{\lambda\mu}$ to $\mathfrak{D}(X^\pm)$ may be regarded as generalized functions in the region $P > -1$ of the Euclidean space $\mathbb{R}^{n-1}$. By (2.2) they look like this (recall that $dx = |x_n|^{-1} dx_1 \cdots dx_{n-1}$):

$$H_{\lambda\mu}^+ = \left\{ \sin\left(\frac{p}{2} + \lambda\right)\pi \cdot P_+^\lambda + \sin\frac{p\pi}{2} \cdot P_-^\lambda \right\} s_{\lambda+\mu}^+,$$

$$H_{\lambda\mu}^- = \left\{ \sin\left(\frac{p}{2} + \mu\right)\pi \cdot P_+^{-\mu} + \sin\frac{p\pi}{2} \cdot P_-^{-\mu} \right\} s_{\lambda+\mu}^-,$$

where

$$s_\alpha^\pm = 2^\alpha(1 + \sqrt{1+P})^{\mp\alpha}(1+P)^{-1/2};$$

these two last-mentioned functions are infinitely differentiable for $P > -1$ and do not vanish. It follows from Chapter IV of [1] that the generalized functions $H_{\lambda\mu}^+$ and $H_{\lambda\mu}^-$ have analytical continuations in λ and μ; the former is entire in μ and meromorphic in λ, while the latter is meromorphic in μ and entire in λ. The case of special interest for us is $\mu = \nu$, $\lambda = -\nu - 1$. In the region $\dot{X}$ the generalized function $H_{-\nu-1,\nu}$ coincides with the ordinary function $\mathring{H}_{-\nu-1,\nu}(x_n)$. Reading Chapter IV of [1] and recalling the definitions given in that book of the generalized functions connected with the quadratic form P, we find the following expressions for the restrictions of the generalized function $H_{-\nu-1,\nu}$ to the spaces $\mathfrak{D}(X^{\pm})$:

$$\frac{H^{+}_{-\nu-1,\nu}}{s^{+}_{-1}}=\begin{cases}(-1)^{q/2}P_{+}^{-\nu-1}, & p\equiv q\equiv 0,\\ \frac{\pi}{2}\Omega_{n-1}\delta(x')+(-1)^{(p-1)/2}P_{-}^{-\nu-1}, & p\equiv q\equiv 1,\\ \frac{\pi}{2}\Omega_{n-1}\delta(x')-\frac{\pi}{\Gamma(\nu+1)}[(-1)^{p/2}\delta_1(P)+\pi^{\nu+1}\delta(x')], & p\equiv 0,\ q\equiv 1,\\ \frac{\pi}{4}\Omega_{n-1}\delta(x')+(-1)^{q/2}P^{-\nu-1}, & p\equiv 1,\ q\equiv 0,\end{cases} \tag{2.5}$$

$$\frac{H^{-}_{-\nu-1,\nu}}{s^{-}_{-1}}=\begin{cases}(-1)^{q/2}P_{+}^{-\nu}, & p\equiv q\equiv 0,\\ (-1)^{(p-1)/2}P_{-}^{-\nu}, & p\equiv q\equiv 1,\\ (-1)^{p/2}\frac{\pi}{\Gamma(\nu)}\delta_1^{(\nu-1)}(P), & p\equiv 0,\ q\equiv 1,\\ (-1)^{(q-2)/2}P^{-\nu}, & p\equiv 1,\ q\equiv 0.\end{cases} \tag{2.6}$$

We proceed to derive the main formula of this section, (2.18). We need a series of lemmas.

LEMMA 2.1. *Under the conditions* (2.3) *one can invert the order of integration with respect to c and w in* (2.4). *In other words,*

$$(H_{\lambda\mu}, f)=\int_{-\infty}^{\infty}dw\int_{-\infty}^{\infty}g_{\mu}(c, w)\,h_{\lambda\mu}(a(c, w))\,\overline{\mathring{f}(c)}\,dc.$$

The proof is analogous to that of Lemma 1.1.

LEMMA 2.2. *If*

$$\operatorname{Re}\lambda>-{}^1/_2,\quad \operatorname{Re}(\lambda-\mu-\sigma)<0,\quad \operatorname{Re}(\lambda+\mu+\sigma)<0 \tag{2.7}$$

then

$$h_{\lambda\mu}(a)=\int_{-\infty}^{\infty}K(\lambda, \mu, \sigma)\,a^{\sigma}d\rho,\quad \rho=\operatorname{Im}\sigma,\quad a>0, \tag{2.8}$$

where

$$K(\lambda, \mu, \sigma)=2^{2\lambda}\pi^{-3/2}\cos(\mu+\sigma)\,\pi\,\frac{\Gamma(-\lambda-\mu-\sigma)\,\Gamma(-\lambda+\mu+\sigma)\,\Gamma({}^1/_2-\mu)}{\Gamma(-\lambda)\,\Gamma(-\lambda-\mu)},$$

and the integral converges uniformly in a on any closed interval not containing $a=1$.

On setting $a=e^{u}$ the lemma follows from the Fourier inversion formula (see [10], 1.9). The expression for K is obtained with the help of [12], 1.5(25).

LEMMA 2.3. *Assume* (2.7) *and, furthermore,* $\operatorname{Re}\lambda>-\frac{1}{4}$. *Then, for any function* $f\in\mathfrak{D}(X)$ *and each real number* $w\neq\pm 1$,

$$\int_{R^{(i)}}h_{\lambda\mu}(a(c, w))\,\overline{\mathring{f}(c)}\,dc=\int_{-\infty}^{\infty}K(\lambda, \mu, \sigma)\,d\rho\int_{R^{(i)}}[a(c, w)]_{+}^{\sigma}\,\overline{\mathring{f}(c)}\,dc \tag{2.9}$$

where $a(c, w)$ *is as in* (1.4) *and* $i=1, 2, 3$.

PROOF. Let $i = 1$. We fix $w \neq \pm 1$. In (2.8) we set $a = a(c, w)$. We multiply both sides by $\mathring{f}(c)$ and integrate over $c > 1$. We must show that in the right member we may invert the integrations with respect to ρ and c. By the previous lemma we may do so for integrals over sets of c on which $a \geqslant \delta$ and $|a - 1| \geqslant \delta$ for some $\delta > 0$. It remains to show that in the integral

$$\int_{-\infty}^{\infty} K(\lambda, \mu, \sigma) G_\sigma^{(1)}(w, \delta)\, d\rho,$$

where

$$G_\sigma^{(1)}(w, \delta) = \int_{\substack{c \geqslant 1 \\ a \geqslant \delta,\ |a-1| \geqslant \delta}} [a(c, w)]_+^\sigma \overline{\mathring{f}(c)}\, dc,$$

we may let $\delta \to 0$ under the integral sign if $\operatorname{Re} \lambda > -\frac{1}{4}$. As

$$K(\lambda, \mu, \sigma) \sim \text{const}\, |\rho|^{-1-2\operatorname{Re}\lambda}, \quad \rho \to \infty, \tag{2.10}$$

by [12], 1.18(6), we need only show (see [13], Proposition 45) that, as $\rho \to \infty$, $G_\sigma^{(1)}(w, \delta) = O(|\rho|^{-1/2})$ uniformly in δ, $0 < \delta < \delta_0$. We write

$$G_\sigma^{(1)}(w, \delta) = G_\sigma^{(1)}(w) - M_\sigma(w, \delta) - N_\sigma(w, \delta),$$

where

$$M_\sigma(w, \delta) = \int_1^\infty a_+^\sigma \overline{\mathring{f}(c)}\, Y(\delta - a)\, dc, \qquad N_\sigma(w, \delta) = \int_1^\infty a_+^\sigma \overline{\mathring{f}(c)}\, Y(\delta - |a - 1|)\, dc,$$

Y denoting the Heaviside function. By Lemma 1.2, $G_\sigma^{(1)}(w) = O(|\rho|^{-1/2})$ independently of δ. Both the functions M and N are $O(\rho^{-1})$ uniformly in δ, $0 < \delta < \delta_0$. This we see by passing in the integrals to the variable a and integrating by parts. For $i = 2$ or 3 the proof is analogous.

LEMMA 2.4. *If*

$$-1 < \operatorname{Re} \sigma < 0, \quad -1 < \operatorname{Re}(\sigma + 2\mu) < 0, \quad \operatorname{Re}(\lambda - \sigma - \mu) < 0, \quad \operatorname{Re} \lambda > -1, \tag{2.11}$$

then for every function $f \in \mathfrak{D}(X)$

$$(H_{\lambda\mu}, f) = \int_{-\infty}^{\infty} K(\lambda, \mu, \sigma)\, d_{\sigma\mu}^{-1} (E_{\sigma\mu}, f)\, d\rho \tag{2.12}$$

(*for* $d_{\sigma\mu}$ *see* (1.2)).

PROOF. On both sides of (2.9) we multiply by the value of g_μ (see (1.3)) in the integrals over $\mathbb{R}^{(i)}$ and sum for $i = 1, 2, 3$; we obtain (see (1.7))

$$\int_{-\infty}^{\infty} h_{\lambda\mu}(a(c, w))\, g_\mu(c, w) \mathring{f}(c)\, dc = \int_{-\infty}^{\infty} K(\lambda, \mu, \sigma)\, G_{\sigma\mu}(w)\, d\rho.$$

Now we integrate with respect to w. On the left-hand side the result is $(H_{\lambda\mu}, f)$ by Lemma 2.1; on the right-hand side the result is

$$\int_{-\infty}^{\infty} dw \int_{-\infty}^{\infty} K(\lambda, \mu, \sigma) G_{\sigma\mu}(w)\, d\rho.$$

This iterated integral converges absolutely by Lemma 1.3, in view of (2.10). We may invert the order of integration if in addition to the conditions of Lemmas 1.3 and 2.3 we also have $\operatorname{Re}\mu > -\frac{1}{2}$, $\operatorname{Re}\lambda > -\frac{1}{4}$ and $-1 < \operatorname{Re}(\sigma + 2\mu) < 0$. Under all these conditions, inverting leads to the required formula (2.12) by Lemma 1.1. By analyticity we may extend (2.12) to the bigger region indicated in Lemma 2.4. In fact, considered in its own right the integral of (2.12) converges absolutely even under the conditions

$$\operatorname{Re}\lambda > -1, \quad (\sigma, \mu) \in \mathfrak{B} \tag{2.13}$$

by (2.10), the asymptotics $d_{\sigma\mu} \sim \text{const}\, e^{\pi|\rho|}|\rho|^{2\mu}$ and Theorem 1.4. The integrand may have a singularity if one of the numbers $-\lambda - \mu - \sigma$, $-\lambda + \mu + \sigma$, $\sigma + 1$ or $-\sigma - 2\mu$ is equal to $0, -1, -2, \ldots$. So (2.12) extends to the region (2.11), which does not contain these singularities. Lemma 2.4 is proved.

As our next step we extend both sides of (2.12) in μ to the point $\mu = \nu$. It is convenient to first rewrite the kernel in the integral. As $E_{\sigma\mu}$ and $d_{\sigma\mu}$ remain the same if we change σ into $-\sigma - 2\mu - 1$, we have

$$H_{\lambda\mu} = \int_{-\infty}^{\infty} L(\lambda, \mu, \sigma) E_{\sigma\mu}\, d\rho, \tag{2.14}$$

where

$$\begin{aligned} L(\lambda, \mu, \sigma) &= \frac{1}{2}[K(\lambda, \mu, \sigma) + K(\lambda, \mu, -\sigma - 2\mu - 1)]\, d_{\sigma\mu}^{-1} \\ &= -2^{2\lambda+\mu}\left(\mu + \sigma + \frac{1}{2}\right) \cos(\mu + \sigma)\pi \\ &\quad \times \frac{\Gamma(\sigma+1)\Gamma(-\sigma-2\mu)\Gamma(-\lambda+\mu+\sigma)\Gamma(-\lambda-\mu-\sigma-1)}{\pi\Gamma(-\lambda)\Gamma(-\lambda-\mu)}. \end{aligned}$$

LEMMA 2.5. *If* $-1 < \operatorname{Re}\lambda < -\frac{1}{2}$, *then*

$$H_{\lambda\nu} = \sum_{r=-[\nu]}^{-1} k_{r\lambda} \left.\frac{E_{r\mu}}{\sin \mu\pi}\right|_{\mu=\nu} + \int_{-\infty}^{\infty} L(\lambda, \nu, \rho) E_{\sigma\nu}\, d\rho,$$

where $\sigma = -n' + i\rho$ *and*

$$k_{r\lambda} = 2^{\nu+2\lambda-1}(-1)^{r+1} \frac{(r+\nu+{}^1\!/_2)\Gamma(-\lambda-\nu-r-1)\Gamma(-\lambda+\nu+r)}{\Gamma(-r)\Gamma(n-2+r)\Gamma(-\lambda)\Gamma(-\lambda-\nu)}.$$

PROOF. Let

$$\operatorname{Re}(\sigma + \mu + {}^1\!/_2) = 0; \tag{2.15}$$

then the conditions (2.11) become

$$-1 < \operatorname{Re}\lambda < -{}^1\!/_2, \quad -{}^1\!/_2 < \operatorname{Re}\mu < {}^1\!/_2, \tag{2.16}$$

and (2.14) may be rewritten as

$$H_{\lambda\mu}=\frac{1}{i}\int_{C_\mu} L(\lambda,\mu,\sigma)E_{\sigma\mu}d\sigma, \tag{2.17}$$

where C_μ is the line in the σ-plane given by (2.15), with σ running upwards. Here the integral converges absolutely under looser conditions than (2.16), namely, $-1<\operatorname{Re}\lambda<-\frac{1}{2}<\operatorname{Re}\mu<[\nu]+1$; these conditions are equivalent to (2.13) together with (2.15). We fix λ and extend both sides of (2.17) in μ to the point $\mu=\nu$. As μ moves from the region $-\frac{1}{2}<\operatorname{Re}\mu<\frac{1}{2}$ into the point ν, singular points of the integrand cross the line C_μ. The points $\sigma=r$ cross from left to right and the points $\sigma=-r-2\mu-1$ cross from right to left; here $r=-1,-2,\ldots,r>-n'$. Note that if n is even and $\mu=\nu$, then the point $\sigma=-n'$ is not singular. From every pair of points $\sigma=r$, $\sigma=-r-2\mu-1$ the right-hand side of (2.17) receives the contribution

$$2\pi\left(\operatorname*{Res}_{\sigma=r} LE-\operatorname*{Res}_{\sigma=-r-2\mu-1} LE\right)=4\pi\operatorname*{Res}_{\sigma=r} LE.$$

This residue is not hard to compute, because $E_{\sigma\mu}$ is entire in σ. The lemma is proved.

THEOREM 2.6. *The generalized function $H_{-\nu-1,\nu}$ has the following expansion with respect to spherical functions*:

$$H_{-\nu-1,\nu}=\sum_{r=-[\nu]}^{-1} k_{r,-\nu-1}\frac{(-1)^{r+1}}{j_r\gamma_r}\Phi_r$$

$$+\int_{-\infty}^{\infty} L(-\nu-1,\nu,\sigma)\sum_{\varepsilon=0,1}\alpha(\rho,\varepsilon)\Phi_{\rho,\varepsilon}d\rho, \tag{2.18}$$

where $\sigma=-n'+i\rho$, explicit expressions for $k_{r\lambda}$ and $L(\lambda,\nu,\sigma)$ are given above, $\alpha(\rho,\varepsilon)$ is as in (1.2), *and j_r and γ_r are as in* §2 *of* [3].

PROOF. We use (1.10) and (1.11) to pass from $E_{\sigma\mu}$ to spherical functions in the expansion of Lemma 2.5. By Theorem 5.1 below the equation obtained extends to all λ for which the left-hand side exists, and in particular to $\lambda=-\nu-1$.

§3. The principal sum expansion

It is the aim of this section to prove Theorem 3.1. We consider the generalized function S_μ on the hyperboloid X. It depends on the complex parameter μ. It is defined with the aid (see [3], §4) of the function $\mathring{S}_\mu(c)$ of a real variable:

$$\mathring{S}_\mu(c)=\begin{cases}\cos\frac{q\pi}{2}(c-1)^{-\mu-1}(c+1)^{-\mu}, & c>1,\\ \cos\left(\frac{q}{2}-\mu-1\right)\pi\cdot(1-c)^{-\mu-1}(c+1)^{-\mu}, & -1<c<1,\\ \cos\left(\frac{q}{2}-2\mu-1\right)\pi\cdot(1-c)^{-\mu-1}(-1-c)^{-\mu}, & c<-1;\end{cases}$$

it is locally integrable for $\operatorname{Re}\mu<0$. The generalized function S_μ depends analytically on μ in the region $\operatorname{Re}\mu<0$ and extends as a meromorphic function to the whole μ-plane. From [1] it follows that S_μ has no singularity for $\mu=\nu$. We just write down, following [1], the expression of S_ν on X^+ and X^- and compare them with the expressions (2.5) and (2.6) of the generalized function $H_{-\nu-1,\nu}$ to convince ourselves that

$$2H_{-\nu-1,\nu} - S_\nu = \pi\Omega_{n-1}\delta, \tag{3.1}$$

where δ denotes the delta function concentrated in the point $x^0 = (0, \ldots, 0, 1)$ on X:

$$(\delta, f) = \overline{f(x^0)}.$$

THEOREM 3.1. *The generalized function S_ν has the following expansion with respect to the spherical functions Φ_r of the (first) discrete series:*

$$S_\nu = -\frac{2^{-\nu-1}\pi}{\Gamma(\nu+1)} \sum_{r=0}^{\infty} (2r+n-2)\frac{1}{j_r\gamma_r}\Phi_r.$$

The proof starts with a pair of lemmas. We consider the generalized function $V_{r\mu}$ on the hyperboloid X. It depends on two parameters, the integer $r = 0, 1, 2, \ldots$ and the complex number μ. It is expressed through the generalized functions $\varphi_{\sigma\mu}^{(i)}$ and $\psi_{\sigma\mu}^{(i)}$ (see [3], §5):

$$V_{r\mu} = \cos\frac{q\pi}{2}\tilde{\psi}_{r\mu}^{(1)} + \cos\frac{q\pi}{2}\tilde{\psi}_{r\mu}^{(2)} + \cos\left(\frac{q}{2} - 2\mu - 1\right)\pi \cdot (-1)^r \tilde{\psi}_{r\mu}^{(3)} + \frac{\pi}{2}\sin\frac{q\pi}{2}\varphi_{r\mu}^{(2)}; \tag{3.2}$$

here we have introduced the abbreviation

$$\tilde{\psi}_{\sigma\mu}^{(i)} = \Gamma(\sigma + 2\mu + 1)\psi_{\sigma\mu}^{(i)}.$$

From §5 of [3] it follows that the function $V_{r\mu}$ is locally integrable for $\operatorname{Re}\mu < 1$ and extends analytically to the μ-plane, the only possible singular points being $\mu = 1, 2 \ldots$ and $\mu = \nu, \nu + 1, \ldots$. It turns out that at the points $\mu = \nu, \nu + 1$ the generalized function $V_{r\mu}$ has no singularities. At the indicated points the singularities of the terms in (3.2) cancel out mutually. Writing down the expression for $V_{r\nu}$, we see that this function is proportional to the spherical function Φ_r (see [3], §7):

$$V_{r\nu} = -\frac{\pi}{2j_r\gamma_r}\Phi_r. \tag{3.3}$$

LEMMA 3.2. *For all complex numbers $z \neq \pm 1$ and all $n = 0, 1, \ldots$ the following estimate for the Legendre function of the second kind holds:*

$$|Q_n(z)| \leqslant |z+\zeta|^{-n-1}\left(1 + \ln 2\left|1 + \frac{z}{\zeta}\right|\right),$$

where $\zeta = \sqrt{z^2 - 1}$; outside the interval $[-1, 1]$ we take for ζ the branch that is positive for $z > 1$; within the interval we take either of the two branches of the root. Moreover, for real c, $-1 < c < 1$, we have the sharper estimate

$$|Q_n(c)| \leqslant B\left(\frac{1}{2}, \frac{n}{2}\right)(1 - c^2)^{-1/2}.$$

PROOF. The first inequality is obtained from Jacobi's integral representation (see [12], 3.7(2)); we replace $|z + \zeta \operatorname{ch} t|$ by $|z + \zeta|$ for $0 < t < \tau$ and by $2^{-1}e^t|\zeta|$ for $t > \tau$, where $\tau = \ln 2|1 + z/\zeta|$. The second inequality is obtained from the same representation with the aid of the inequality $|c + i\sqrt{1 - c^2} \operatorname{ch} t|^2 > 1 + (1 - c^2)t^2$ for $-1 < c < 1$.

LEMMA 3.3. *For any* $f \in \mathfrak{D}(X)$

$$\lim_{r\to\infty} (V_{r,\nu+1}, f) = 0.$$

PROOF. If $\operatorname{Re} \mu < \nu + 2$, then for every $f \in \mathfrak{D}(X)$

$$(V_{r\mu}, f) = (\mathring{V}_{r+j,\mu-j}, \mathring{f}^{(j)}), \quad j = 0, 1, \ldots, \tag{3.4}$$

so Lemma 3.3 reduces to the assertion

$$\lim_{r\to\infty} (\mathring{V}_{r\varkappa}, g) = 0,$$

where $\varkappa = \nu + 1 - [n']$ and $g = \mathring{f}^{([n'])}$. Note that $\varkappa = \frac{1}{2}$ or 1 if n is even or odd, respectively. It follows from §4 of [3] that the function $g(c)$ has compact support and is infinitely differentiable for $c \neq \pm 1$, and for $c = 1$ it has the asymptotic expansion

$$g(c) \sim \sum_{k=0}^{\infty} \alpha_k (c - 1)^k + \xi(c) \sum_{k=0}^{\infty} \beta_k (c - 1)^k,$$

where

$$\xi(c) = \begin{cases} (c - 1)_-^{-1/2}, & p \equiv q \equiv 0, \\ (c - 1)_+^{-1/2}, & p \equiv q \equiv 1, \\ Y(c - 1), & p \equiv 0,\ q \equiv 1, \\ \ln |c - 1|, & p \equiv 1,\ q \equiv 0. \end{cases}$$

An analogous formula holds for $c = -1$.

Recalling the behavior of $\varphi_{\sigma\mu}(c)$ and $\psi_{\sigma\mu}(c)$ near the points $c = \pm 1$ (see [3], §5), we see that the value $v_{r\mu} = (\mathring{V}_{r\mu}, g)$ is given, for $\operatorname{Re} \mu < \varkappa$, by the absolutely convergent integral

$$v_{r\mu} = \int_{-\infty}^{\infty} \mathring{V}_{r\mu}(c) \overline{g(c)}\, dc.$$

Let us extend $v_{r\mu}$ to the region $\operatorname{Re} \mu < \varkappa + 1$; it contains the point we need, namely $\mu = \varkappa$.

Suppose first that $\operatorname{supp} \mathring{f} \subset (0, 2)$. Then the formula

$$v_{r\mu} = \int_0^2 \mathring{V}_{r\mu}(c) g_1(c)\, dc + \beta_0 w_{r\mu} \tag{3.5}$$

will accomplish the required extension. Here $w_{r\mu}$ is defined for $\operatorname{Re}\mu < \kappa$ by the integral

$$w_{r\mu} = \int_0^2 \overset{\circ}{V}_{r\mu}(c)\,\xi(c)\,dc,$$

and it extends by analyticity to the remaining values of μ; moreover, we have set $g_1(c) = \overline{g(c)} - \bar{\beta}_0\xi(c)$. Now

$$w_{r\kappa} = \begin{cases} (-1)^{1+q/2}\, 2\sqrt{\pi}\left\{\sum_{k=0}^{r} \dfrac{\sin (2k+1)\frac{\pi}{4}}{2k+1} - \dfrac{\pi}{4}\right\}, & p \equiv q \equiv 0, \\ 0, \ q \equiv 1, & \\ (-1)^{q/2}\left\{\displaystyle\int_0^2 Q_{r+1}(c)\frac{dc}{c-1} - \frac{\pi^2}{4}\right\}, & p \equiv 1, \ q \equiv 0; \end{cases}$$

the last integral is to be understood in the sense of Cauchy's principal value. Let us show that $\lim_{r\to\infty} w_{r\kappa} = 0$. In fact, for $q \equiv 1$ this is trivial. For $p \equiv q \equiv 0$ it follows from the equation

$$\pi/4 = \sum_{k=0}^{\infty} (2k+1)^{-1} \sin [(2k+1)\,\pi/4].$$

For $p \equiv 1$ and $q \equiv 0$ we pass to complex integration,

$$w_{r,1} = (-1)^{q/2}\left\{\int_{L^+} Q_{r+1}(z)\frac{dz}{z-1} + \int_{L^-} Q_{r+1}(z)\frac{dz}{z-1}\right\},$$

where $L^{\pm}$ are the semicircles $|z-1| = 1$, $\operatorname{Im} z \gtrless 0$, z running from left to right. Now we apply Lemma 3.2.

It remains to show that for $\mu = \kappa$ the first term on the right-hand side of (3.5), which we will henceforth call η_r, tends to 0 as $r \to \infty$. For even n the Legendre functions of which $\overset{\circ}{V}_{r,1/2}(c)$ is composed have expressions in terms of elementary functions. If $p \equiv q \equiv 0$, then

$$\overset{\circ}{V}_{r,1/2}(c) = \begin{cases} (-1)^{q/2}\sqrt{\dfrac{\pi}{2}}\,\dfrac{e^{-(r+1)t}}{\operatorname{sh} t}, & c = \operatorname{ch} t, \\ (-1)^{1+q/2}\sqrt{\dfrac{\pi}{2}}\,\dfrac{\sin (r+1)t}{\sin t}, & c = \cos t, \end{cases}$$

and if $p \equiv q \equiv 1$ then

$$\overset{\circ}{V}_{r,1/2}(c) = \begin{cases} 0, \ |c| > 1, & \\ (-1)^{(q-1)/2}\sqrt{\dfrac{\pi}{2}}\,\dfrac{\cos (r+1)t}{\sin t}, & c = \cos t. \end{cases}$$

Thus, for even n the task reduces to showing that the integrals

$$\int_0^{\operatorname{arcch} 2} e^{-(r+1)t} g_1(\operatorname{ch} t)\,dt, \qquad \int_0^{\pi/2} e^{i(r+1)t} g_1(\cos t)\,dt$$

tend to 0 as $r \to \infty$. This is an easy consequence of the continuity and boundedness of g_1.

For odd n we use the relation

$$\left(\frac{d}{dc}\right)\mathring{\Gamma}_{r\mu} = -\mathring{\Gamma}_{r-1,\,\mu+1}$$

and write η_r in the form

$$\eta_r = \int_0^2 \mathring{V}_{r+1,\,0}(c)\, g_1'(c)\, dc.$$

If $p \equiv 0$ and $q \equiv 1$, then

$$\eta_r = \frac{\pi}{2}(-1)^{(q-1)/2}\int_0^1 P_{r+1}(c)\, g_1'(c)\, dc.$$

We replace the Legendre polynomial with its expression by the Mehler-Dirichlet formula (see [12], 3.7(27)) and switch the order of integration. This is legitimate because $g_1'(c)$ is infinitely differentiable on [0, 1]. We obtain

$$\eta_r = \int_0^\pi \gamma(v)\cos\left(r+\frac{3}{2}\right)v\,dv,$$

where $\gamma(v)$ is continuous on $[0, \pi]$. Consequently, $\eta_r \to 0$. If $p \equiv 1$ and $q \equiv 0$, then $\mathring{V}_{r+1,0}(c) = (-1)^{q/2}Q_{r+1}(c)$, and Lemma 3.2 yields $\eta_r \to 0$.

In the case $\operatorname{supp}\mathring{f} \subset (-2, 0)$ the assertion of the lemma reduces to what we have just seen, because $\hat{V}_{r,\nu+1} = (-1)^{p+r}V_{r,\nu+1}$ (for the operator ^ see [3], §3). If, finally, $\operatorname{supp}\mathring{f}$ does not contain the points $c = \pm 1$, then the proof is analogous to the one we just gave of the fact that $\eta_r \to 0$. As any function in $\mathfrak{D}(X)$ can be written as a sum of functions of the three types considered, Lemma 3.3 is proved.

We now finish the proof of Theorem 3.1. By (3.3) we must show that

$$S_\nu = \frac{1}{2^\nu\Gamma(\nu+1)}\sum_{r=0}^\infty (2r+2\nu+1)V_{r\nu}. \tag{3.6}$$

We write $\tilde{\psi}_{\sigma\mu}(z) = (z^2-1)^{-\mu/2}Q^\mu_{\sigma+\mu}(z)$ for complex z not lying on the ray $(-\infty, 1]$. For these z, from [12], 3.9(3), we easily deduce the formula

$$(2\sigma+2\mu+1)\tilde{\psi}_{\sigma\mu}(z) = \tilde{\psi}_{\sigma-2,\,\mu+1}(z) - \tilde{\psi}_{\sigma,\,\mu+1}(z).$$

Setting $\sigma = r = 0, 1, 2, \ldots$ and summing over $r = 0, 1, \ldots, N$, we obtain

$$\sum_{r=0}^N (2r+2\mu+1)\tilde{\psi}_{r\mu}(z) = \tilde{\psi}_{-2,\,\mu+1}(z) + \tilde{\psi}_{-1,\,\mu+1}(z) - \tilde{\psi}_{N-1,\,\mu+1}(z) - \tilde{\psi}_{N,\,\mu+1}(z).$$

The first two terms on the right-hand side contribute (see [12], 3.2(5)) $2^\mu\Gamma(\mu+1) \times (z-1)^{-\mu-1}(z+1)^{-\mu}$. We use the formulas 3.4(9) and 3.3(12) of [12] to pass to the ray $(-\infty, 1]$, and recall the generalized functions S_μ and $V_{r\mu}$; for $\operatorname{Re}\mu < 0$ we obtain the expansion

$$2^\mu\Gamma(\mu+1)S_\mu = \sum_{r=0}^N (2r+2\mu+1)V_{r\mu} + V_{N-1,\,\mu+1} + V_{N,\,\mu+1}.$$

This extends by analyticity to the point $\mu = \nu$. Now let $N \to \infty$. By Lemma 3.3 we obtain (3.6).

§4. Proof of Theorems 0.1 and 0.2

LEMMA 4.1. *Let δ be the delta function on the hyperboloid X concentrated at $x^0 = (0, \ldots, 0, 1)$. In terms of the spherical functions Φ_r and $\Phi_{\rho,\varepsilon}$ the following expansion obtains*:

$$\delta = \sum_{r=-[\nu]}^{\infty} \omega_r j_r^{-1} \Phi_r + \int_{-\infty}^{\infty} \sum_{\varepsilon=0,1} \omega(\rho, \varepsilon) \Phi_{\rho,\varepsilon} d\rho, \tag{4.1}$$

where ω_r and $\omega(\rho, \varepsilon)$ are given by (0.2) *and* (0.3); *for j_r see* [3], §2.

This follows from (3.1) and Theorems 2.6 and 3.1.

Let us take an $f \in \mathfrak{D}(X)$ and apply (4.1) to the translate $U(g)f$. Writing $x = x^0 g$, we obtain

$$\overline{f(x)} = \sum_{r=-[\nu]}^{\infty} \frac{\omega_r}{j_r} (\Phi_r, U(g) f) + \int_{-\infty}^{\infty} \sum_{\varepsilon=0,1} \omega(\rho, \varepsilon) (\Phi_{\rho,\varepsilon}, U(g) f) \, d\rho. \tag{4.2}$$

Note that, considered as functions of g, the terms of the sum and the integrand depend only on the coset $x = G_0 g$, because the spherical functions are G_0-invariant. From (4.2) Theorem 0.1 follows with the aid of several facts set forth in [3], namely, the definition of the spherical functions (§7) and the formulas (2.8), (3.9) and (1.4).

LEMMA 4.2. *For every function $f \in \mathfrak{D}(X)$ the sum and integral in* (4.2) *converge absolutely and uniformly on any compact set $B \subset X$.*

PROOF. We embed X continuously in G, $x \mapsto g_x$. We fix the function $f \in \mathfrak{D}(X)$ and the compact set $B \subset X$. The union of the supports of the functions $U(g_x)f$, $x \in B$, is again a compact set $Q \subset X$.

In proving the assertion of the lemma, let us first treat the sum. It is enough to consider the sum

$$\sum_{r=1}^{\infty} (2r + n - 2) (V_{r\nu}, U(g) f),$$

whose terms differ by a common factor from the terms appearing in the sum of (4.2) when $r \geq 1$. Let a be a number satisfying $Q \subset \{|x_n| \leq a\}$. Let the space $\mathring{\mathfrak{D}}_a$ and the norm $\|\cdot\|_m$, $m = [\nu]$, on it be as in [3], §4. Now the Laplace operator Δ on $\mathfrak{D}(X)$ is continuous, and so is the averaging map $\mathfrak{D}(X) \to \mathring{\mathfrak{D}}$. Thus, for every $k = 0, 1, 2, \ldots$ the numerical function $s_k(x) = \|(U(g_x)\Delta^k f)^0\|_m$ on B is continuous. Using Theorem 5.7 below, we obtain the estimate

$$|(V_{r\nu}, U(g) f)| \leq K_a r^{-2k} \max s_k, \quad x^0 g \in B,$$

which yields our assertion for the sum when $k \geq 2$.

Let us consider the integral in (4.2). For every $k = 0, 1, \ldots,$ we consider the function $n_k(x) = N(\Delta^k U(g_x)f) = N(U(g_x)\Delta^k f)$ on X, where N is a functional as in Theorem 5.1 below. This function is continuous on X because N is continuous. Let

/ 7

$\nu_k = \max_{x\in B} n_k(x)$. Theorem 5.1 shows that the function $C_Q\nu_k\omega(\rho, \varepsilon)|u(\chi)|^2|\sigma|^{-2k}$ is, uniformly in $x = x^0g \in B$, a majorant of the integrand in (4.2). For $\rho \to \infty$ this majorant behaves asymptotically like $\mathrm{const}|\rho|^{n-2-2k}$. It remains to take k so that $n - 2 - 2k < -1$, or $k > (n - 1)/2$. Lemma 4.2 is proved.

We multiply both sides of (4.2) by $f(x)$ and integrate with respect to x. As f has compact support, Lemma 4.2 shows that the integration on the right-hand side may be performed before the summation with respect to r and the integration with respect to ρ. By what we have shown in [3] at the end of §7 we obtain Theorem 0.2.

§5. Estimates for the spherical functions

In this section we derive estimates for the spherical functions $\Phi_{\rho,\varepsilon}$ and Φ_r of the continuous and (first) discrete series, respectively (Theorems 5.1 and 5.7). These estimates were used in the proofs of Lemma 4.2 and Theorem 2.6. Instead of Φ_r it is more convenient to deal with the generalized function $V_{r\nu}$, which differs from it by a factor (see (3.3)).

THEOREM 5.1. *For every compact set Q on the hyperboloid X there is a number C_Q such that for every function $f \in \mathfrak{D}(X)$ with support in Q*

$$|(\Phi_{\rho,\varepsilon}, f)| \leqslant C_Q |u(\chi)|^2 N(\Delta^k f)|\sigma|^{-2k}, \quad |\rho| > R, \tag{5.1}$$

where $\chi = (\sigma, \varepsilon)$, $\sigma = -n' + i\rho$, k is a nonnegative integer, N is a certain positive continuous functional on $\mathfrak{D}(X)$, R is a certain number and Δ is the Laplace operator on X (for $u(\chi)$ see [3], (2.3)).

PROOF. It is enough to establish (5.1) for $k = 0$, i.e.

$$|(\Phi_{\rho,\varepsilon}, f)| \leqslant C_Q |u(\chi)|^2 N(f) \tag{5.2}$$

because $(\Phi_{\rho,\varepsilon}, \Delta f) = |\sigma|^2(\Phi_{\rho,\varepsilon}, f)$ (see [3], (6.3)).

Recall (see [3], §7) that the spherical function $\Phi_{\rho,\varepsilon}$ is defined thus:

$$(\Phi_{\rho,\varepsilon}, f) = (\theta_\chi, F_\chi f), \quad \chi = (-n' + i\rho, \varepsilon). \tag{5.3}$$

So let us first estimate the generalized function θ_χ on S (see [3], §2). We fix γ, $0 < \gamma < 1$. We consider the following functionals l_m, $m = 0, 1, \ldots,$ on $\mathfrak{D}(S)$: we define $l_m(\varphi)$ to be the largest of the numbers

$$\max_{|t|\leqslant 1} |\varphi^\#(t)|, \quad \max_{|t|\leqslant\gamma} \left|\left(\frac{d}{dt}\right)^k \varphi^\#(t)\right|, \quad k = 0, 1, \ldots, m$$

(for the mapping $^\#$ see [3], §2). These functionals are continuous on $\mathfrak{D}(S)$, the mapping $^\#$ being continuous.

LEMMA 5.2. *For every function $\varphi \in \mathfrak{D}_\varepsilon(S)$ and every number $m = 0, 1, \ldots$*

$$|(\theta_\chi, \varphi)| \leqslant C_m l_m(\varphi)|u(\chi)|, \quad |\rho| \geqslant \rho_m,$$

where $\chi = (\sigma, \varepsilon)$, $\rho = \operatorname{Im}\sigma$, and the constants C_m and ρ_m depend only on $\operatorname{Re}\sigma$.

PROOF. Here we need a somewhat more general formula than (2.4) of **[3]**; for $\operatorname{Re}\sigma > -m - 1$ we have the following representation of the value of (θ_χ, φ) by an absolutely convergent integral:([1])

$$(\theta_\chi, \varphi) = \Omega_p \Omega_{q-1} \left[\int_{-\gamma}^{\gamma} \theta_\chi(t) \left\{ \varphi^\#(t) - \sum_{k=0}^{m-1} \frac{1}{k!} t^k \alpha_k \right\} (1-t^2)^{(q-3)/2} dt \right.$$
$$+ \frac{1}{2} u(\chi) \sum_{k=0}^{m-1} [1 + (-1)^{\varepsilon+k}] \frac{1}{k!} B_{\gamma^2}\left(\frac{\sigma+k+1}{2}, \frac{q-1}{2}\right) \alpha_k$$
$$\left. + \int_{|t|\geqslant\gamma} \theta_\chi(t)\, \varphi^\#(t) (1-t^2)^{(q-3)/2} dt \right], \tag{5.4}$$

where $\alpha_k = (d/dt)^k \varphi^\#(0)$ and $B_a(\lambda, \mu)$ is the incomplete beta function (see [12], 2.5.3). It is easy to show that $B_a(\lambda, \mu)$ is bounded for $\operatorname{Im}\lambda$ large:

$$|B_a(\lambda, \mu)| \leqslant C, \quad |\operatorname{Im}\lambda| \geqslant A, \tag{5.5}$$

where C and A depend on $\operatorname{Re}\lambda$ (for fixed μ and a, $0 < a < 1$). The quantity in curly brackets in (5.4) is equal to

$$\frac{1}{m!} t^m \left(\frac{d}{dt}\right)^m \varphi^\#(t'),$$

where t' lies between 0 and t. From this and (5.5) the lemma follows.

Further, keeping in mind (5.3), let us estimate $l_m(\varphi)$ where $\varphi = F_\chi f$. Let L denote the element $L_{n-1,n}$ of the Lie algebra of the group G (see [3], §1). We consider the average with respect to the subgroup K_0 (see [3], §2) on the manifold S and on the hyperboloid X:

$$\tilde{\varphi}(s) = \int_{K_0} \varphi(sk_0)\, dk_0, \quad \tilde{f}(x) = \int_{K_0} f(xk_0)\, dk_0.$$

LEMMA 5.3. *If $\varphi = F_\chi f$, then for any m*

$$l_m(\varphi) \leqslant C'_m \max_{0\leqslant j\leqslant m} \max_{s\in S} |F_\chi U(L^j) \tilde{f}(s)|,$$

where C'_m is a constant.

PROOF. The mapping F_χ commutes with averaging over K_0, so $\tilde{\varphi} = F_\chi \tilde{f}$. To both sides of this equation we apply the Lie operator L at the point $s_t = (1, 0, \ldots, 0, \sqrt{1-t^2}, t)$, and on the right we shift F_χ to the front:

$$T_\chi(L)\tilde{\varphi}(s_t) = F_\chi U(L) \tilde{f}(s_t).$$

It is not hard to verify that the left-hand side of this equation is equal to

$$\sqrt{1-t^2}\, (d/dt)\, \varphi^\#(t),$$

so

$$\left(\frac{d}{dt}\right)^k \varphi^\#(t) = \sum_{j=1}^{k} (1-t^2)^{-k+j/2} P_{k-j}(t)\, F_\chi U(L^j) \tilde{f}(s_t),$$

([1])Formula (2.4) of [3] is obtained on setting $\gamma = 1$.

where the $P_i(t)$, $i = 0, 1, \ldots,$ are polynomials. From this the assertion of the lemma follows.

Finally, we estimate the Fourier component $F_\chi f$ of f. We consider the Radon transform of functions on the hyperboloid. The Radon transform of a function $f \in \mathfrak{D}(X)$ is the function $\check{f}$ on $\mathbb{R} \times S$ given by

$$\check{f}(b, s) = \int_X f(x)\,\delta([x, s] - b)\,dx.$$

The following two lemmas are easy to prove.

LEMMA 5.4. *For every compact set $Q \subset X$ there is an interval $E \subset \mathbb{R}$ such that* $\operatorname{supp} \check{f} \subset E \times S$ *whenever* $\operatorname{supp} f \subset Q$.

LEMMA 5.5. *The mapping $f \mapsto \check{f}$ is a continuous mapping $\mathfrak{D}(X) \to \mathfrak{D}(\mathbb{R} \times S)$.*

If v is a generalized function on $\mathbb{R}$, then the mapping

$$f \mapsto \check{v}f(s) = (v(b), \check{f}(b, s)) \tag{5.6}$$

takes $\mathfrak{D}(X)$ into $\mathfrak{D}(S)$. If v has parity ε, then $\check{v}f$ also has parity ε and thus falls within $\mathfrak{D}_\varepsilon(S)$. In particular, if for v we take the generalized function $\theta_\chi(b)$ (see [3], (2.2)), then (5.6) will be the mapping F_χ of §3 in [3]. Consequently, for $\operatorname{Re}\sigma > -m - 1$, $m = 0, 1, \ldots,$ the Fourier component $F_\chi f$ of f can be written as an absolutely convergent integral:

$$F_\chi f(s) = \frac{u(\chi)(-1)^m}{(\sigma+1)\ldots(\sigma+m)} \int_{-\infty}^{\infty} \{b_+^{\sigma+m} + (-1)^{\varepsilon+m} b_-^{\sigma+m}\} \left(\frac{\partial}{\partial b}\right)^m \check{f}(b, s)\,db. \tag{5.7}$$

Moreover, from Lemma 5.5 it follows that the functional

$$M_k(f) = \max_{b,s} \left|\left(\frac{\partial}{\partial b}\right)^k \check{f}(b, s)\right|$$

is continuous on $\mathfrak{D}(X)$. (5.7) enables us to give the required estimate of the Fourier component.

LEMMA 5.6. *If* $\operatorname{supp} f$ *is contained in the compact set $Q \subset X$, then for every* $m = 0, 1, \ldots$

$$\max_{s \in S} |F_\chi f(s)| \leqslant \frac{|u(\chi)|}{|\sigma+1|\ldots|\sigma+m|} M_m(f)\,A(Q, m, \operatorname{Re}\sigma).$$

(Here the last factor is $\int_E |b|^{m+\operatorname{Re}\sigma}db$, where E is the interval corresponding to Q by Lemma 5.4).

Now we are in a position to finish the proof of Theorem 5.1. Let $\sigma = -n' + i\rho$, and take m so that $-n' > -m - 1$ or, equivalently, $m > n' - 1$. We estimate (5.3) according to Lemmas 5.2, 5.3 and 5.6, in that order. Writing $R = \rho_m$ and $N(f) = \max_{0 \leqslant j \leqslant m} M_m(U)L^j(\tilde{f})$, we obtain (5.2). Theorem 5.1 is proved.

THEOREM 5.7. *For every real number $a > 1$ there exists a constant K_a such that for every function $f \in \mathfrak{D}(X)$ with support in the set $\{|x_n| \leqslant a\}$*

$$|(V_{r\nu}, f)| \leqslant K_a r^{-2k} \|(\Delta^k f)^0\|_m, \quad r = 1, 2, \ldots,$$

where $m = [\nu]$ and the norm $\|\cdot\|_m$ is defined as in §4 *of* [3].

PROOF. It is enough to establish the inequality for $k = 0$, i.e. $|(V_{r\nu}, f)| \leqslant K_a \|\mathring{f}\|_m$, because $\Delta V_{r\nu} = r(2 - n - r)V_{r\nu}$. First we estimate $\mathring{V}_{r,\nu-m}(c)$. Using the explicit expressions for it that we have mentioned in §3, Lemma 3.2, and the inequality 8.917(5) of [11], we find that the function has, uniformly in $r = 0, 1, \ldots,$ a locally integrable majorant: $|\mathring{V}_{r,\nu-m}(c)| \leqslant M(c)$. For $M(c)$ we may choose, e.g., $A + B|c^2 - 1|^{-1/2}$, where A and B are suitable constants. Now in virtue of (3.4) we have

$$|(V_{r\nu}, f)| = \left| \int_{-a}^{a} \mathring{V}_{r+m,\,\nu-m}(c) \overline{\mathring{f}^{(m)}(c)}\, dc \right|$$

$$\leqslant \int_{-a}^{a} M(c) |\{\mathring{f}(c) - b_0\gamma(c) - b_0'\gamma(-c)\}^{(m)}|\, dc + |b_0| \int_{-a}^{a} M(c) |\gamma^{(m)}(c)|\, dc$$

$$+ |b_0'| \int_{-a}^{a} M(c) |\gamma^{(m)}(-c)|\, dc \leqslant K_a \|\mathring{f}\|_m,$$

where b_0, b_0' and $\gamma(c)$ are taken from §4 of [3], and K_a is the largest of the three numbers

$$\int_{-a}^{a} M(c)\, dc, \quad \int_{-a}^{a} M(c) |\gamma^{(m)}(c)|\, dc, \quad \int_{-a}^{a} M(c) |\gamma^{(m)}(-c)|\, dc.$$

BIBLIOGRAPHY

1. I. M. Gel'fand and M. I. Graev, *An application of the horisphere method to the spectral analysis of functions in real and imaginary Lobachevsky spaces*, Trudy Moskov. Mat. Obšč. **11** (1962), 243–308. (Russian)

2. I. M. Gel'fand, M. I. Graev and N. Ja. Vilenkin, *Generalized functions*. Vol. 5: *Integral geometry and representation theory*, Fizmatgiz, Moscow, 1962; English transl., Academic Press, 1966.

3. V. F. Molčanov, *Spherical functions on hyperboloids*, Mat. Sb. **99** (**141**) (1976), 139–161; English transl. in Math. USSR Sb. **28** (1976).

4. ______, *Harmonic analysis on an hyperboloid of one sheet*, Dokl. Akad. Nauk SSSR **171** (1966), 794–797; English transl. in Soviet Math. Dokl. **7** (1966).

5. ______, *An analog of the Plancherel formula for hyperboloids*, Dokl. Akad. Nauk SSSR **183** (1968), 288–291; English transl. in Soviet Math. Dokl. **9** (1968).

6. Tokuro Shintani, *On the decomposition of regular representation of the Lorentz group on a hyperboloid of one sheet*, Proc. Japan. Acad. **43** (1967), 1–5.

7. N. Limić, J. Niederle and R. Rączka, *Eigenfunction expansions associated with the second-order invariant operator on hyperboloids and cones*. III, J. Mathematical Phys. **8** (1967), 1079–1093.

8. Robert S. Strichartz, *Harmonic analysis on hyperboloids*, J. Functional Analysis **12** (1973), 341–383.

9. V. F. Molčanov, *Representations of pseudo-orthogonal groups associated with a cone*, Mat. Sb. **81** (**123**) (1970), 358–375; English transl. in Math. USSR Sb. **10** (1970).

10. E. C. Titchmarsh, *Introduction to the theory of Fourier integrals*, Clarendon Press, Oxford, 1937.

11. I. S. Gradšteĭn and I. M. Ryžik, *Tables of integrals, series and products*, 4th ed., Fizmatgiz, Moscow, 1963; English transl., Academic Press, 1965.

12. A. Erdélyi et al., *Higher transcendental functions*. Vol. 1, McGraw-Hill, 1953.

13. Laurent Schwartz, *Méthodes mathématiques pour les sciences physiques*, Hermann, Paris, 1961; English transl., Addison-Wesley, Reading, Mass., 1966.

Translated by B. ROIDER

Труды Матем. Инст. Стеклов.
147(1980)

Proceedings of the Steklov
Institute of Mathematics
1981, Issue 2

A FUNCTIONAL MODEL OF PERTURBATION THEORY AND ITS APPLICATION TO SCATTERING THEORY

UDC 517.9

S. N. NABOKO

ABSTRACT. A new model of the perturbation theory of nonselfadjoint operators in Hilbert space is constructed. Formulas for functions of a certain class of operators which are close to selfadjoint are obtained within the framework of a model representation. Conditions for the existence and completeness of wave operators for a pair of operators of this class are established, and explicit formulas for the wave operators and the scattering operator are also obtained.

Bibliography: 18 titles.

This work is devoted to the construction of a new model of the perturbation theory of operators in Hilbert space and some of its applications to scattering theory. This model is closely related to the well-known functional model of Sz.-Nagy and Foiaş for dissipative, completely nonselfadjoint operators [1] and contains it as a special case. In the author's papers [2] and [3] similar constructions were used in the study of completely nonselfadjoint, nondissipative operators.

In the present paper the model is considered for an entire class (see (1), below) of operators $L^{(\kappa)}$ which are "close" to selfadjoint operators; this makes it possible to apply it to problems of perturbation theory both for nonselfadjoint and selfadjoint operators. Even in the latter case the basic ideas and computations of the theory of completely nonselfadjoint operators are used in an essential way. In particular, formulas are given for computing the resolvent and other functions of all operators $L^{(\kappa)}$ in the same model representation. Here κ is a bounded operator in an auxiliary Hilbert space E and plays the role of a parameter. This makes it possible to obtain explicit formulas for the wave operators for the pair $(A, L^{(\kappa)})$, where $A = L^{(\kappa)}|_{\kappa=0}$, and the scattering operator in the model representation by using the characteristic function S of (3) of the dissipative operator $L^{\|} \equiv L^{(\kappa)}|_{\kappa=iI}$. Consideration of questions of the existence and completeness of wave operators (both for nuclear and smooth perturbations) makes it possible to complement the results of Sahnovič [4]–[6] in this area.

The basic results of the paper were announced in [7].

§0. Notation

In a Hilbert space H we consider a class of operators of the form

$$L^{(\kappa)} = A + V^{(\kappa)}, \tag{1}$$

1980 *Mathematics Subject Classification.* Primary 47A45, 47A55, 47A40.

where A is a selfadjoint operator in H and the "perturbation" $V^{(\kappa)}$ admits the factorization

$$V^{(\kappa)}=\alpha\kappa\alpha/2. \tag{1'}$$

Here $\alpha \geqslant 0$ is a nonnegative operator in H and κ is a bounded operator in the subspace $E \overset{\text{def}}{=} \overline{\mathfrak{R}(\alpha)} \subseteq H$ generated by the range $\mathfrak{R}(\alpha)$ of the operator α which plays the role of a parameter in this decomposition. In order that the expression $A + V^{(\kappa)}$ be meaningful, we impose the condition that $V^{(\kappa)}$ be subordinate to the operator A. Let $\alpha^2/2$ be (A)-bounded with relative bound less than 1, i.e. (see [8]) the domains of the operators satisfy $D(A) \subset D(\alpha^2)$ and for some a and b $(a < 1)$

$$\left\|\frac{\alpha^2}{2}u\right\|\leqslant a\|Au\|+b\|u\|,\quad u\in D(A). \tag{2}$$

The operator $L^{(\kappa)}$ is well defined(¹) for $\|\kappa\| < 1/a$ in terms of a quadratic form (see [8]), and in the cases of interest to us (see below) the operator α commutes with κ; then $D(L^{(\kappa)})$ coincides with $D(A)$.

A nonselfadjoint operator L of "general form" with imaginary part subordinate to its real part can be written in the form (1): $L = A + iV \equiv A + i\alpha J\alpha/2 = L^{(\kappa)}|_{\kappa=iJ}$, where $\alpha \overset{\text{def}}{=} (2|V|)^{1/2}$ and $J \overset{\text{def}}{=} \operatorname{sign} V$, by using the polar decomposition $V = |V|J$ of the selfadjoint operator V. Together with L there are selfadjoint operators $A = L^{(\kappa)}|_{\kappa=0}$ and $A + V = L^{(\kappa)}|_{\kappa=J}$ among the operators $L^{(\kappa)}$ as well as the maximal dissipative operator $L^{\|} \overset{\text{def}}{=} A + i\alpha^2/2 = L^{(\kappa)}|_{\kappa=iI}$ and its adjoint $L^{-\|} \overset{\text{def}}{=} (L^{\|})^* = A - i\alpha^2/2 = L^{(\kappa)}|_{\kappa=-iI}$. In addition to the conditions imposed above, we shall assume that the normal spectrum of each operator $L^{(\kappa)}$ is discrete.

Since the functional model of Sz.-Nagy and Foiaş [1] for dissipative completely nonselfadjoint operators will be used below, we require that $L^{\|}$ be completely nonselfadjoint, i.e. that it not have reducing, selfadjoint parts. We shall show that the requirement of complete nonselfadjointness of $L^{\|}$ is not restrictive from the viewpoint of perturbation theory. Namely, there is the following result which generalizes the theorem on the decomposition of an operator into selfadjoint and completely nonselfadjoint parts (see [2] and [9]).

PROPOSITION 1. *The subspace $H_1 \subseteq H$ which is reducing for all operators $L^{(\kappa)}$ on which $L^{\|}$ induces a completely nonselfadjoint dissipative operator has the form*

$$H_1=\bigvee_{(\lambda\notin\sigma(L^{\|}))}(L^{\|}-\lambda)^{-1}\mathcal{R}(\alpha)=\bigvee_{(\lambda\notin\sigma(L^{(\kappa)}))}(L^{(\kappa)}-\lambda)^{-1}\mathcal{R}(\alpha),$$

while the restriction to the subspace $H_2 \overset{\text{def}}{=} H \ominus H_1$ of all operators $L^{(\kappa)}$ coincides with the selfadjoint operator $A|H_2$ $(L^{(\kappa)}|H_2 = A|H_2)$.

PROOF. On the basis of the Hilbert identity (if condition (2) is satisfied and $\|\kappa\| < 1/a$ the operators $\alpha(L^{(\kappa)} - \lambda)^{-1}$ and $(L^{\|} - \lambda)^{-1}\alpha$ are bounded off the spectrum)

$$(L^{(\kappa)}-\lambda)^{-1}=(L^{\|}-\lambda)^{-1}+((L^{\|}-\lambda)^{-1}\alpha)\left(\frac{i-\kappa}{2}\right)(\alpha(L^{(\kappa)}-\lambda)^{-1}),$$

(¹) Its resolvent will be constructed below (see (6)) explicitly in the model representation.

we obtain the inclusion

$$(L^{(\varkappa)}-\lambda)^{-1}\mathcal{R}(\alpha)\subseteq(L^{\|}-\lambda)^{-1}\mathcal{R}(\alpha)$$

for all λ, $\lambda\notin\sigma(L^{(\kappa)})$, $\lambda\notin(L^{\|})$. By changing the roles of $L^{(\kappa)}$ and $L^{\|}$ in the Hilbert identity we obtain the reverse inclusion, i.e. $(L^{(\kappa)}-\lambda)^{-1}\mathcal{R}(\alpha)=(L^{\|}-\lambda)^{-1}\mathcal{R}(\alpha)$, and hence

$$\bigvee_{(\lambda\notin\sigma(L^{(\varkappa)}))}(L^{(\varkappa)}-\lambda)^{-1}\mathcal{R}(\alpha)=\bigvee_{(\lambda\notin\sigma(L^{\|}))}(L^{\|}-\lambda)^{-1}\mathcal{R}(\alpha)\equiv H_1.$$

The discreteness of the nonreal spectrum of the operators $L^{(\kappa)}$ and $L^{\|}$ has been used here. If we set $\kappa=-iI$, then the last equality gives

$$H_1=\bigvee_{(\lambda\notin\sigma(L^{-\|}))}((L^{\|})^*-\lambda)^{-1}\mathcal{R}(\alpha).$$

Hence H_1 is a reducing subspace for $L^{\|}$ on which it induces a completely nonselfadjoint operator (see Lemma 0 of [2]). Similarly, passing from κ to the adjoint operator κ^*, we prove the reducibility of H_1 for all (admissible) operators $L^{(\kappa)}$.

Finally, the equality $L^{(\kappa)}|H_2=A|H_2$ follows from the fact that the condition of orthogonality to H_1 implies the equality $\alpha(L^{(\kappa)}-\lambda)^{-1}|H_2=0$, $\lambda\notin\sigma(L^{(\kappa)})$. Again, by the Hilbert identity,

$$\begin{aligned}(L^{(\varkappa)}-\lambda)^{-1}|H_2&=(A-\lambda)^{-1}|H_2-((A-\lambda)^{-1}\alpha)\frac{\varkappa}{2}(\alpha(L^{(\varkappa)}-\lambda)^{-1})|H_2\\&=(A-\lambda)^{-1}|H_2.\end{aligned}$$

§1. Explicit construction of the model of Sz.-Nagy and Foiaş for dissipative operators

The present section is devoted to the construction of the model of Sz.-Nagy and Foiaş for dissipative, completely nonselfadjoint operators in explicit form. This makes it possible to utilize it for the model representation of the operators $L^{(\kappa)}$ for all admissible values of κ simultaneously, including the selfadjoint operators A and $A+V$. In the latter case the model representation has little in common with the spectral theorem for selfadjoint operators, but is convenient for problems in the theory of perturbations of selfadjoint operators and, in particular, for problems of scattering theory. Here, as follows from Proposition 1, the requirement of complete nonselfadjointness of $L^{\|}$ corresponds to the orthogonal separation of the maximal reducing subspace for A and $A+V$ on which these operators coincide.

We note that for the dissipative Schrödinger operator an "explicit" form of the model was constructed by Pavlov [10]. The construction presented in this section makes essential use of the basic features of the symmetric spectral representation of the selfadjoint dilation of the dissipative Schrödinger operator in [10]. The next theorem makes the construction possible.

THEOREM 1. *Suppose that the maximal dissipative operator $L^{\|}\equiv A+i\alpha^2/2$ satisfies condition* (2); *then*

$$\|\alpha(L^{\|}-\lambda)^{-1}u\|_{H_2^-(E)}\leqslant\sqrt{2\pi}\,\|u\|,\qquad u\in H,$$

where $\alpha(L^{\|}-\lambda)^{-1}u$ *is considered as an analytic function of* λ (Im $\lambda<0$) *with values in the Hilbert space* E.(²)

PROOF. We make use of the fact that for the maximal dissipative operator $L^{\|}$ there exists a selfadjoint dilation [1], i.e. an operator $\mathfrak{L}$ in a larger Hilbert space $\mathcal{H}$ with $\mathfrak{L}^*=\mathfrak{L}$ such that $(L^{\|}-\lambda)^{-1}=P_H(\mathfrak{L}-\lambda)^{-1}|H$, Im $\lambda<0$. Here P_H is the orthogonal projection of $\mathcal{H}$ onto H. Passing to adjoint operators, we obtain the equality

$$(L^{-\|}-\lambda)^{-1}=P_H(\mathfrak{L}-\lambda)^{-1}|H, \quad \operatorname{Im}\lambda>0.$$

Let $\varepsilon>0$; then

$$\begin{aligned}\int_{-\infty}^{\infty}\|\alpha(L^{\|}-\sigma+i\varepsilon)^{-1}u\|^2\,d\sigma &= \int_{-\infty}^{\infty}\big((L^{-\|}-\sigma-i\varepsilon)^{-1}\alpha^2(L^{\|}-\sigma+i\varepsilon)^{-1}u,u\big)\,d\sigma\\ &=\frac{1}{i}\int_{-\infty}^{\infty}\big((L^{-\|}-\sigma-i\varepsilon)^{-1}\big[(L^{\|}-\sigma+i\varepsilon)-(L^{-\|}-\sigma-i\varepsilon)-2i\varepsilon\big]\\ &\quad\times(L^{\|}-\sigma+i\varepsilon)^{-1}u,u\big)\,d\sigma\\ &=\frac{1}{i}\int_{-\infty}^{\infty}\big(\big((L^{-\|}-\sigma-i\varepsilon)^{-1}-(L^{\|}-\sigma+i\varepsilon)^{-1}\big)u,u\big)\,d\sigma\\ &\quad-2\varepsilon\int_{-\infty}^{\infty}\|(L^{\|}-\sigma+i\varepsilon)^{-1}u\|^2\,d\sigma\\ &=\frac{1}{i}\int_{-\infty}^{\infty}\big(\big[(\mathfrak{L}-\sigma-i\varepsilon)^{-1}-(\mathfrak{L}-\sigma+i\varepsilon)^{-1}\big]u,u\big)_{\mathcal{H}}\,d\sigma\\ &\quad-2\varepsilon\int_{-\infty}^{\infty}\|(L^{\|}-\sigma+i\varepsilon)^{-1}u\|^2\,d\sigma.\end{aligned}$$

We denote by E_t, $t\in\mathbb{R}$, the resolution of the identity of the selfadjoint operator $\mathfrak{L}$, and use Fubini's theorem:

$$\begin{aligned}&\frac{1}{i}\int_{-\infty}^{\infty}\big(\big[(\mathfrak{L}-\sigma-i\varepsilon)^{-1}-(\mathfrak{L}-\sigma+i\varepsilon)^{-1}\big]u,u\big)_{\mathcal{H}}\,d\sigma\\ &\quad=\frac{1}{i}\int_{-\infty}^{\infty}\left\{\int_{-\infty}^{\infty}\big[(t-\sigma-i\varepsilon)^{-1}-(t-\sigma+i\varepsilon)^{-1}\big]\,d(E_tu,u)_{\mathcal{H}}\right\}d\sigma\\ &\quad=\int_{-\infty}^{\infty}\left\{\frac{1}{i}\int_{-\infty}^{\infty}2i\varepsilon\big[(t-\sigma)^2+\varepsilon^2\big]^{-1}\,d\sigma\right\}d(E_tu,u)_{\mathcal{H}}\\ &\quad=2\pi\int_{-\infty}^{\infty}d(E_tu,u)_{\mathcal{H}}=2\pi\|u\|^2.\end{aligned}$$

We thus have the identity

$$\int_{-\infty}^{\infty}\|\alpha(L^{\|}-\sigma+i\varepsilon)^{-1}u\|^2\,d\sigma=2\pi\|u\|^2-2\varepsilon\int_{-\infty}^{\infty}\|(L^{\|}-\sigma+i\varepsilon)^{-1}u\|^2\,d\sigma,$$

(²) The symbol $H_2^{(\mp)}(E)$ denotes the Hardy class of analytic functions f in the upper (lower) half plane with values in the Hilbert space E such that

$$\sup_{\varepsilon>0}\int_{-\infty}^{\infty}\|f(\sigma\underset{(-)}{+}i\varepsilon)\|^2\,d\sigma\equiv\|f\|_{H_2^{(\mp)}(E)}<\infty.$$

The class of these functions will henceforth be identified with the class of boundary values of these functions on the real axis which forms a subspace of $L_2(E)$, the Hilbert space of E-valued functions which are square summable on the axis $(-\infty,\infty)$ (see [1] and [11]).

which implies the inequality

$$\int_{-\infty}^{\infty}\|\alpha(L^{\|}-\sigma+i\varepsilon)^{-1}u\|^2\,d\sigma\leqslant 2\pi\|u\|^2.$$

We note that along the way we have proved the estimate

$$\int_{-\infty}^{\infty}\|(L^{\|}-\sigma+i\varepsilon)^{-1}u\|^2\,d\sigma\leqslant\frac{\pi}{\varepsilon}\|u\|^2.$$

By the definition of the norm in the Hardy class, we have

$$\|\alpha(L^{\|}-\lambda)^{-1}u\|^2_{H_2^-(E)}=\sup_{\varepsilon>0}\int_{-\infty}^{\infty}\|\alpha(L^{\|}-\sigma+i\varepsilon)^{-1}u\|^2\,d\sigma\leqslant 2\pi\|u\|^2.$$

Construction of the spectral representation of a selfadjoint dilation of the dissipative operator $L^{\|}$. It is known [1] that for the dissipative operator $L^{\|}$ there exists a functional model. We briefly describe it, following [12].

The characteristic function S of the operator $L^{\|}$ is the contractive, analytic, operator-valued function $S(\lambda)\colon E\to E$ defined for $\operatorname{Im}\lambda>0$ by

$$S(\lambda)=I+i\alpha((L')^*-\lambda)^{-1}\alpha,\quad \operatorname{Im}\lambda>0. \tag{3}$$

In the case of an unbounded operator α the function $S(\lambda)$ is first defined by (3) on the manifold $E\cap D(\alpha)$ (here $S(\lambda)\colon D(\alpha)\to D(\alpha)$) and is then extended by continuity to the whole space E. The operator $S(\lambda)$ is invertible in the half plane $\operatorname{Im}\lambda>0$ everywhere with the exception of points of the spectrum of the operator $L^{\|}$. Formula (3) makes it possible to consider $S(\lambda)$ for $\operatorname{Im}\lambda<0$ with $S(\bar\lambda)=(S^*(\lambda))^{-1}$. Finally, $S(\lambda)$ possesses boundary values on the real axis in the strong sense: $S(k)\equiv S(k+i0)$, $k\in\mathbb{R}$.

As the model space $\mathcal{H}$ we take $\mathcal{H}\overset{\text{def}}{=}L_2\left(\begin{smallmatrix} I & S^*\\ S & I\end{smallmatrix}\right)$, the Hilbert space of two-component functions $(\tilde g,g)$ on the axis ($\tilde g(k), g(k)\in E$, $k\in\mathbb{R}$) with metric

$$\left(\begin{pmatrix}\tilde g\\ g\end{pmatrix},\begin{pmatrix}\tilde g\\ g\end{pmatrix}\right)=\int_{-\infty}^{\infty}\left(\begin{pmatrix} I & S^*(k)\\ S(k) & I\end{pmatrix}\begin{pmatrix}\tilde g(k)\\ g(k)\end{pmatrix},\begin{pmatrix}\tilde g(k)\\ g(k)\end{pmatrix}\right)_{E\oplus E}d_k.$$

It is assumed that the set of two-component functions has been factored by the set of elements with norm equal to zero.

The subspaces in $\mathcal{H}$

$$D_-=\begin{pmatrix}0\\ H_2^-(E)\end{pmatrix},\quad D_+=\begin{pmatrix}H_2^+(E)\\ 0\end{pmatrix}$$

are respectively the incoming and outgoing subspaces [13] for the unitary group of translations in $\mathcal{H}$:

$$U_t(\tilde g,g)(k)=\exp(ikt)(\tilde g(k),g(k)),\qquad t\in\mathbb{R},\ k\in\mathbb{R}.$$

This form of the model constructed in [12] differs from the usual model of Sz.-Nagy and Foiaş only in the symmetry with respect to the subspaces $D_\pm$ which is convenient for our purposes. Let $K\overset{\text{def}}{=}\mathcal{H}\ominus[D_-\oplus D_+]$ ($K=\{(\tilde g,g)\in\mathcal{H}\colon \tilde g+S^*g\in H_2^-(E),\ S\tilde g+g\in H_2^+(E)\}$), and let P_K be the orthogonal projection of $\mathcal{H}$ onto K:

$$P_K\begin{pmatrix}\tilde g\\ g\end{pmatrix}=\begin{pmatrix}\tilde g-P_+(\tilde g+S^*g)\\ g-P_-(S\tilde g+g)\end{pmatrix}, \tag{4}$$

where $P_\pm$ are orthogonal projections of $L_2(E)$ onto $H_2^\pm(E)$. The following assertion holds [1], [13]: *the operator $L^\parallel$ is unitarily equivalent to the generator of the semigroup $Z_t = P_K U_t|K$, $t \geqslant 0$, acting in the space K.* The latter is equivalent to the following: *the operator $(L^\parallel - \lambda_0)^{-1}$ is unitarily equivalent to the operator $P_K(k - \lambda_0)^{-1}|K$ for all λ_0*, $\operatorname{Im}\lambda_0 < 0$; that is, the operator of multiplication by k in $\mathcal{H}$ serves as the (minimal) selfadjoint dilation of the operator $L^\parallel$.

The remainder of this section is devoted to finding explicit formulas for passing from the Hilbert space H to the space K whereby the operator $L^\parallel$ goes over into the model operator. To this end we first describe the selfadjoint dilation of a dissipative operator constructed in [10]. We use the same notation for the objects unitarily equivalent to the $(H, D_\pm)$ introduced earlier and hope that this leads to no confusion.

The operator $\mathfrak{L}$, which is a selfadjoint dilation of the operator $L^\parallel$, acts in the Hilbert space

$$\mathcal{H} = L_2((-\infty, 0); E) \oplus H \oplus L_2((0, \infty); E) \equiv D_- \oplus H \oplus D_+.$$

The space H is imbedded in $\mathcal{H}$ as a subspace in a natural way: $H \equiv 0 \oplus H \oplus 0$. On its domain

$$D(\mathfrak{L}) = \{(v_-, u, v_+): u \in D(A), v_- \in W_2^1((-\infty, 0); E), \\ v_+ \in W_2^1((0, \infty); E), v_+(0) - v_-(0) = i\alpha u\}$$

it is given by the formula

$$\mathfrak{L}\begin{pmatrix} v_- \\ u \\ v_+ \end{pmatrix} \begin{pmatrix} -\dfrac{1}{i}\dfrac{d}{d\xi}v_-(\xi) \\ Au + (\alpha/2)[v_+(0) + v_-(0)] \\ -\dfrac{1}{i}\dfrac{d}{d\xi}v_+(\xi) \end{pmatrix}.$$

We note that the boundary values $v_\pm(0) = v_\pm(\xi)|_{\xi=0}$ exist because $v_\pm$ belong to the Sobolev spaces. It can be verified [10] that $\mathfrak{L}^* = \mathfrak{L}$ and $(L^\parallel - \lambda_0)^{-1} = P_H(\mathfrak{L} - \lambda_0)^{-1}|H$, $\operatorname{Im}\lambda_0 < 0$, where P_H is the orthogonal projection onto $0 \oplus H \oplus 0$ in $\mathcal{H}$. Moreover, this dilation is minimal, i.e. $\mathcal{H} = \bigvee_{t\in\mathbf{R}} \exp(i\mathfrak{L}t)H$.

From the general theory of Sz.-Nagy and Foiaş it follows that there exists a spectral representation of the operator $\mathfrak{L}$ in which it acts as the operator of multiplication by k in the space $L_2(\begin{smallmatrix} I & S^* \\ S & I \end{smallmatrix})$. A concrete manner of passing to this spectral representation will be indicated below.

We introduce the following two bounded mappings of $\mathcal{H}$ into $L_2(E)$:

$$\mathfrak{F}_+: (v_-, u, v_+) \to -(1/\sqrt{2\pi})\alpha(L^\parallel - k + i0)^{-1}u + S^*(k)\hat{v}_-(k) + \hat{v}_+(k),$$

$$\mathfrak{F}_-: (v_-, u, v_+) \to -(1/\sqrt{2\pi})\alpha(L^{-\parallel} - k - i0)^{-1}u + \hat{v}_-(k) + S(k)\hat{v}_+(k).$$

The boundary values in the strong sense on the real axis $\alpha(L^{\pm\parallel} - k \pm i0)^{-1}u$ exist and belong to $H_2^\mp(E)$ because $\alpha(L^{\pm\parallel} - \lambda)^{-1}u$ belongs to the Hardy class $H_2^\mp(E)$ (Theorem 1), while $\hat{v}_\pm = (1/\sqrt{2\pi})\int_{\mathbf{R}} \exp(ik\xi)v_\pm(\xi)\,d\xi$ is the Fourier transform of the function $v_\pm(\xi)$ extended by zero to the complementary semiaxis. Under the mapping $\mathfrak{F}_{+(-)}$ the outgoing (incoming) subspace $D_{+(-)}$ is mapped onto $H_2^{+(-)}(E)$ (the Paley-Wiener theorem [14]).

The mappings $\mathfrak{F}_{\pm}$ are used in constructing the spectral representation of $\mathfrak{L}$ in the following manner.

THEOREM 2. *There exists an isometric mapping of* $\mathcal{H} = L_2((-\infty, 0); E) \oplus H \oplus L_2((0, \infty); E)$ *onto* $L_2\left(\begin{smallmatrix} I & S^* \\ S & I \end{smallmatrix}\right)$ *satisfying the conditions*

$$\begin{aligned} \tilde{g} + S^*g &= \mathfrak{F}_+(v_-, u, v_+), \\ S\tilde{g} + g &= \mathfrak{F}_-(v_-, u, v_+), \end{aligned} \tag{5}$$

under which $\mathfrak{L}$ *goes over into the operator of multiplication by the real variable* k, *while the subspace* $0 \oplus H \oplus 0$ *is mapped onto* K.

The proof is direct computation. From the properties of the norm in $L_2\left(\begin{smallmatrix} I & S^* \\ S & I \end{smallmatrix}\right)$ it follows easily that

$$\|(\tilde{g}, g)\|_{\mathcal{H}} \geqslant \max\{\|\tilde{g} + S^*g\|, \|S\tilde{g} + g\|\}$$

on the one hand,(3) and, on the other hand, the element $(\tilde{g}, g)$ for which $\tilde{g} + S^*g = 0$ and $S\tilde{g} + g = 0$ is equivalent to zero. Therefore, the mapping satisfying conditions (5) is unique. The proof of its existence is broken into three lemmas. According to Proposition 1, linear combinations of elements of the form $(L^{-\|} - \lambda_0)^{-1}\alpha v$, $v \in E$, λ_0 a regular point of $L^{-\|}$, are dense in H (it follows from condition (2) that the operator $(L^{-\|} - \lambda_0)^{-1}\alpha$ is bounded). We first give our mapping onto elements of this form.

LEMMA 1. *The mapping*

$$\begin{aligned} &\left(v_-, \left(L^{-\|} - \lambda_0\right)^{-1}\alpha v, v_+\right) \\ &\qquad \to \left(\hat{v}_+ + (i/\sqrt{2\pi})(k - \lambda_0)^{-1}v;\ \hat{v}_- - (i\sqrt{2\pi})(k - \lambda_0)^{-1}S(\lambda_0)v\right) \end{aligned}$$

satisfies condition (5).

PROOF. The lemma is verified by direct computation:

$$\begin{aligned} S\tilde{g} + g &= S\hat{v}_+ + (i/\sqrt{2\pi})(k - \lambda_0)^{-1}Sv + \hat{v}_- - (i/\sqrt{2\pi})(k - \lambda_0)^{-1}S(\lambda_0)v \\ &= \hat{v} + S\hat{v}_+ + (i\sqrt{2\pi})(k - \lambda_0)^{-1}(S(k) - S(\lambda_0))v \\ &= \hat{v}_- + S\hat{v}_+ + (i/\sqrt{2\pi})(k - \lambda_0)^{-1}\left(i\alpha(L^{-\|} - k - i0)^{-1}\alpha - i\alpha\left(L^{-\|} - \lambda_0\right)^{-1}\alpha\right)v \\ &= \hat{v}_- + S\hat{v}_+ - (1/\sqrt{2\pi})(k - \lambda_0)^{-1}\left(\alpha(L^{-\|} - k - i0)^{-1}\right) \\ &\qquad \times \left((L^{-\|} - \lambda_0) - (L^{-\|} - k - i0)\right)\left(L^{-\|} - \lambda_0\right)^{-1}\alpha v \\ &= \hat{v}_- + S\hat{v}_+ - (1/\sqrt{2\pi})\left(\alpha(L^{-\|} - k - i0)^{-1}\right)\left(L^{-\|} - \lambda_0\right)^{-1}\alpha v \\ &= \mathfrak{F}_-\left(v_-, \left(L^{-\|} - \lambda_0\right)^{-1}\alpha v, v_+\right). \end{aligned}$$

The other equality of (5) is verified similarly.

The mapping given in Lemma 1 can be extended by linearity to linear combinations of elements of the form $(L^{-\|} - \lambda_i)^{-1}\alpha v_i$, $\operatorname{Im} \lambda_i \neq 0$, $i = 0, 1, 2, \ldots$. This extension is unique because of the uniqueness of the mapping satisfying conditions (5).

(3) Unless otherwise specified, the norm and inner product of single-component functions are always understood to be those of $L_2(E)$.

LEMMA 2. *The mapping* Φ *described in Lemma 2 extended by linearity to elements of the form* $\Sigma_i(L^{-\|} - \lambda_i)^{-1}\alpha v_i$ *is isometric.*

PROOF. Because of the bilinearity of the inner product, it suffices to verify that the inner product is preserved only for the two elements $(L^{-\|} - \lambda_0)^{-1}\alpha v$ and $(L^{-\|} - \lambda_0')^{-1}\alpha v'$. There here arise four different situations corresponding to the cases $\pm \operatorname{Im} \lambda_0 > 0$, $\pm \operatorname{Im} \lambda_0' > 0$, which are considered in the same way. We consider only the case $\operatorname{Im} \lambda_0 > 0$, $\operatorname{Im} \lambda_0' < 0$. We compute the inner product of the images of the elements

$$\left(v_-, \sqrt{2\pi}\,(L^{-\|} - \lambda_0)^{-1}\alpha v, v_+\right) \quad \text{and} \quad \left(v', \sqrt{2\pi}\,(L^{-\|} - \lambda_0')^{-1}\alpha v', v_+'\right)$$

as follows:

$$\begin{aligned}
&\left(\hat{v}_+ + i(k - \lambda_0)^{-1}v + S^*\hat{v}_- - iS^*(k - \lambda_0)^{-1}S(\lambda_0)v,\ \hat{v}_+' + i(k - \lambda_0')^{-1}v'\right)\\
&\quad + \left(S\hat{v}_+ + iS(k - \lambda_0)^{-1}v + \hat{v}_- - i(k - \lambda_0)^{-1}S(\lambda_0)v,\ \hat{v}_-' - i(k - \lambda_0')^{-1}S(\lambda_0')v'\right)\\
&= \left(\hat{v}_+, \hat{v}_+' + i(k - \lambda_0')^{-1}v'\right)\\
&\quad + \left(i(S(k) - S(\lambda_0))(k - \lambda_0)^{-1}\,v + S\hat{v}_+, -i(k - \lambda_0')^{-1}S(\lambda_0')v'\right) + (\hat{v}_-, \hat{v}_-')\\
&= (\hat{v}_+, \hat{v}_+') + (\hat{v}_-, \hat{v}_-') + \left(\hat{v}_+, i(I - S^*S(\lambda_0'))(k - \lambda_0')^{-1}v'\right)\\
&\quad - \left((S(k) - S(\lambda_0))(k - \lambda_0)^{-1}v, S(\lambda_0')(k - \lambda_0')^{-1}v'\right)\\
&= (\hat{v}_+, \hat{v}_+') + (\hat{v}_-, \hat{v}_-') - \left((S(k) - S(\lambda_0))(k - \lambda_0)^{-1}v, S(\lambda_0')(k - \lambda_0)^{-1}v'\right)
\end{aligned}$$

Here we have used the identity $S(\lambda_0') = (S^*(\overline{\lambda_0'}))^{-1}$, by virtue of which

$$I - S^*(k)S(\lambda_0')(k - \lambda_0')^{-1}v' \in H_2^-(E).$$

Further, evaluation of the integral by residues gives

$$\begin{aligned}
&-\left((S(k) - S(\lambda_0))(k - \lambda_0)^{-1}v, S(\lambda_0')(k - \lambda_0')v'\right)\\
&= -2\pi i\left((S(\overline{\lambda_0'}) - S(\lambda_0))\left(\overline{\lambda_0'} - \lambda_0\right)^{-1}v, S(\lambda_0')v'\right)_E\\
&= -2\pi i\left(\overline{\lambda_0'} - \lambda_0\right)^{-1}\left((I - S^*(\lambda_0')S(\lambda_0))v, v'\right)_E\\
&= -2\pi i\left(\overline{\lambda_0'} - \lambda_0\right)^{-1}\Big(\Big(-i\alpha(L^{-\|} - \lambda_0)^{-1}\alpha + i\alpha\left(L^{\|} - \overline{\lambda_0'}\right)^{-1}\alpha + i\alpha\left(L^{\|} - \overline{\lambda_0'}\right)^{-1}\\
&\qquad \times i\alpha^2(L^{-\|} - \lambda_0)^{-1}\alpha\Big)v, v'\Big)_E\\
&= -2\pi i\left(\overline{\lambda_0'} - \lambda_0\right)^{-1}\Big(\Big(-i\alpha(L^{-\|} - \lambda_0)^{-1}\alpha + i\alpha\left(L^{\|} - \overline{\lambda_0'}\right)^{-1}\alpha\\
&\qquad + i\alpha\left(L^{\|} - \overline{\lambda_0'}\right)^{-1}\left[\left(L^{\|} - \overline{\lambda_0'}\right) - (L^{-\|} - \lambda_0) + \left(\overline{\lambda_0'} - \lambda_0\right)\right](L^{-\|} - \lambda_0)^{-1}\alpha\Big)v, v'\Big)_E\\
&= 2\pi\left(\alpha\left(L^{\|} - \overline{\lambda_0'}\right)^{-1}(L^{-\|} - \lambda_0)^{-1}\alpha v, v'\right)_E\\
&= 2\pi\left((L^{-\|} - \lambda_0)^{-1}\alpha v, (L^{-\|} - \lambda_0')^{-1}\alpha v'\right)_E.
\end{aligned}$$

From this and the above computations it follows that the inner product is preserved under our mapping.

Since the mapping Φ is isometric, it can be extended by continuity to the whole space $\mathcal{H}$, because, according to Proposition 1, elements of the form

$$\left(v_-, \sum_i (L^{-\|} - \lambda_i)^{-1}\alpha v_i, v_+\right)$$

form a dense set in $\mathcal{H}$. We shall verify that under this mapping $\mathfrak{L}$ goes over into the operator of multiplication by k. It is more convenient to verify this assertion for the resolvents $(\mathfrak{L} - \lambda_0)^{-1}$ and $(k - \lambda_0)^{-1}$.

LEMMA 3. *Under the mapping* Φ,

$$\Phi \circ (\mathfrak{L} - \lambda_0)^{-1} = (k - \lambda_0)^{-1} \circ \Phi, \quad \operatorname{Im} \lambda_0 \neq 0.$$

PROOF. It suffices to prove that $\mathcal{F}_\pm(\mathfrak{L} - \lambda_0)^{-1} = (k - \lambda_0)^{-1}\mathcal{F}_\pm$. To be specific, we assume that $\operatorname{Im} \lambda_0 < 0$, and we verify this equality for $\mathcal{F}_+$. Let $(\mathfrak{L} - \lambda_0)^{-1}(v_-, u, v_+) \overset{\text{def}}{=} (v'_-, u', v'_+)$; from the definition of $\mathfrak{L}$ it then follows easily that

$$\hat{v}'_+ = (k - \lambda_0)^{-1}\left(\hat{v}_+(k) - (1/i\sqrt{2\pi})v'_+(0)\right) \in H_2^+(E)$$

and

$$\hat{v}'_- = (k - \lambda_0)^{-1}\left(\hat{v}_-(k) + (1/i\sqrt{2\pi})v'_-(0)\right) \in H_2^-(E),$$

while

$$\hat{v}_-(\lambda_0) = -(1/i\sqrt{2\pi})v'_-(0), \qquad v'_+(0) - v'_-(0) = i\alpha u'$$

(for the details see [2], where an analogous computation is made). Moreover,

$$(A - \lambda_0)u' + (\alpha/2)[v'_+(0) + v'_-(0)] = u,$$

and hence

$$u' = \left(L^{\|} - \lambda_0\right)^{-1}(u - \alpha v'_-(0)).$$

Using these relations, we arrive at the required result:

$$\begin{aligned}
\mathcal{F}_+(v'_-, u', v'_+) &= -(1/\sqrt{2\pi})\alpha(L^{\|} + k + i0)^{-1}u' + S^*(k)\hat{v}'_-(k) + \hat{v}'_+(k) \\
&= -(1/\sqrt{2\pi})\alpha(L^{\|} - k + i0)^{-1}\left(L^{\|} - \lambda_0\right)^{-1}(u - \alpha v'_-(0)) \\
&\quad + (k - \lambda_0)^{-1}\left(S^*(k)\hat{v}_-(k)\right. \\
&\quad \left. + (1/i\sqrt{2\pi})S^*(k)v'_-(0) + \hat{v}_+(k) - (1/i\sqrt{2\pi})v'_+(0)\right) \\
&= (k - \lambda_0)^{-1}\mathcal{F}_+(v_-, u, v_+) + (k - \lambda_0)^{-1}(1/i\sqrt{2\pi}) \\
&\quad \times \left(S^*(k)v'_-(0) - v'_+(0) + i\alpha\left(L^{\|} - \lambda_0\right)^{-1}(u - \alpha v'_-(0))\right. \\
&\quad \left. + i\alpha(L^{\|} - k + i0)^{-1}\alpha v'_-(0)\right) \\
&= (k - \lambda_0)^{-1}\mathcal{F}_+(v_-, u, v_+) + (k - \lambda_0)^{-1}(1/i\sqrt{2\pi}) \\
&\quad \times (v'_-(0) - v'_+(0) + i\alpha u') \\
&= (k - \lambda_0)^{-1}\mathcal{F}_+(v_-, u, v_+).
\end{aligned}$$

Verification of the identity $\mathcal{F}_-(\mathfrak{L} - \lambda_0)^{-1} = (k - \lambda_0)^{-1}\mathcal{F}_-$ basically coincides with the proof just presented.

To complete the proof of Theorem 2, it now suffices to show that the range of the mapping Φ is the entire space $L_2\left(\begin{smallmatrix} I & S^* \\ S & I \end{smallmatrix}\right)$. From the definition of Φ it follows that

$$\Phi\left(L_2((-\infty, 0); E) \oplus 0 \oplus L_2((0, \infty); E)\right) = (H_2^+(E); H_2^-(E)).$$

By applying Lemma 3, we extend this set to

$$\bigvee_{(\operatorname{Im}\lambda_0',\ \lambda_0'\neq 0)} \left((k-\lambda_0)^{-1}H_2^+(E), (k-\lambda_0')^{-1}H_2^-(E)\right) = \left(\overline{L_2(E); L_2(E)}\right).$$

It is possible to write the closed linear span, since the range of an isometric mapping is closed. Further, $(\overline{L_2(E); L_2(E)})$ coincides with the entire space. Finally, the equality $\Phi(0 \oplus H \oplus 0) = K$ follows immediately from the fact that Φ is isometric and the obvious properties

$$\Phi(L_2((-\infty, 0);\ E) \oplus 0 \oplus 0) = (0;\ H_2^-(E)),$$
$$\Phi(0 \oplus 0 \oplus L_2((0,\ \infty);\ E)) = (H_2^+(E);\ 0).$$

The proof of Theorem 2 is complete.

§2. The functional model for a class of operators "close" to selfadjoint operators

The formulas for passing to the spectral representation of the selfadjoint dilation of the dissipative operator $L^{\parallel}$ indicated in Theorem 2 make it possible to compute the action of the operators $L^{(\kappa)}$ in the model representation. We preserve the notation $\mathcal{H}$ for the model space $L_2\begin{pmatrix} I & S^* \\ S & I\end{pmatrix}$.

THEOREM 3. *In the model representation $\mathcal{H}$ the following formulas for the action of the resolvent of the operator $L^{(\kappa)}$ hold (λ_0 is a regular point of $L^{(\kappa)}$, and $(\tilde{g}, g) \in K$):*

1) *for* $\operatorname{Im}\lambda_0 < 0$

$$(L^{(\kappa)} - \lambda_0)^{-1}\begin{pmatrix}\tilde{g} \\ g\end{pmatrix} = P_K \frac{1}{(k-\lambda_0)}\begin{bmatrix} \tilde{g} \\ g - \frac{1+i\kappa}{2}\left[I + \left(S^*(\bar{\lambda}_0) - I\right)\frac{1+i\kappa}{2}\right]^{-1}(\tilde{g} + S^*g)(\lambda_0)\end{bmatrix}, \tag{6}$$

2) *for* $\operatorname{Im}\lambda_0 > 0$

$$(L^{(\kappa)} - \lambda_0)^{-1}\begin{pmatrix}\tilde{g} \\ g\end{pmatrix} = P_K \frac{1}{(k-\lambda_0)}\begin{bmatrix} \tilde{g} - \frac{1-i\kappa}{2}\left[I + (S(\lambda_0) - I)\frac{1-i\kappa}{2}\right]^{-1}(S\tilde{g} + g)(\lambda_0) \\ g\end{bmatrix},$$

where $(\tilde{g} + S^*g)(\lambda_0)$ denotes the value at the point λ_0 of the analytic extension of the function $\tilde{g} + S^*g \in H_2^-(E)$ to the lower half plane ($(S\tilde{g} + g)(\lambda_0)$ has the similar meaning, with $H_2^-(E)$ replaced by $H_2^+(E)$).

PROOF. We note first of all that the inverse operators in the above formulas exist because of the following identities:

$$\left(I + (S^*(\bar{\lambda}_0) - I)\frac{1+i\kappa}{2}\right)^{-1} = I + i\alpha(L^{(\kappa)} - \lambda_0)^{-1}\alpha\left(\frac{1+i\kappa}{2}\right),$$
$$\left(I + (S(\lambda_0) - I)\frac{1-i\kappa}{2}\right)^{-1} = I - i\alpha(L^{(\kappa)} - \lambda_0)^{-1}\alpha\left(\frac{1-i\kappa}{2}\right).$$

These are easily proved on the basis of the Hilbert identity for the resolvents. Let $\operatorname{Im}\lambda_0 < 0$. To prove the first of the formulas, we consider the preimage $(0, u, 0)$, $u \in H$, of the element $(\tilde{g}, g) \in K$ under the mapping Φ. According to Theorem 2 (below the elements of $H_2^{-(+)}(E)$ are represented by analytic functions in the half plane $\operatorname{Im}\lambda < 0$ ($\operatorname{Im}\lambda > 0$)),

$$\begin{aligned}\sqrt{2\pi}\,(\tilde{g} + S^* g)(\lambda) &= -\alpha(L^{\|} - \lambda)^{-1} u \\ &= \alpha(L^{\|} - \lambda)^{-1}\big[(L^{\|} - \lambda) - (L^{(\kappa)} - \lambda)\big](L^{(\kappa)} - \lambda)^{-1}u - \alpha(L^{(\kappa)} - \lambda)^{-1}u \\ &= -\left(I - i\alpha(L^{\|} - \lambda)^{-1}\alpha\frac{1 + i\kappa}{2}\right)\alpha(L^{(\kappa)} - \lambda)^{-1}u \\ &= -\left(I + (S^*(\bar{\lambda}) - I)\frac{1 + i\kappa}{2}\right)\alpha(L^{(\kappa)} - \lambda)^{-1}u \\ &\overset{\text{def}}{=} -\Theta'_\kappa(\lambda)\alpha(L^{(\kappa)} - \lambda)^{-1}u\end{aligned}$$

for $\operatorname{Im}\lambda < 0$, while for $\operatorname{Im}\lambda > 0$

$$\begin{aligned}\sqrt{2\pi}\,(S\tilde{g} + g)(\lambda) &= -\alpha(L^{-\|} - \lambda)^{-1}u \\ &= \alpha(L^{-\|} - \lambda)^{-1}\big[(L^{-\|} - \lambda) - (L^{(\kappa)} - \lambda)\big](L^{(\kappa)} - \lambda)^{-1}u - \alpha(L^{(\kappa)} - \lambda)^{-1}u \\ &= -\left(I + i\alpha(L^{-\|} - \lambda)^{-1}\alpha\frac{1 - i\kappa}{2}\right)\alpha(L^{(\kappa)} - \lambda)^{-1}u \\ &= -\left(I + (S(\lambda) - I)\frac{1 - i\kappa}{2}\right)\alpha(L^{(\kappa)} - \lambda)^{-1}u \\ &\overset{\text{def}}{=} -\Theta_\kappa(\lambda)\alpha(L^{(\kappa)} - \lambda)^{-1}u.\end{aligned}$$

We shall prove that the element $\Phi(0, (L^{(\kappa)} - \lambda_0)^{-1}u, 0)$, $\operatorname{Im}\lambda_0 < 0$, coincides with the element indicated in Theorem 3. It suffices to verify the corresponding equalities (5) for $\mathfrak{F}_\pm(0, (L^{(\kappa)} - \lambda_0)^{-1}u, 0)$. We have

$$\begin{aligned}&\alpha(L^{\|} - \lambda)^{-1}(L^{(\kappa)} - \lambda_0)u = \Theta'_\kappa(\lambda)\alpha(L^{(\kappa)} - \lambda)^{-1}(L^{(\kappa)} - \lambda_0)^{-1}u \\ &= (\lambda - \lambda_0)^{-1}\Theta'_\kappa(\lambda)\alpha(L^{(\kappa)} - \lambda)^{-1}u - (\lambda - \lambda_0)^{-1}\Theta'_\kappa(\lambda)\alpha(L^{(\kappa)} - \lambda_0)^{-1}u \\ &= (\lambda - \lambda_0)^{-1}\alpha(L^{\|} - \lambda)^{-1}u - (\lambda - \lambda_0)^{-1}\Theta'_\kappa(\lambda)(\Theta'_\kappa(\lambda_0))^{-1}\alpha(L^{\|} - \lambda_0)^{-1}u \\ &= -(\lambda - \lambda_0)^{-1}\sqrt{2\pi}\,(\tilde{g} + S^*g)(\lambda) + (\lambda - \lambda_0)^{-1}\sqrt{2\pi}\,\Theta'_\kappa(\lambda)(\Theta'_\kappa(\lambda_0))^{-1}(\tilde{g} + S^*g)(\lambda_0).\end{aligned}$$

Passing to the boundary values on the real axis, we obtain

$$\begin{aligned}\left(\mathfrak{F}_+\left(0, (L^{(\kappa)} - \lambda_0)^{-1}u, 0\right)\right)(k) &= (k - \lambda_0)^{-1}(\tilde{g} + S^*g)(k) \\ &\quad - (k - \lambda_0)^{-1}\Theta'_\kappa(k)(\Theta'_\kappa(\lambda_0))^{-1}(\tilde{g} + S^*g)(\lambda_0),\end{aligned}$$

where $\Theta'_\kappa(k) \equiv \Theta'_\kappa(k - i0)$ is the boundary value in the strong sense of the analytic, operator-valued function introduced above. On the other hand, by (6) $\mathfrak{F}_+(0, (L^{(\kappa)} - \lambda_0)^{-1}u, 0)$ must coincide with the following expression (see (4)):

$$(k-\lambda_0)^{-1}\tilde g - P_+\Big[(k-\lambda_0)^{-1}(\tilde g + S^*g) - (k-\lambda_0)^{-1}S^*\frac{1+i\kappa}{2}(\Theta'_\kappa(\lambda_0))^{-1}$$

$$\times(\tilde g + S^*g)(\lambda_0)\Big] + S^*\Big[(k-\lambda_0)^{-1}g - \frac{1+i\kappa}{2}(\Theta'_\kappa(\lambda_0))^{-1}(k-\lambda_0)^{-1}(\tilde g + S^*g)(\lambda_0)$$

$$- P_-\Big((k-\lambda_0)^{-1}(S\tilde g + g) - \frac{1+i\kappa}{2}(\Theta'_\kappa(\lambda_0))^{-1}(k-\lambda_0)^{-1}(\tilde g + S^*g)(\lambda_0)\Big)\Big]$$

$$= (k-\lambda_0)^{-1}(\tilde g + S^*g) - (k-\lambda_0)^{-1}(\tilde g + S^*g)(\lambda_0) + (k-\lambda_0)^{-1}S^*(\lambda_0)\frac{1+i\kappa}{2}$$

$$\times(\Theta'_\kappa(\lambda_0))^{-1}(\tilde g + S^*g)(\lambda_0) - S^*\frac{1+i\kappa}{2}(\Theta'_\kappa(\lambda_0))^{-1}(k-\lambda_0)^{-1}(\tilde g + S^*g)(\lambda_0)$$

$$= (k-\lambda_0)^{-1}(\tilde g + S^*g) - \Big[\Theta'_\kappa(\lambda_0) - S^*(\bar\lambda_0)\frac{1+i\kappa}{2} + S^*\frac{1+i\kappa}{2}\Big](\Theta'_\kappa(\lambda_0))^{-1}$$

$$\times(k-\lambda_0)^{-1}(\tilde g + S^*g)(\lambda_0)$$

$$= (k-\lambda_0)^{-1}(\tilde g + S^*g) - \Theta'_\kappa(k)(\Theta'_\kappa(\lambda_0))^{-1}(k-\lambda_0)^{-1}(\tilde g + S^*g)(\lambda_0).$$

The required identity has been proved. We have here used simple formulas for the computation of $P_\pm$ of a meromorphic function having a pole of first order in the corresponding half plane. To complete the proof of assertion 1) of Theorem 3, it remains to verify the analogous assertion for $\mathcal{F}_-$. Here we use the identity

$$\alpha(L^{-1}-\lambda)^{-1}(L^{(\varkappa)}-\lambda_0)^{-1} = (\lambda-\lambda_0)^{-1}\alpha(L^{-1}-\lambda)^{-1}$$

$$-(\lambda-\lambda_0)^{-1}\Theta_\varkappa(\lambda)\alpha(L^{(\varkappa)}-\lambda_0)^{-1}$$

and its corollary

$$\big(\mathcal{F}_-(0,(L^{(\kappa)}-\lambda_0)^{-1}u,0)\big)(k) = (k-\lambda_0)^{-1}(S\tilde g + g)(k)$$

$$-(k-\lambda_0)^{-1}\Theta_\kappa(k)(\Theta'_\kappa(\lambda_0))^{-1}(\tilde g + S^*g)(\lambda_0).$$

Verification of the second of conditions (5) is accomplished by the following computations:

$$(k-\lambda_0)^{-1}(S\tilde g + g) - SP_+\Big[(k-\lambda_0)^{-1}(\tilde g + S^*g) - S^*\frac{1+i\kappa}{2}(\Theta'_\kappa(\lambda_0))^{-1}(k-\lambda_0)^{-1}$$

$$\times(\tilde g + S^*g)(\lambda_0)\Big]$$

$$-\frac{1+i\kappa}{2}(\Theta'_\kappa(\lambda_0))^{-1}(k-\lambda_0)^{-1}(\tilde g + S^*g)(\lambda_0)$$

$$= (k-\lambda_0)^{-1}(S\tilde g + g) - S(k-\lambda_0)^{-1}(\tilde g + S^*g)(\lambda_0)$$

$$+ S(k)S^*(\bar\lambda_0)\frac{1+i\kappa}{2}(\Theta'_\kappa(\lambda_0))^{-1}(k-\lambda_0)^{-1}(\tilde g + S^*g)(\lambda_0)$$

$$-\frac{1+i\kappa}{2}(\Theta'_\kappa(\lambda_0))^{-1}(k-\lambda_0)^{-1}(\tilde g + S^*g)(\lambda_0)$$

$$= (k-\lambda_0)^{-1}(S\tilde g + g) - \Big[S(k)\Theta'_\kappa(\lambda_0) - S(k)S^*(\bar\lambda_0)\frac{1+i\kappa}{2} + \frac{1+i\kappa}{2}\Big]$$

$$+ (\Theta'_\kappa(\lambda_0))^{-1}(k-\lambda_0)^{-1}(\tilde g + S^*g)(\lambda_0)$$

$$= (k-\lambda_0)^{-1}(S\tilde g + g) - \Theta_\kappa(k)(\Theta'_\kappa(\lambda_0))^{-1}(k-\lambda_0)^{-1}(\tilde g + S^*g)(\lambda_0).$$

Since the proof of assertion 2) of Theorem 3 is analogous to the foregoing, we omit it.

REMARKS. 1. Theorem 3 for $\kappa = iJ$ gives formulas for computing the resolvents $(L - \lambda_0)^{-1}$, $\operatorname{Im}\lambda_0 \neq 0$, of an, in general, nondissipative operator L on the whole space K (in contrast to Theorem 1 of [2]). In this case Θ_κ and Θ'_κ coincide with the divisors of the characteristic function of the operator L in its triangular factorization (see [2] and [3]).

2. For $\kappa = iI$ and $\operatorname{Im}\lambda_0 < 0$ equation (6) coincides with the well-known formula for the selfadjoint dilation of the dissipative operator $L^{\|}$.

3. The operator-valued functions

$$\Theta_\varkappa(\lambda) = I + (S(\lambda) - I)\frac{1 - i\varkappa}{2} \quad \text{and} \quad \Theta_{\varkappa'}(\lambda) = I + (S^*(\bar{\lambda}) - I)\frac{1 + i\varkappa}{2},$$

which are analytic and bounded in their respective half planes, may be considered a generalization of the characteristic function of a completely nonselfadjoint operator. In particular, this pertains to the selfadjoint operators A and $A + V$ ($\kappa = 0, J$). Such operator-valued functions arise naturally in problems of perturbation theory. Here by appealing to the characteristic functions of a dissipative operator it is possible to deal not with a meromorphic function but rather with an operator-valued function which is analytic and bounded in a half plane without imposing additional restrictions on the perturbation V. An important consequence of this is the possibility of considering the boundary values of these functions on the real axis.

§3. Computation of the resolvent of the operator $L^{(\kappa)}$ on "smooth" vectors

It is known [1], [12] that the absolutely continuous subspace of the dissipative, completely nonselfadjoint operator $L^{\|}$ coincides with the closure of the set of vectors in K of the form $P_K(\tilde{g}, -S\tilde{g})$. It is shown in [2] that this set coincides with the collection of images in K under the mapping Φ of vectors $u \in H$ such that $\alpha(L^{\|} - \lambda)^{-1}u \in H_2^+(E)$. Here $\alpha(L^{\|} - \lambda)^{-1}u$ is considered as an analytic vector-valued function of λ ($\operatorname{Im}\lambda > 0$) with values in the auxiliary Hilbert space E. The analytic description of a dense set in the absolutely continuous subspace of the operator $L^{\|}$ made it possible to determine the absolutely continuous subspace N_e of the, in general, nondissipative operator L in an analogous fashion [3]:

$$N_e \overset{\text{def}}{=} \overline{\tilde{N}_e},$$

where

$$\tilde{N}_e = \{u \in H\colon \chi_+\alpha(L - \lambda)^{-1}u \in H_2^+(E),$$
$$\operatorname{Im}\lambda > 0;\ \chi_-\alpha(L - \lambda)^{-1}u \in H_2^-(E),\ \operatorname{Im}\lambda < 0\},$$

and $\chi_\pm \overset{\text{def}}{=} (I \pm J)/2$ are projections in E onto the positive and negative parts of the spectrum of V. In the model space K the linear manifold $\tilde{N}_e$ can be described as follows:

$$\tilde{N}_e = \{P_K(\tilde{g}, g)\colon \chi_-(\tilde{g} + S^*g) + \chi_+(S\tilde{g} + g) = 0\}$$

(see [2], Theorem 1). The next theorem is a direct generalization of these results for the operators $L^{(\kappa)}$ (in particular, for the selfadjoint operators A and $A + V$).

THEOREM 4. *The following three assertions are equivalent*:

I) $(L^{(\kappa)}-\lambda_0)^{-1}P_K\binom{\tilde g}{g} = P_K\frac{1}{(k-\lambda_0)}\binom{\tilde g}{g}$ *for all* λ_0, $\operatorname{Im}\lambda_0<0$ $(\operatorname{Im}\lambda_0>0)$;

$$\text{II)}\quad P_{-(+)}\left(\frac{1+i\kappa}{2}(\tilde g+S^*g)+\frac{1-i\kappa}{2}(S\tilde g+g)\right)=0; \tag{7}$$

III) $P_K\binom{\tilde g}{g}$ *is the image of a vector* $u\in H$ *such that*

$$\left(I\underset{(-)}{+}i\kappa\right)\alpha\left(L^{(\kappa)}-\lambda_0\right)^{-1}u\in H_2^{(\mp)}(E)$$

respectively; *here* $(I\underset{(-)}{+}i\kappa)\alpha(L^{(\kappa)}-\lambda_0)^{-1}u$ *is considered as an analytic vector-valued function of* λ_0, $\operatorname{Im}\lambda_0<0$ $(\operatorname{Im}\lambda_0>0)$, *with values in the Hilbert space* E.(4)

PROOF. II) $\Rightarrow$ I). We use assertion 1) of Theorem 3:

$$(L^{(\kappa)}-\lambda_0)^{-1}P_K\binom{\tilde g}{g}=(L^{(\kappa)}-\lambda_0)^{-1}\begin{pmatrix}\tilde g-P_+(\tilde g+S^*g)\\ g-P_-(S\tilde g+g)\end{pmatrix}$$

$$=P_K\frac{1}{(k-\lambda_0)}\begin{pmatrix}\tilde g-P_+(\tilde g+S^*g)\\ g-P_-(S\tilde g+g)-\frac{1+i\kappa}{2}\left[I+(S^*(\bar\lambda_0)-I)\frac{1+i\kappa}{2}\right]^{-1}\\ \times\left(P_-(\tilde g+S^*g)(\lambda_0)-S^*(\bar\lambda_0)P_-(S\tilde g+g)(\lambda_0)\right)\end{pmatrix}.$$

We transform the last expression by using condition II) and the identity

$$\frac{1+i\kappa}{2}\left[I+(S^*(\bar\lambda_0)-I)\frac{1+i\kappa}{2}\right]^{-1}=\frac{1+i\kappa}{2}\left(I+i\alpha(L^{(\kappa)}-\lambda_0)^{-1}\alpha\frac{1+i\kappa}{2}\right)$$
$$=\left(I+\frac{1+i\kappa}{2}i\alpha(L^{(\kappa)}-\lambda_0)^{-1}\alpha\right)\frac{1+i\kappa}{2}=\left[I+\frac{1+i\kappa}{2}(S^*(\bar\lambda_0)-I)\right]^{-1}\frac{1+i\kappa}{2}.$$

We obtain

$$P_-(S\tilde g+g)(\lambda_0)+\left[I+\frac{1+i\kappa}{2}(S^*(\bar\lambda_0)-I)\right]^{-1}\frac{1+i\kappa}{2}\Big(P_-(\tilde g+S^*g)(\lambda_0)$$
$$-S^*(\bar\lambda_0)P_-(S\tilde g+g)(\lambda_0)\Big)$$

$$=\left[I+\frac{1+i\kappa}{2}(S^*(\bar\lambda_0)-I)\right]^{-1}\times\left[\left(I+\frac{1+i\kappa}{2}(S^*(\bar\lambda_0)-I)\right)P_-(S\tilde g+g)(\lambda_0)\right.$$
$$\left.+\frac{1+i\kappa}{2}P_-(\tilde g+S^*g)(\lambda_0)-\frac{1+i\kappa}{2}S^*(\bar\lambda_0)P_-(S\tilde g+g)(\lambda_0)\right]$$

$$=\left[I+\frac{1+i\kappa}{2}(S^*(\bar\lambda_0)-I)\right]^{-1}$$
$$\times\left[\frac{1-i\kappa}{2}P_-(S\tilde g+g)(\lambda_0)+\frac{1+i\kappa}{2}P_-(\tilde g+S^*g)(\lambda_0)\right]$$

(4) Vectors satisfying condition III) are called "smooth" below. We note that this condition is close to the concept of the relative smoothness of a perturbation in the sense of Kato [15].

Since

$$(k-\lambda_0)^{-1}P_+(\tilde g + S^*g) \in H_2^+(E),$$
$$(k-\lambda_0)^{-1}(P_-(S\tilde g + \tilde g) - P_-(S\tilde g + g)(\lambda_0)) \in H_2^-(E),$$

it follows that

$$\begin{aligned}(L^{(\kappa)}-\lambda_0)^{-1}P_K\begin{pmatrix}\tilde g\\ g\end{pmatrix} &= P_K\frac{1}{k-\lambda_0}\begin{pmatrix}\tilde g - P_+(\tilde g + S^*g)\\ g-(P_-(S\tilde g+g)-P_-(S\tilde g+g)(\lambda_0))\end{pmatrix}\\ &= P_K\frac{1}{k-\lambda_0}\begin{pmatrix}\tilde g\\ g\end{pmatrix}.\end{aligned}$$

I) $\Rightarrow$ II). We introduce the notation

$$\begin{aligned}c(\lambda_0) &\overset{\text{def}}{=} \left(I+\frac{1+i\kappa}{2}(S^*(\bar\lambda_0)-I)\right)^{-1}\\ &\times\left[\frac{1-i\kappa}{2}P_-(S\tilde g+g)(\lambda_0)+\frac{1+i\kappa}{2}P_-(\tilde g+S^*g)(\lambda_0)\right]\in E.\end{aligned}$$

The preceding computation shows that under condition I)

$$\begin{aligned}(L^{(\kappa)}-\lambda_0)^{-1}P_K\begin{pmatrix}\tilde g\\ g\end{pmatrix} - P_K\frac{1}{(k-\lambda_0)}\begin{pmatrix}\tilde g\\ g\end{pmatrix} &= P_K\begin{pmatrix}0\\ -c(\lambda_0)(k-\lambda_0)^{-1}\end{pmatrix}\\ &= \begin{bmatrix}S^*(\lambda_0)c(\lambda_0)(k-\lambda_0)^{-1}\\ -c(\lambda_0)(k-\lambda_0)^{-1}\end{bmatrix}=0.\end{aligned}$$

Hence, for all λ_0, $\operatorname{Im}\lambda_0<0$,

$$\left(S^*(k)-S^*(\bar\lambda_0)\right)c(\lambda_0)(k-\lambda_0)^{-1}=0,$$

and therefore

$$S^*(k)c(\lambda_0)=S^*(\bar\lambda_0)c(\lambda_0),\qquad k\in R.$$

From this, because of the simplicity of the operator $L^{\|}$, it follows that $c(\lambda_0)\equiv 0$, $\operatorname{Im}\lambda_0<0$ (for details see the proof of Lemma 4 of [2]). This leads to the required equality

$$\frac{1-i\kappa}{2}P_-(S\tilde g+g)+\frac{1+i\kappa}{2}P_-(\tilde g+S^*g)=0.$$

Strictly speaking, this argument is valid only for λ_0 which are regular points of the operator $L^{(\kappa)}$. By the analyticity of $P_-(S\tilde g+g)$ and $P_-(\tilde g+S^*g)$ it is also valid for all λ_0, $\operatorname{Im}\lambda_0<0$.

II) $\Rightarrow$ III). Let $u\overset{\text{def}}{=}\Phi^{-1}P_K\binom{\tilde g}{g}$, where the vector $\binom{\tilde g}{g}$ satisfies (7). From the identity

$$\begin{aligned}\alpha(L^{(\kappa)}-\lambda_0)^{-1}u &= (\Theta_\kappa'(\lambda_0))^{-1}\alpha(L^{\|}-\lambda_0)^{-1}u\\ &= (\Theta_\kappa'(\lambda_0))^{-1}(-\sqrt{2\pi})\left(P_-(\tilde g+S^*g)(\lambda_0)-S^*(\bar\lambda_0)P_-(S\tilde g+g)(\lambda_0)\right),\end{aligned}$$

which is easily obtained from (5), it follows that (see above)

$$(-\sqrt{2\pi})^{-1}\frac{1+i\kappa}{2}\alpha(L^{(\kappa)}-\lambda_0)^{-1}u=\frac{1+i\kappa}{2}\left[I+(S^*(\bar\lambda_0)-I)\frac{1+i\kappa}{2}\right]^{-1}$$
$$\times\left(P_-(\tilde g+S^*g)(\lambda_0)-S^*(\bar\lambda_0)P_-(S\tilde g+g)(\lambda_0)\right)$$
$$=\left[I+\frac{1+i\kappa}{2}(S^*(\bar\lambda_0)-I)\right]^{-1}\left[-\frac{1-i\kappa}{2}-\frac{1+i\kappa}{2}S^*(\bar\lambda_0)\right]P_-(S\tilde g+g)(\lambda_0)$$
$$=-P_-(S\tilde g+g)(\lambda_0).$$

Thus

$$\frac{1+i\kappa}{2}\alpha(L^{(\kappa)}-\lambda_0)^{-1}u=\sqrt{2\pi}\,P_-(S\tilde g+g)(\lambda_0)\in H_2^-(E).$$

III) $\Rightarrow$ II). The assertion is that for $(1+i\kappa)\alpha(L^{(\kappa)}-\lambda_0)^{-1}u\in H_2^-(E)$ there exists a vector $(\tilde g,g)\in\mathcal{H}$ satisfying condition (7) such that $P_K(\tilde g,g)=\Phi u$. The mapping Φ here is given by Lemmas 2 and 3. We set $(\tilde g,g)\overset{\text{def}}{=}\Phi u+(0,h^-)$, where $h^-\in H_2^-(E)$. Then $P_K(\tilde g,g)=\Phi u$, and we shall find the required value of h^- in the course of the direct computation ($\Phi u\overset{\text{def}}{=}(\tilde g_0,g_0)\in K$):

$$P_-\left(\frac{1+i\kappa}{2}(\tilde g_0+S^*(g_0+h^-))+\frac{1-i\kappa}{2}(S\tilde g_0+g_0+h^-)\right)$$
$$=P_-\frac{1+i\kappa}{2}(\tilde g_0+S^*g_0)-P_-\frac{1-i\kappa}{2}(S\tilde g_0+g_0)+\left(\frac{1+i\kappa}{2}S^*h^-+\frac{1-i\kappa}{2}h^-\right)$$
$$=\frac{1+i\kappa}{2}(\tilde g_0+S^*g_0)+\left[I+\frac{1+i\kappa}{2}(S^*-I)\right]h^-$$
$$=\frac{1+i\kappa}{2}(-\sqrt{2\pi})^{-1}\alpha(L^{\parallel}-k+i0)^{-1}u+\left[I+\frac{1+i\kappa}{2}(S^*-I)\right]h^-.$$

Making use of the analytic continuation of functions in $H_2^-(E)$ from the real axis to the lower half plane, we consider the following expression for $\operatorname{Im}\lambda_0<0$:

$$\frac{1+i\kappa}{2}(-\sqrt{2\pi})^{-1}\alpha(L^{\parallel}-\lambda_0)^{-1}u+\left(I+\frac{1+i\kappa}{2}(S^*(\bar\lambda_0)-I)\right)h^-(\lambda_0)$$
$$=\frac{1+i\kappa}{2}(-\sqrt{2\pi})^{-1}\left(I+(S^*(\bar\lambda_0)-I)\frac{1+i\kappa}{2}\right)\alpha(L^{(\kappa)}-\lambda_0)^{-1}u$$
$$+\left(I+\frac{1+i\kappa}{2}(S^*(\bar\lambda_0)-I)\right)h^-(\lambda_0)$$
$$=\left(I+\frac{1+i\kappa}{2}(S^*(\bar\lambda_0)-I)\right)\left[h^-(\lambda_0)+\frac{1+i\kappa}{2}(-\sqrt{2\pi})^{-1}\alpha(L^{(\kappa)}-\lambda_0)^{-1}u\right].$$

Condition (7) is satisfied if we set

$$h^-(\lambda_0)=\frac{1+i\kappa}{2}(\sqrt{2\pi})^{-1}\alpha(L^{(\kappa)}-\lambda_0)^{-1}u.$$

Here the inclusion $h^-\in H_2^-(E)$ holds because of the validity of assertion III).

Finally, the second version, $\operatorname{Im}\lambda_0>0$, can be considered in a way analogous to the foregoing.

Combining the cases $\operatorname{Im}\lambda_0<0$ and $\operatorname{Im}\lambda_0>0$ of Theorem 4 we obtain a corollary.

COROLLARY 1. *The linear manifold*

$$\tilde N_e^{(\kappa)} \overset{\text{def}}{=} \{u \in H\colon (1+i\kappa)\alpha(L^{(\kappa)}-\lambda_0)^{-1}u \in H_2^-(E),\ \operatorname{Im}\lambda_0<0;$$
$$(1-i\kappa)\alpha(L^{(\kappa)}-\lambda_0)^{-1}u \in H_2^+(E),\ \operatorname{Im}\lambda_0>0\}$$

in the model representation $\mathcal{H}$ *consists of elements*

$$\left\{P_K\begin{pmatrix}\tilde g\\ g\end{pmatrix}\colon \frac{1+i\kappa}{2}(\tilde g + S^* g) + \frac{1-i\kappa}{2}(S\tilde g + g) = 0\right\}. \tag{8}$$

For these elements of K, *and only these, the following formula holds for all* λ_0, $\operatorname{Im}\lambda_0 \neq 0$:

$$(L^{(\kappa)}-\lambda_0)^{-1}P_K\begin{pmatrix}\tilde g\\ g\end{pmatrix} = P_K\frac{1}{k-\lambda_0}\begin{pmatrix}\tilde g\\ g\end{pmatrix}. \tag{9}$$

PROOF. By Theorem 4 it remains to show that under the condition $u \in \tilde N_e^{(\kappa)}$ there exists a vector $(\tilde g, g) \in \mathcal{H}$ satisfying the relation

$$((1+i\kappa)/2)(\tilde g + S^* g) + ((1-i\kappa)/2)(S\tilde g + g) = 0$$

and such that $P_K(\tilde g, g) = \Phi u$. Theorem 4 provides us with a pair of vectors $(\tilde g, g)$ and $(\tilde g', g')$ such that

$$P_K(\tilde g, g) = P_K(\tilde g', g') = \Phi u,$$
$$P_-\left(\frac{1+i\kappa}{2}(\tilde g + S^* g) + \frac{1-i\kappa}{2}(S\tilde g + g)\right) = 0,$$
$$P_+\left(\frac{1+i\kappa}{2}(\tilde g' + S^* g') + \frac{1-i\kappa}{2}(S\tilde g' + g')\right) = 0.$$

Since $P_K(\tilde g - \tilde g', g - g') = 0$, it follows that $\tilde g - \tilde g' \overset{\text{def}}{=} h^+ \in H_2^+(E)$ and $g - g' \overset{\text{def}}{=} h^- \in H_2^-(E)$. The verification that for $\Phi u = P_K(\tilde g', g)$ equation (8) is satisfied reduces to the computation

$$P_-\left(\frac{1+i\kappa}{2}(\tilde g' + S^* g) + \frac{1-i\kappa}{2}(S\tilde g' + g)\right)$$
$$= P_-\left(\frac{1+i\kappa}{2}(\tilde g' + h^+ + S^* g) + \frac{1-i\kappa}{2}(S(\tilde g' + h^+) + g)\right)$$
$$= P_-\left(\frac{1+i\kappa}{2}(\tilde g + S^* g) + \frac{1+i\kappa}{2}(S\tilde g + g)\right) = 0,$$

$$P_+\left(\frac{1+i\kappa}{2}(\tilde g' + S^* g) + \frac{1-i\kappa}{2}(S\tilde g' + g)\right)$$
$$= P_+\left(\frac{1+i\kappa}{2}(\tilde g' + S^*(g - h^-)) + \frac{1-i\kappa}{2}(S\tilde g' + g - h^-)\right)$$
$$= P_+\left(\frac{1+i\kappa}{2}(\tilde g' + S^* g') + \frac{1-i\kappa}{2}(S\tilde g' + g')\right) = 0.$$

REMARKS. 1. The invariance of the subspaces

$$N_e^{(\varkappa)} \overset{\text{def}}{=} \overline{\tilde N_e^{(\varkappa)}} \quad\text{and}\quad N_{\substack{+\\(-)}}^{(\varkappa)} \overset{\text{def}}{=} \overline{\tilde N_{\substack{+\\(-)}}^{(\varkappa)}},$$

where

$$\tilde N_{\substack{+\\(-)}}^{(\kappa)} \overset{\text{def}}{=} \{u \in H\colon (1\overset{-}{(+)}\, i\kappa)\alpha(L^{(\kappa)}-\lambda_0)^{-1}u \in H_2^{\overset{+}{(-)}}(E)\},$$

with respect to $(L^{(\kappa)} - \lambda_0)^{-1}$, $\operatorname{Im}\lambda_0 \neq 0$, follows directly from Theorem 4 and Corollary 1. It is not hard to show (see [16]) that $N_\pm^{(\kappa)}$ and $N_e^{(\kappa)} \equiv \tilde{N}_+^{(\kappa)} \cap N_-^{(\kappa)}$ are actually regular invariant subspaces, i.e.

$$\overline{(L^{(\varkappa)} - \lambda_0)^{-1} N_\pm^{(\varkappa)}} = N_\pm^{(\varkappa)}, \quad \overline{(L^{(\varkappa)} - \lambda_0)^{-1} N_e^{(\varkappa)}} = N_e^{(\varkappa)}$$

for all λ_0, $\operatorname{Im}\lambda_0 \neq 0$. Here $(L^{(\kappa)} - \lambda_0)^{-1}|N_{\underset{(-)}{+}}^{(\kappa)}$ is analytic in the upper (lower) half plane, and estimates of the norm of the resolvent restricted to the regular invariant subspace $N_{\underset{(-)}{+}}^{(\kappa)}$ hold,

$$\|(L^{(\varkappa)} - \lambda_0)^{-1} | N_+^{(\varkappa)}\| \leqslant \frac{1}{\operatorname{Im}\lambda_0}\left(1 + \frac{1}{2}\left\|\frac{1 - i\varkappa}{2}\right\| \|(\tilde{\Theta}_\varkappa)_e^{-1}(\lambda_0)\|\right), \quad \operatorname{Im}\lambda_0 > 0,$$

$$\|(L^{(\varkappa)} - \lambda_0)^{-1} | N_-^{(\varkappa)}\| \leqslant \frac{1}{(|\operatorname{Im}\lambda_0|)}\left(1 + \frac{1}{2}\left\|\frac{1 + i\varkappa}{2}\right\| \|(\tilde{\Theta}'_\varkappa)_e^{-1}(\lambda_0)\|\right), \quad \operatorname{Im}\lambda_0 < 0,$$

where in the notation the canonical factorization of the bounded, analytic, operator-valued functions

$$\Theta_\varkappa = (\Theta_\varkappa)_i (\Theta_\varkappa)_e, \quad \Theta'_\varkappa = (\Theta'_\varkappa)_i (\Theta'_\varkappa)_e$$

into a product of inner and outer factors has been used [2]. The proof of these inequalities for $\kappa = iJ$ in [16] carries over to the general case without major changes.

2. Similar inequalities are valid also for the estimate of the norm of the resolvent on the whole space (Corollary of Theorem 3):

$$\|(L^{(\varkappa)} - \lambda_0)^{-1}\| \leqslant \frac{1}{\operatorname{Im}\lambda_0}\left(1 + \frac{1}{2}\left\|\frac{1 - i\varkappa}{2}\right\| \|(\Theta_\varkappa(\lambda_0))^{-1}\|\right), \quad \operatorname{Im}\lambda_0 > 0,$$

$$\|(L^{(\varkappa)} - \lambda_0)^{-1}\| \leqslant \frac{1}{(|\operatorname{Im}\lambda_0|)}\left(1 + \frac{1}{2}\left\|\frac{1 + i\varkappa}{2}\right\| \|(\Theta'_\varkappa(\lambda_0))^{-1}\|\right), \quad \operatorname{Im}\lambda_0 < 0.$$

3. The condition $(1(\overset{+}{-})\, i\kappa)\alpha(L^{(\kappa)} - \lambda_0)^{-1}u \in H_2^{(\mp)}(E)$ in Theorem 4 can be replaced by the equivalent condition $\alpha(L^{(\kappa)} - \lambda_0)^{-1}u \in H_2^{(+)}(E)$. This follows from the identity

$$S(\lambda_0)\left(\frac{1 - i\varkappa}{2}\right)\alpha(L^{(\varkappa)} - \lambda_0)^{-1} = (I + i\alpha(L^{-\|} - \lambda_0)^{-1}\alpha)\frac{1 - i\varkappa}{2}\alpha(L^{(\varkappa)} - \lambda_0)^{-1}$$

$$= \frac{1 - i\varkappa}{2}\alpha(L^{(\varkappa)} - \lambda_0)^{-1} + \alpha(L^{-\|} - \lambda_0)^{-1}[(L^{(\varkappa)} - \lambda_0) - (L^{-\|} - \lambda_0)](L^{(\varkappa)} - \lambda_0)^{-1}$$

$$= \frac{1 - i\varkappa}{2}\alpha(L^{(\varkappa)} - \lambda_0)^{-1} + \alpha(L^{-\|} - \lambda_0)^{-1} - \alpha(L^{(\varkappa)} - \lambda_0)^{-1}$$

$$= \alpha(L^{-\|} - \lambda_0)^{-1} - \frac{1 + i\varkappa}{2}\alpha(L^{(\varkappa)} - \lambda_0)^{-1},$$

whence

$$\alpha(L^{(\varkappa)} - \lambda_0)^{-1} = (I + S(\lambda_0))\frac{1 - i\varkappa}{2}\alpha(L^{(\varkappa)} - \lambda_0)^{-1} + \alpha(L^{-\|} - \lambda_0)^{-1}.$$

It remains to make use of Theorem 1, according to which $\alpha(L^{-\|} - \lambda_0)^{-1}u \in H_2^+(E)$ for all $u \in H$. For $\operatorname{Im}\lambda_0 < 0$ one can proceed in analogous fashion.

§4. Formulas for the wave operators

This section is devoted to studying wave operators for nonselfadjoint operators. The principal tool of the investigation is the functional model in the concrete form in which it was constructed in preceding sections. In terms of this model it is possible to calculate the action of the wave operators for the pair $(L^{(\kappa)}, A)$ and to obtain certain "explicit" formulas for them which will be interpreted in the next section. Here the case of selfadjoint operators ($\kappa = J$) is not excluded.

Construction of functions of the operator $L^{(\kappa)}$. Formula (9) makes it possible to consider in the model representation the action of the function $(k - \lambda_0)^{-1}$, $\operatorname{Im} \lambda_0 \neq 0$, of the operator $L^{(\kappa)}|N_e^{(\kappa)}$ restricted to the regular invariant subspace $N_e^{(\kappa)}$. We recall that $N_e^{(\kappa)}$ for $\kappa = iJ$ coincides with the absolutely continuous subspace of L. Conditions under which $N_e^{(\kappa)}$ coincides with the absolutely continuous subspace of $L^{(\kappa)}$ will also be discussed below.

Other functions of the operator $L^{(\kappa)}$, in particular $\exp(ikt)$, $t \in \mathbb{R}$, can be considered in a similar way. To avoid an accumulation of symbols, we denote the function $\exp(ikt)$ of the operator $L^{(\kappa)}|N_e^{(\kappa)}$ by $\exp(iL^{(\kappa)}t)$, implying hereby that the operator $\exp(iL^{(\kappa)}t)$ acts in $N_e^{(\kappa)}$.

PROPOSITION 2. *Let* $P_K\binom{\tilde g}{g} \in \tilde N_e^{(\kappa)}$; *then*

$$\exp(iL^{(\kappa)}t)P_K\begin{pmatrix}\tilde g\\ g\end{pmatrix} = P_K \exp(ikt)\begin{pmatrix}\tilde g\\ g\end{pmatrix}, \qquad t \in \mathbb{R}.$$

PROOF. To define the, in general, unbounded operator $\exp(iL^{(\kappa)}t)$ we use the formula [8]

$$\exp(iL^{(\kappa)}t) = s - \lim_{n\to+\infty} (I - iL^{(\kappa)}t/n)^{-n}.$$

It is not hard to show that this limit exists on $\tilde N_e^{(\kappa)}$ and is equal to

$$\exp(iL^{(\kappa)}t)P_K\begin{pmatrix}\tilde g\\ g\end{pmatrix} = P_K \exp(ikt)\begin{pmatrix}\tilde g\\ g\end{pmatrix}, \qquad t \in \mathbb{R},\ P_K\begin{pmatrix}\tilde g\\ g\end{pmatrix} \in \tilde N_e^{(\kappa)}.$$

For this it is necessary to note that

$$|(1 - ikt/n)^{-n}| \leqslant 1, \qquad (1 - ikt/n)^{-n} \underset{n\to\infty}{\longrightarrow} \exp(ikt)$$

for all k and t, and to use the Lebesgue theorem on passing to the limit. Moreover, it is easy to see that the function $\exp(ikt)$ thus defined satisfies the operator differential equation $du(t)/dt = L^{(\kappa)}u(t)$ in the following sense ($\operatorname{Im} \lambda_0 \neq 0$):

$$\frac{d}{dt}(L^{(\kappa)} - \lambda_0)^{-1} \exp(iL^{(\kappa)}t)P_K\begin{pmatrix}\tilde g\\ g\end{pmatrix} = iL^{(\kappa)}(L^{(\kappa)} - \lambda_0)^{-1}P_K\begin{pmatrix}\tilde g\\ g\end{pmatrix}.$$

The operators $\exp(iL^{(\kappa)}t)$, $t \in \mathbb{R}$, form a quasibounded group if V is a bounded operator [8]. The next lemma gives another boundedness condition.

LEMMA 4. *The operators* $\exp(iL^{(\kappa)}t)$, $t \geqslant 0$, *form a bounded semigroup under the condition*

$$\sup_{k\in\mathbb{R}} \|(\Theta'_\kappa(k))^{-1}\| < \infty \qquad \Big(\sup_{k\in\mathbb{R}} \|(\Theta_\kappa(k))^{-1}\| < \infty\Big).$$

PROOF. Let $t \geqslant 0$; then for $P_K(\tilde g, g) \in \tilde N_e^{(\kappa)}$

$$\left\|\exp(iL^{(\kappa)}t)P_K\begin{pmatrix}\tilde g\\ g\end{pmatrix}\right\| = \left\|P_K\exp(ikt)\begin{pmatrix}\tilde g\\ g\end{pmatrix}\right\| \leqslant \left\|P_K\begin{pmatrix}\tilde g\\ g\end{pmatrix}\right\| + \left\|P_K\exp(ikt)(I-P_K)\begin{pmatrix}\tilde g\\ g\end{pmatrix}\right\|$$

$$= \left\|P_K\begin{pmatrix}\tilde g\\ g\end{pmatrix}\right\| + \left\|P_K\begin{pmatrix}0\\ \exp(ikt)P_-(S\tilde g+g)\end{pmatrix}\right\| \leqslant \left\|P_K\begin{pmatrix}\tilde g\\ g\end{pmatrix}\right\| + \|-P_-(S\tilde g+g)\|_{L_2(E)}.$$

On the other hand, let $\Phi u = P_K(\tilde g, g)$; then as in the proof of Theorem 4 we obtain

$$P_-(S\tilde g+g)(k) = (-\sqrt{2\pi})^{-1}\frac{1+i\kappa}{2}\left(\alpha(L^{(\kappa)}-k+i0)^{-1}u\right)$$
$$= (-\sqrt{2\pi})^{-1}\frac{1+i\kappa}{2}(\Theta'_\kappa(k))^{-1}\left(\alpha(L^{\|}-k+i0)^{-1}u\right),$$

whence by Theorem 1

$$\|P_-(S\tilde g+g)\| \leqslant \left\|\frac{1+i\kappa}{2}\right\|\left(\sup_{k\in\mathbf{R}}\|(\Theta'_\kappa(k))^{-1}\|\right)\|u\|.$$

Finally,

$$\left\|P_K\exp(ikt)\begin{pmatrix}\tilde g\\ g\end{pmatrix}\right\|_{\mathcal{K}} \leqslant \left\|P_K\begin{pmatrix}\tilde g\\ g\end{pmatrix}\right\|_{\mathcal{K}} + \left\|\frac{1+i\kappa}{2}\right\|\left(\sup_{k\in\mathbf{R}}\|(\Theta'_\kappa(k))^{-1}\|\right)\left\|P_K\begin{pmatrix}\tilde g\\ g\end{pmatrix}\right\|_{\mathcal{K}},$$

and hence

$$\|\exp(iL^{(\kappa)}t)\| \leqslant 1 + \left\|\frac{1+i\kappa}{2}\right\|\left(\sup_{k\in\mathbf{R}}\|(\Theta'_\kappa(k))^{-1}\|\right), \qquad t \geqslant 0.$$

Similarly,

$$\|\exp(iL^{(\kappa)}t)\| \leqslant 1 + \left\|\frac{1-i\kappa}{2}\right\|\left(\sup_{k\in\mathbf{R}}\|(\Theta_\kappa(k))^{-1}\|\right)$$

for $t \leqslant 0$.

We note that in the hypotheses of the lemma conditions are imposed essentially on the outer components of the operator-valued functions Θ_κ and Θ'_κ.

Calculation of the wave operators. Proposition 2 gives formulas for computing the action of the operators $\exp(iL^{(\kappa)}t)$ on a distinguished linear manifold that depends on κ. This makes it possible to compare the asymptotic behavior as $t \to \pm\infty$ of $\exp(iL^{(\kappa)}t)$ and $\exp(iAt)$ and thus to find formulas for computing wave operators for the pair $(L^{(\kappa)}, A)$. Indeed, let $P_K(\tilde g, g) \in \tilde N_e^{(\kappa)}$, i.e.

$$\frac{1+i\kappa}{2}(\tilde g + S^*g) + \frac{1-i\kappa}{2}(S\tilde g + g) = 0.$$

We suppose that there exists a vector $P_K(\tilde g', g')$ with the same element g, $(\tilde g, g) \in \mathcal{K}$, belonging to the linear manifold $\tilde N_e^0 \equiv \tilde N_e^{(\kappa)}|_{\kappa=0}$ such that $(\tilde g' + S^*g) + (S\tilde g' + g) = 0$. Then

$$\left\|\exp(-iL^{(\kappa)}t)P_K\begin{pmatrix}\tilde g\\ g\end{pmatrix} - \exp(-iAt)P_K\begin{pmatrix}\tilde g'\\ g\end{pmatrix}\right\|$$

$$= \left\|P_K\exp(-ikt)\begin{pmatrix}\tilde g\\ g\end{pmatrix} - P_K\exp(-ikt)\begin{pmatrix}\tilde g'\\ g\end{pmatrix}\right\| = \left\|P_K\begin{pmatrix}\exp(-ikt)(\tilde g' - g)\\ 0\end{pmatrix}\right\|$$

$$= \left\|P_K\begin{pmatrix}P_-\exp(-ikt)(\tilde g - \tilde g')\\ 0\end{pmatrix}\right\| \leqslant \|P_-\exp(-ikt)(\tilde g - \tilde g')\|_{L_2(E)}.$$

We observe that $(\tilde{g} - \tilde{g}') \in L_2(E)$, since $(\tilde{g} - \tilde{g}', 0) = (\tilde{g}, g) - (\tilde{g}', g) \in \mathcal{H}$. Further, $\exp(ikt) \in H^+_\infty$ for $t \leqslant 0$. Obviously, for $t \to -\infty$

$$\|P_- \exp(-ikt)(\tilde{g} - \tilde{g}')\|_{L_2(E)} = \left(\int_{-\infty}^{t} \|\widehat{(\tilde{g} - \tilde{g}')}(\lambda)\|_E^2 \, d\lambda\right)^{1/2} \to 0,$$

where the Fourier transform of the function $(\tilde{g} - \tilde{g}') \in L_2(E)$ is denoted by $\widehat{(\tilde{g} - \tilde{g}')}(\lambda)$.

Thus,

$$\left\|\exp(-iL^{(\kappa)}t)P_K\begin{pmatrix}\tilde{g}\\ g\end{pmatrix} - \exp(-iAt)P_K\begin{pmatrix}\tilde{g}'\\ g\end{pmatrix}\right\| \underset{t\to-\infty}{\rightarrow} 0,$$

which makes it possible to define the, in general, unbounded wave operator $W_-(A, L^{(\kappa)})$:

$$W_-(A, L^{(\kappa)})P_K\begin{pmatrix}\tilde{g}\\ g\end{pmatrix} = P_K\begin{pmatrix}\tilde{g}'\\ g\end{pmatrix} \equiv P_K\begin{pmatrix}-(I + S)^{-1}(I + S^*)g\\ g\end{pmatrix}. \tag{10}$$

The operator $W_-(A, L^{(\kappa)})$ is defined for those elements $P_K(\tilde{g}, g) \in \tilde{N}_e^{(\kappa)}$ for which $P_K(\tilde{g}', g) \in \tilde{N}_e^0$ for some $\tilde{g}'$ which enables us to attribute meaning to the expression $(I + S)^{-1}(I + S^*)g$. The latter is obtained after formal solution of the equation $(\tilde{g}' + S^*g) + (S\tilde{g}' + g) = 0$ for g'. In this case it is obvious that

$$W_-(A, L^{(\kappa)})P_K\begin{pmatrix}\tilde{g}\\ g\end{pmatrix} = \lim_{t\to-\infty} \exp(iAt)\cdot\exp(-iL^{(\kappa)}t)P_K\begin{pmatrix}\tilde{g}\\ g\end{pmatrix}.$$

Similar formal expressions can easily be obtained for the other wave operators $W_+(A, L^{(\kappa)})$ and $W_\pm(L^{(\kappa)}, A)$ as well. For $P_K(\tilde{g}, g) \in \tilde{N}_e^{(\kappa)}$

$$W_+(A, L^{(\kappa)})P_K\begin{pmatrix}\tilde{g}\\ g\end{pmatrix} = \lim_{t\to+\infty} \exp(iAt)\cdot\exp(-iL^{(\kappa)}t)P_K\begin{pmatrix}\tilde{g}\\ g\end{pmatrix}$$
$$= P_K\begin{pmatrix}\tilde{g}\\ -(I + S^*)^{-1}(I + S)\tilde{g}\end{pmatrix}. \tag{11}$$

For $P_K(\tilde{g}, g) \in \tilde{N}_e^0$

$$W_-(L^{(\kappa)}, A)P_K\begin{pmatrix}\tilde{g}\\ g\end{pmatrix}$$
$$= P_K\begin{bmatrix}-\left(I + \frac{1 - i\kappa}{2}(S - I)\right)^{-1}\left(I + \frac{1 + i\kappa}{2}(S^* - I)\right)g\\ g\end{bmatrix}, \tag{12}$$

$$W_+(L^{(\kappa)}, A)P_K\begin{pmatrix}\tilde{g}\\ g\end{pmatrix}$$
$$= P_K\begin{bmatrix}\tilde{g}\\ -\left(I + \frac{1 + i\kappa}{2}(S^* - I)\right)^{-1}\left(I + \frac{1 - i\kappa}{2}(S - I)\right)\tilde{g}\end{bmatrix}. \tag{13}$$

The operators $W_\pm(L^{(\kappa)}, A)$ are defined by a comparison of the asymptotic behavior of

$$\left\|\exp(-iuL^{(\kappa)}t)\,W_\pm(L^{(\kappa)}, A)P_K\begin{pmatrix}\tilde{g}\\ g\end{pmatrix} - \exp(-iAt)P_K\begin{pmatrix}\tilde{g}\\ g\end{pmatrix}\right\| \underset{t\to\pm\infty}{\rightarrow} 0$$

(here $P_K(\tilde{g}, g) \in \tilde{N}_e^0$). It is possible to interpret the expressions (10)–(13) in the same manner as for the operator $W_-(A, L^{(\kappa)})$.

The following sections are devoted to a correct definition of the wave operators under certain restrictions on the perturbation based on (10)–(13). Conditions that they be bounded and also the condition of the denseness of N_e^0 in the absolutely continuous subspace H_a of the selfadjoint operator A will be studied.

§5. The scalar multiple of operator-valued functions and conditions for the existence and completeness of the wave operators

The wave operators $W_\pm(A, L^{(\kappa)})$ and $W_\pm(L^{(\kappa)}, A)$ can be constructed directly on the basis of formulas (10)–(13) if it is possible to attribute meaning to the expressions

$$(I + ((1 - i\kappa)/2)(S(k) - I))^{-1}, \qquad (I + ((1 + i\kappa)/2)(S^*(k) - I))^{-1}$$

for almost all k on the real axis. A sufficient condition for this is that the operator-valued functions (the inversion of which is required in (10)–(13)) have a scalar multiple.([5]) We recall that a scalar multiple of an operator-valued function $Q(\lambda)$: $E \to E$ which is bounded and analytic in the half plane $\operatorname{Im} \lambda > 0$ ($\operatorname{Im} \lambda < 0$) is a bounded, analytic, scalar-valued function $d(\lambda) \not\equiv 0$ such that

$$\Omega(\lambda) Q(\lambda) = Q(\lambda) \Omega(\lambda) = d(\lambda) I, \quad \operatorname{Im} \lambda > 0 \,(\operatorname{Im} \lambda < 0),$$

where $\Omega(\lambda)$ is a bounded, analytic, operator-valued function in the half plane $\operatorname{Im} \lambda > 0$ ($\operatorname{Im} \lambda < 0$). The definition of a scalar multiple given by Sz.-Nagy and Foiaş [1] for contractive operator-valued functions is here extended in a natural way to a bounded operator-valued function $\|Q(\lambda)\|$, $\|\Omega(\lambda)\| \leqslant C < \infty$. The latter frees us from having to divide by a normalizing factor.

The following proposition gives a simple condition for the existence of a scalar multiple for a contractive analytic function (E is an auxiliary Hilbert space).

PROPOSITION 3. *Let $Q(\lambda)$ be a contractive* ($\|Q(\lambda)\| \leqslant 1$) *operator-valued function* $Q(\lambda)$: $E \to E$ *analytic in the half plane* $\operatorname{Im} \lambda > 0$ *and such that* $Q(\lambda) - I \in \sigma_1$ *for* $\operatorname{Im} \lambda > 0$. *If $Q(\lambda)$ is boundedly invertible at at least one point λ_0,* $\operatorname{Im} \lambda_0 > 0$, *then it possesses a scalar multiple.*

PROOF. The verification of the assertion essentially repeats the proof of a theorem of Sz.-Nagy and Foiaş on the existence of a scalar multiple for the characteristic function of a weak contraction [1], of which it is a simple generalization.

We denote by P_n the orthogonal projection onto the subspace $\bigvee_1^n e_k$, where $\{e_k\}_1^\infty$ is an orthonormal basis in E (under the assumption that $\dim E = \infty$). Let $P_n^\perp \overset{\text{def}}{=} I - P_n$; then for the contraction $P_n Q(\lambda) P_n$ in the finite-dimensional subspace $(P_n E)$ we have the identity ($\operatorname{Im} \lambda > 0$)

$$(P_n Q(\lambda) P_n) \Omega_n(\lambda) = \Omega_n(\lambda) (P_n Q(\lambda) P_n) = \det(P_n Q(\lambda) P_n) I_n, \tag{14}$$

([5]) It can be shown that if the operator-valued functions Θ_κ, Θ'_κ and $(I + S)$ have scalar multiples, then the wave operators for the pair $(L^{(\kappa)}, A)$ exist and are complete.

where I_n is the identity operator in (P_nE) and

$$\Omega_n(\lambda) \overset{\text{def}}{=} (P_nQ(\lambda)P_n)^{-1} \det (P_nQ(\lambda)P_n).$$

By analyticity, either $d_n(\lambda) \overset{\text{def}}{=} \det(P_nQ(\lambda)P_n)$ is different from zero for almost all λ, or it vanishes identically. In the first case $\Omega_n(\lambda)$ is defined in correspondence with the formula presented above, and in the second case we set $\Omega_n(\lambda) = 0$. We note that $|d_n(\lambda)| \leqslant 1$ and $\|\Omega_n(\lambda)\| \leqslant 1$. Indeed, for any contraction B ($\|B\| \leqslant 1$) in (P_nE) we have

$$\|Bu\| \geqslant \Big(\min_i s_i(B)\Big)\|u\| \geqslant \Big(\prod_i s_i(B)\Big)\|u\| = \det(|B|)\|u\| = |\det B|\,\|u\|,$$

where $\{s_i(B)\}$ are the singular numbers of B. Hence $\|B^{-1}\| \cdot |\det B| \leqslant 1$. It remains to set $B = P_nQ(\lambda)P_n$. The identity (14) can be rewritten for operators on the entire space E:

$$\Omega_n'(\lambda)(P_n^{\perp} + P_nQ(\lambda)P_n) = (P_n^{\perp} + P_nQ(\lambda)P_n)\Omega_n'(\lambda) = d_n(\lambda)I,$$
$$\Omega_n'(\lambda) \overset{\text{def}}{=} \Omega_n(\lambda)P_n + d_n(\lambda)P_n^{\perp}, \quad \|\Omega_n'(\lambda)\| \leqslant 1, \quad \operatorname{Im}\lambda > 0.$$

By the Vitali-Montel theorem there exists a subsequence $n_i \to \infty$ such that $d_{n_i}(\lambda) \to d(\lambda)$ and $\Omega_{n_i}'(\lambda)$ converges weakly to the analytic operator-valued function $\Omega(\lambda)$. Here $|d(\lambda)| \leqslant 1$ and $\|\Omega(\lambda)\| \leqslant 1$, $\operatorname{Im}\lambda > 0$. Since $P_n^{\perp} + P_nQ(\lambda)P_n \to Q(\lambda)$ strongly, while $(\Omega_n'(\lambda))^*$ converges to $\Omega^*(\lambda)$ weakly, in the equation

$$\Omega_{n_i}'(\lambda)(P_{n_i}^{\perp} + P_{n_i}Q(\lambda)P_{n_i}) = d_{n_i}(\lambda)I$$

it is possible to pass to the limit as $n_i \to \infty$. In summary, we obtain $\Omega(\lambda)Q(\lambda) = d(\lambda)I$. From the bounded invertibility of $Q(\lambda)$ at λ_0, and hence in some neighborhood, we obtain the equality $Q(\lambda)\Omega(\lambda) = d(\lambda)I$ for λ belonging to this neighborhood. Analyticity in λ leads to the identity $Q(\lambda)\Omega(\lambda) = d(\lambda)I$ in the entire half plane.

The existence of a scalar multiple will be proved if it is shown that $d(\lambda) \not\equiv 0$. Since $d(\lambda)$ is a divisor of the operator $Q(\lambda) \equiv I + A(\lambda)$, $A(\lambda) \in \sigma_1$ (see [17]), it follows that $d(\lambda) = \prod_i(1 + \lambda_i(A(\lambda)))$, where $\lambda_i(A(\lambda))$ are the eigenvalues of the operator $A(\lambda)$. Noting the invertibility of $Q(\lambda_0)$, we obtain

$$|d(\lambda_0)| \geqslant C \prod_{(|\lambda_i|<1)} (1 - |\lambda_i(A(\lambda_0))|) \geqslant C\exp\Big(-\sum_i |\lambda_i(A(\lambda_0))|\Big)$$
$$\geqslant C\exp\Big(-\sum_i s_i(A(\lambda_0))\Big) = C\exp(-\|A(\lambda_0)\|_{\sigma_1}) > 0,$$

where

$$C = \prod_{(|\lambda_i|\geqslant 1)} |(1 + \lambda_i(A(\lambda_0)))| \neq 0.$$

In this estimate the Weyl inequality for the singular numbers $s_i(A(\lambda_0))$ of the operator $A(\lambda_0)$ has been applied [17].

We note that the condition $\|Q(\lambda)\| \leqslant 1$ is essential, since we do not require the uniform boundedness of $\|A(\lambda_0)\|_{\sigma_1}$. However, this assertion can be extended to operator-valued functions of triangular form $(\chi_- + Q(\lambda)\chi_+)$ which are, in general, not contractive ($\chi_\pm = (I \pm J)/2$, $J^* = J$, $J^2 = I$, $J\colon E \to E$).

PROPOSITION 4. *Let $Q(\lambda)$ be a contractive, analytic, operator-valued function such that $Q(\lambda) - I \in \sigma_1$; then $(\chi_- + Q(\lambda)\chi_+)$ has a scalar multiple under the condition of invertibility at at least one point λ_0,* $\operatorname{Im}\lambda_0 > 0$.

PROOF. We consider the contractive operator-valued function $\chi_+ Q(\lambda)\chi_+$ acting in the space $(\chi_+ E)$. Since $\chi_+ Q(\lambda)\chi_+$ is boundedly invertible at λ_0 (it is easy to see this by direct computation; see also [21]), all the hypotheses of Proposition 3 are satisfied for it. Hence $\chi_+ Q(\lambda)\chi_+$ possesses a scalar multiple $d_+(\lambda)$:

$$\Omega_+(\lambda)(\chi_+ Q(\lambda)\chi_+) = (\chi_+ Q(\lambda)\chi_+)\Omega_+(\lambda) = d_+(\lambda)\chi_+.$$

These equalities can be rewritten in the following form:

$$\begin{aligned}(d_+(\lambda)\chi_- + (I - \chi_- Q(\lambda)\chi_+)\Omega_+(\lambda)\chi_+)(\chi_- + Q(\lambda)\chi_+)\\ = (\chi_- + Q(\lambda)\chi_+)(d_+(\lambda)\chi_- + (I - \chi_- Q(\lambda)\chi_+)\Omega_+(\lambda)\chi_+) = d_+(\lambda) I,\end{aligned}$$

from which it follows immediately that $(\chi_- + Q(\lambda)\chi_+)$ has a scalar multiple equal to $d_+(\lambda)$.

N_e^0 and the absolutely continuous subspace of the operator A. The interrelation of the linear manifold $\tilde{N}_e^0$ and the absolutely continuous subspace $H_a \equiv H_a(A)$ of the selfadjoint operator A will be studied below.

LEMMA 5. $\tilde{N}_e^0 \subseteq H_a(A)$.

PROOF. We first verify that the Cauchy transform of the complex measure $d\mu(t)$ on $\mathbb{R}$ ($\hat{\mu}(z) \stackrel{\text{def}}{=} \int_{\mathbb{R}} (t - z)^{-1}\, d\mu(t)$) satisfies one of the conditions $\hat{\mu}(z) \in H_2^{(\pm)}$ for $\operatorname{Im} z > 0$, $\operatorname{Im} z < 0$ respectively only if $d\mu(t)$ is absolutely continuous with respect to Lebesgue measure on the line. This assertion is a direct corollary of the theorem of F. and M. Riesz (see [11]). Indeed, the condition $\hat{\mu}(z) \in H_2^-$ implies the existence of a function $f_-(t)$, $f_- \in H_2^-$, such that

$$\int_{\mathbb{R}} (t - z)^{-1}\{f_-(t)\,dt - d\mu(t)\} = 0, \quad \operatorname{Im} z < 0.$$

Evaluating this expression at some point z_0, $\operatorname{Im} z_0 < 0$, we obtain

$$\int_{\mathbb{R}} (t - z)^{-1} d\tilde{\mu}(t) \equiv \int_{\mathbb{R}} (t - z)^{-1}(t - z_0)^{-1}(f_-(t)\,dt - d\mu(t)) = 0, \quad \operatorname{Im} z < 0,$$

where $d\tilde{\mu}(t)$ is a complex measure. By the F. and M. Riesz theorem (to apply it we must first carry out the usual transformation from the half plane to the disk) the measure $d\tilde{\mu}(t) \stackrel{\text{def}}{=} (t - z_0)^{-1}(f_-(t)dt - d\mu(t))$ is absolutely continuous with respect to Lebesgue measure on the line. The same obviously holds for the measure $d\mu(t)$. We note that only one of the conditions $\hat{\mu}(z) \in H_2^{(\pm)}$ has been used here.

Let E_t be the resolution of the identity of the selfadjoint operator A. Then

$$(\alpha(A - \lambda)^{-1}u, v) = \int_{\mathbb{R}} (t - \lambda)^{-1} d(E_t u, \alpha v), \qquad u \in H,\ v \in D(\alpha),\ \operatorname{Im}\lambda < 0.$$

From what has been said above it follows that the condition $\alpha(A - \lambda)^{-1}u \in H_2^-(E)$ implies the absolute continuity of the complex measure $d(E_t u, \alpha v)$. We decompose the vector u into absolutely continuous ($P_a u$) and singular ($P_s u$) components. We have

$$d(E_t u, \alpha v) = \left(\frac{d}{dt}(E_t u, \alpha v)\right) dt = \left(\frac{d}{dt}(E_t P_a u, \alpha v)\right) dt.$$

Hence

$$\int_{\mathbf{R}} (t - \lambda)^{-1} d(E_t P_s u, \alpha v) \equiv 0,$$

i.e.

$$((A - \lambda)^{-1} u, \alpha v) = 0, \quad \operatorname{Im} \lambda \neq 0.$$

The last condition means that $P_s u$ is orthogonal to $\bigvee_{\operatorname{Im}\lambda \neq 0} (A - \lambda)^{-1} \alpha D(\alpha)$. Since $H = \bigvee_{\operatorname{Im}\lambda \neq 0} (A - \lambda)^{-1} \alpha D(\alpha)$, by the complete nonselfadjointness of the operator $L^{\|}$, the condition of the lemma $\tilde{N}_e^0 \subseteq H_a$ is satisfied.

For the correctness of the definition of the wave operators given in §4 it is necessary that the stronger condition $N_e^0 = H_a(A)$ be satisfied. The additional requirement of nuclearity of the perturbation $V \in \sigma_1$ is sufficient for this. According to the Rosenblum lemma, in this case $\int_{\mathbf{R}} \|\alpha \exp(iAt) u\|^2 \, dt < \infty$ for the dense set of vectors u in $H_a(A)$ for which

$$\sup_{t \in \mathbf{R}} \left| \frac{d}{dt}(Etu, u) \right| < \infty.$$

By taking the Fourier transform on t of the functions $(\chi_\pm(t))\alpha \exp(iAt)u$ ($\chi_\pm(t)$ is the characteristic function of the interval $(0, \infty)$ $((-\infty, 0))$), we obtain by the Parseval equality

$$\|\alpha(A - \lambda)^{-1} u\|^2_{H_2^-(E)} + \|\alpha(A - \lambda)^{-} u\|^2_{H_2^+(E)} = 2\pi \int_{\mathbf{R}} \|\alpha \exp(iAt) u\|^2 \, dt.$$

Therefore, $u \in \tilde{N}_e^0$ for those vectors u for which $\int_{\mathbf{R}} \|\alpha \exp(iAt) u\|^2 \, dt < \infty$. The equality $N_e^0 = H_a(A)$ under the condition $V \in \sigma_1$ has been proved. Of course, there is no harm in replacing the condition $V \in \sigma_1$ by the condition $\alpha(A - \lambda_0)^{-1} \in \sigma_2$ for some λ_0, $\operatorname{Im} \lambda_0 \neq 0$, i.e.

$$[(L - \lambda_0)^{-1} - (A - \lambda_0)^{-1}] \in \sigma_1.$$

Finally, we note that a detailed analysis of the somewhat stronger condition $\tilde{N}_e^0 = H$ (which is equivalent to the existence of the two estimates $\|\alpha(A - \lambda)^{-1} u\|_{H_2^\pm(E)} \leqslant C_\pm \|u\|$, $u \in H$) is given by Kato [15]. In particular, that paper contains a list of conditions equivalent to the condition $\tilde{N}_e^0 = H_a$.

Theorems for the existence of wave operators. The conditions of the existence of a scalar multiple obtained in Propositions 3 and 4 make it possible to assign a meaning to the formal expressions for the wave operators which we presented above. Of course, because of the unboundedness of the operators, this can be done in the general case only on dense sets. The next theorem is a generalization of the Rosenblum-Kato theorem [8] on nuclear perturbations to the case of nonselfadjoint operators.

THEOREM 5. *Let* $(\kappa - \kappa^*)/2i \equiv \operatorname{Im} \kappa = 0$ *or* J, *and suppose that* $V \in \sigma_1$; *then the wave operators* $W_\pm(A, L^{(\kappa)})$ $(W_\pm(L^{(\kappa)}, A))$ *exist on a dense set in* $N_e^{(\kappa)}$ $(H_a(A))$ *and are given respectively by formulas* (10)–(13).(6) *The range of these operators is a dense set in* $H_a(A)$ $(N_e^{(\kappa)})$.

PROOF. We shall show that the operator-valued functions

$$(I+S(\lambda)),\ \left(I+\frac{1-i\varkappa}{2}(S(\lambda)-I)\right),\ (I+S^*(\bar\lambda)),\ \left(I+\frac{1+i\varkappa}{2}(S^*(\bar\lambda)-I)\right),$$

contained in (10)–(13) possess scalar multiples. The cases $\operatorname{Im}\kappa = 0$ and $\operatorname{Im}\kappa = J$ are considered separately.

I. $\operatorname{Im}\kappa = 0$. Since the operator-valued functions in question are not, in general, contractive, Proposition 3 cannot be applied directly. However, the operator-valued function $(I + ((1 - i\kappa)/2)(S(\lambda) - I))$ can be represented as the product of a contractive operator-valued function and a boundedly invertible function. Namely, we have the identity

$$\left(I+\frac{1-i\varkappa}{2}(S(\lambda)-I)\right)=\left(I+(i\alpha/2)\left(L^{-\shortparallel}+\frac{\alpha\varkappa\alpha}{2}-\lambda\right)^{-1}\alpha\right) \times (I+(\varkappa\alpha/2)(L^{-\shortparallel}-\lambda)^{-1}\alpha),$$

the proof of which is based on the Hilbert identity:

$$\begin{aligned}&\left(I+i(\alpha/2)\left(L^{-\shortparallel}+\frac{\alpha\varkappa\alpha}{2}-\lambda\right)^{-1}\alpha\right)(I+\varkappa(\alpha/2)(L^{-\shortparallel}-\lambda)^{-1}\alpha)\\ &=I+i(\alpha/2)\left(L^{-\shortparallel}+\frac{\alpha\varkappa\alpha}{2}-\lambda\right)^{-1}\alpha+\varkappa(\alpha/2)(L^{-\shortparallel}-\lambda)^{-1}\alpha\\ &+i(\alpha/2)\left(L^{-\shortparallel}+\frac{\alpha\varkappa\alpha}{2}-\lambda\right)^{-1}\left[\left(L^{-\shortparallel}+\frac{\alpha\varkappa\alpha}{2}-\lambda\right)-(L^{-\shortparallel}-\lambda)\right]\\ &\times(L^{-\shortparallel}-\lambda)^{-1}\alpha=I+\varkappa(\alpha/2)(L^{-\shortparallel}-\lambda)^{-1}\alpha+i(\alpha/2)(L^{-\shortparallel}-\lambda)^{-1}\alpha\\ &=I+i\frac{1-i\varkappa}{2}\alpha(L^{-1}-\lambda)^{-1}\alpha=I+\frac{1-i\varkappa}{2}(S(\lambda)-I).\end{aligned}$$

The operator-valued function $I + \kappa(\alpha/2)(L^{-\shortparallel} - \lambda)^{-1}\alpha$ analytic in the half plane $\operatorname{Im}\lambda > 0$ is bounded, i.e. $\|(I + \kappa(\alpha/2)(L^{-\shortparallel} - \lambda)^{-1}\alpha)\| \leqslant 1 + \|\kappa\|$, together with its inverse, for which it is not hard to give the explicit expression

$$(I+\varkappa(\alpha/2)(L^{-\shortparallel}-\lambda)^{-1}\alpha)^{-1}=I-\varkappa(\alpha/2)\left(L^{-\shortparallel}+\frac{\alpha\varkappa\alpha}{2}-\lambda\right)^{-1}\alpha.$$

Indeed,

$$\begin{aligned}&(I+\varkappa(\alpha/2)(L^{-\shortparallel}-\lambda)^{-1}\alpha)\left(I-\varkappa(\alpha/2)\left(L^{-\shortparallel}+\frac{\alpha\varkappa\alpha}{2}-\lambda\right)^{-1}\alpha\right)\\ &=I+\varkappa(\alpha/2)(L^{-\shortparallel}-\lambda)^{-1}\alpha-\varkappa(\alpha/2)\left(L^{-\shortparallel}+\frac{\alpha\varkappa\alpha}{2}-\lambda\right)^{-1}\alpha\\ &-\varkappa(\alpha/2)(L^{-\shortparallel}-\lambda)^{-1}\left[\left(L^{-\shortparallel}+\frac{\alpha\varkappa\alpha}{2}-\lambda\right)-(L^{-\shortparallel}-\lambda)\right]\\ &\times\left(L^{-\shortparallel}+\frac{\alpha\varkappa\alpha}{2}-\lambda\right)^{-1}\alpha=I.\end{aligned}$$

(6) The condition $\operatorname{Im}\kappa = 0$, J makes it possible to consider all the interesting cases $\kappa = 0$, J, iJ.

The verification that $(I - \kappa(\alpha/2)(L^{-\|} + \alpha\kappa\alpha/2 - \lambda)^{-1}\alpha)$ is also a left inverse for $(I + \kappa(\alpha/2)(L^{-\|} - \lambda)^{-1}\alpha)$ is analogous. The boundedness of the operator-valued function $(I - \kappa(\alpha/2)(L^{-\|} + \alpha\kappa\alpha/2 - \lambda)^{-1}\alpha)$, $\operatorname{Im}\lambda > 0$, follows immediately from the fact that $(L^{\|} + \alpha\kappa\alpha/2)$ is a dissipative operator in H with imaginary part $\alpha^2/2$:

$$\left\| I - \kappa(\alpha/2)\left(L^{-\|} + \frac{\alpha\kappa\alpha}{2} - \lambda\right)^{-1}\alpha \right\| \leqslant 1 + \|\kappa\|.$$

The other factor

$$(I + i(\alpha/2)(L^{-\|} + \alpha\kappa\alpha/2 - \lambda)^{-1}\alpha) \equiv (I + S_\kappa(\lambda))/2,$$

where $S_\kappa(\lambda)$ is the characteristic function of the dissipative operator

$$L^{\|} + \alpha\kappa\alpha/2 \equiv (A + \alpha\kappa\alpha/2) + i\alpha^2/2, \quad \|S_\kappa(\lambda)\| \leqslant 1, \quad \operatorname{Im}\lambda > 0),$$

is obviously a contractive, operator-valued function. The condition $V \in \sigma_1$ (i.e. $\alpha \in \sigma_2$, the class of Hilbert-Schmidt operators) implies that $\alpha(L^{-\|} + \alpha\kappa\alpha/2 - \lambda)^{-1}\alpha \in \sigma_1$. All the conditions of Proposition 3 pertaining to the function

$$(I + i(\alpha/2)(L^{-\|} + \alpha\kappa\alpha/2 - \lambda)^{-1}\alpha)$$

are satisfied. It therefore possesses a scalar multiple. The factorization indicated makes it possible to assert the same for the operator-valued function

$$\left(I + \left(\frac{(1 - i\kappa)}{2}\right)(S(\lambda) - I)\right),$$

while its scalar multiple may be taken to coincide with the scalar multiple of

$$(I + i(\alpha/2)(L^{-\|} + \alpha\kappa\alpha/2 - \lambda)^{-1}\alpha).$$

The existence of a scalar multiple for

$$\left(I + \left(\frac{(1 + i\kappa)}{2}\right)(S^*(\bar\lambda) - I)\right), \quad \operatorname{Im}\lambda > 0,$$

is proved in the same way. Finally, $(I + S(\lambda))$ and $(I + S^*(\bar\lambda))$ possess scalar multiples, since $(I + S(\lambda))/2$ and $(I + S^*(\bar\lambda))/2$ are contractive operator-valued functions.

II. $\operatorname{Im}\kappa = J$. An important special case here is $\kappa = iJ$ $(L^{(\kappa)} = L)$. Just as in case I, it is possible to justify the following factorization $(\chi_\pm = (I \pm J)/2$, $\kappa \equiv \kappa_R + iJ$, $\operatorname{Im}\kappa_R = 0)$:

$$\left(I + \frac{1 - i\kappa}{2}(S(\lambda) - I)\right) = \left(I + i\chi_+\alpha\left(L^{-\|} + \frac{\alpha\kappa_R\alpha}{2} - \lambda\right)^{-1}\alpha\right) \times \left(I + \kappa_R(\alpha/2)(L^{-\|} - \lambda)^{-1}\alpha\right).$$

Regarding the function $(I + (\kappa_R\alpha/2)(L^{-\|} - \lambda)^{-1}\alpha)$, it has been shown that it is bounded and boundedly invertible (κ_R is a bounded, selfadjoint operator). In order to apply Proposition 4 to the operator-valued function

$$(I + i\chi_+\alpha(L^{-\|} + \alpha\kappa_R\alpha/2 - \lambda)^{-1}\alpha) \equiv (\chi_- + \chi_+ S_{\kappa_R}(\lambda)),$$

it suffices to go over to the adjoint operator. Here again S_{κ_R} is the characteristic function of the dissipative operator $(L^{\|} + \alpha\kappa_R\alpha/2)$, and hence it is contractive for $\operatorname{Im}\lambda > 0$.

We remark that it is easy to foresee the above factorizations if use is made of the identity

$$(L^{(\varkappa)} - \lambda)^{-1}\alpha\,(I + ((1 - i\varkappa)/2)\,(S(\lambda) - I)) = (L^{-1} - \lambda)^{-1}\alpha.$$

An analogous identity for $\kappa = iJ$ was proved in [2].

Now that the existence of scalar multiples has been proved it is not hard to verify the assertion of the theorem. We do this for one of the wave operators, $W_-(A, L^{(\kappa)})$. All arguments can be carried over without essential change to the other three operators, $W_+(A, L^{(\kappa)})$ and $W_\pm(L^{(\kappa)}, A)$.

It follows from the results of §4 that the wave operator, $W_-(A, L^{(\kappa)})$ is defined (and is given by (10)) for vectors of $\tilde{N}_e^{(\kappa)}$ of the form $P_K\chi_n(k)(\tilde{g}, g)$, where $P_K(\tilde{g}, g) \in \tilde{N}_e^{(\kappa)}$, and the scalar-valued function $\chi_n(k)$ is the characteristic function of the set $\{k \in \mathbb{R}: |d(k)| > 1/n\}$, $n \in \mathbb{N}$. Here $d(\lambda)$ is the scalar multiple of the operator-valued function

$$(I + S(\lambda))\ ((I + S(\lambda))\ \Omega(\lambda) = \Omega(\lambda)(I + S(\lambda)) = d(\lambda)I,\ \operatorname{Im}\lambda > 0).$$

Indeed,

$$\left(-\left[\chi_n(I + S)^{-1}\right](I + S^*)g,\ \chi_n g\right) \in \mathcal{H},$$

and it belongs to $\tilde{N}_e^0$ if $[\chi_n(I + S)^{-1}](k)$ is understood to mean $\chi_n(k)d^{-1}(k + i0) \times \Omega(k + i0)$. Residency in the basic space $\mathcal{H}$ follows from the fact that

$$\begin{aligned}\left(-\left[\chi_n(I + S)^{-1}\right](I + S^*)g,\ \chi_n g\right) &- (\chi_n\tilde{g},\ \chi_n g)\\ &\equiv \left(-\left[\chi_n(I + S)^{-1}\right]((I + S^*)g + (I + S)\tilde{g}),\ 0\right)\\ &= \left(-\left[\chi_n(I + S)^{-1}\right](\tilde{g} + S^*g + S\tilde{g} + g),\ 0\right) \in (L_2(E),\ 0),\end{aligned}$$

while the inclusion in N_e^0 is a result of direct computation. Further, by Fatou's theorem, $\chi_n(k) \to 1$ as $n \to \infty$ for almost all k. The Lebesgue theorem on passage to the limit makes it possible to assert that

$$P_K\left(\chi_n\begin{pmatrix}\tilde{g}\\ g\end{pmatrix}\right) \underset{n\to\infty}{\to} P_K\begin{pmatrix}\tilde{g}\\ g\end{pmatrix}.$$

The domain of $W_-(A, L^{(\kappa)})$ is dense in $\tilde{N}_e^{(\kappa)}$ and hence also in $N_e^{(\kappa)}$. That the range of $W_-(A, L^{(\kappa)})$ is dense in $H_a(A)$ follows from the fact that the domain of the operator $W_-(L^{(\kappa)}, A)$ is dense and the results of §4.

Under additional conditions the wave operators are found to be bounded.[7]

THEOREM 6. *Suppose that the hypotheses of Theorem 5 are satisfied and, moreover, the operators $\Theta_\kappa(k)$ and $\Theta'_\kappa(k)$ are invertible with*

$$\sup_{k\in\mathbb{R}} \|(\Theta_\kappa(k))^{-1}\| < \infty, \quad \sup_{k\in\mathbb{R}} \|(\Theta'_\kappa(k))^{-1}\| < \infty.$$

Then the complete bounded wave operators $W_\pm(L^{(\kappa)}, A)$ and $W_\pm(A, L^{(\kappa)})$ exist.

([7]) We remark that the local wave operators for subsets of R on which $\Theta_\kappa(k)$ and $\Theta'_\kappa(k)$ are uniformly boundedly invertible are already bounded without additional assumptions.

PROOF. By Theorem 5 the wave operators $W_{\pm}(A, L^{(\kappa)})$ ($W_{\pm}(L^{(\kappa)}, A)$) are defined on smooth vectors of a dense set in $N_e^{(\kappa)}$ ($H_a(A)$) and have range dense in $H_a(A)$ ($N_e^{(\kappa)}$). The operators $W_{\pm}(A, L^{(\kappa)})$ and $W_{\pm}(L^{(\kappa)}, A)$ are therefore mutual inverses. It therefore suffices to prove that they are bounded. According to Lemma 4, under our conditions

$$\sup_{t\in\mathbf{R}} \| \exp(iL^{(\kappa)}t) \| \leqslant C < \infty,$$

where for C it is possible to take

$$1 + (1 + \|\kappa\|)/2 \max\left\{ \sup_{k\in\mathbf{R}} \|(\Theta_\kappa(k))^{-1}\|, \sup_{k\in\mathbf{R}} \|(\Theta'_\kappa(k))^{-1}\| \right\}.$$

We recall that the operator $\exp(iL^{(\kappa)}t)$ is here considered on a subspace of smooth vectors of $N_e^{(\kappa)}$ (for $\kappa = J\,(iJ)$, $N_e^{(\kappa)}$ coincides with the absolutely continuous subspace of the operator $A + V(L)$). Since

$$\|\exp(-iAt)u - \exp(-iL^{(\kappa)}t) W_{\pm}(L^{(\kappa)}, A)u\| \underset{t\to\pm\infty}{\to} 0$$

for a dense set of vectors $u \in \tilde{N}_e^0$ in $H_a(A)$, and $\|\exp(-iAt)\| = 1$, while the operators $\exp(iL^{(\kappa)}t)$, $t \in \mathbb{R}$, form a bounded group, it follows that $\|W_{\pm}(L^{(\kappa)}, A)\| \leqslant C$. Similarly, we find that $\|W_{\pm}(A, L^{(\kappa)})\| \leqslant C$; therefore $W_{\pm}(A, L^{(\kappa)})$ ($W_{\pm}(L^{(\kappa)}, A)$) can be extended to bounded complete wave operators.

REMARKS. 1. For the operator L the condition that its characteristic function $\Theta(\lambda) = I + iJ\alpha(L^* - \lambda)^{-1}\alpha$ ($\|\Theta(\lambda)\| \leqslant C$, $\operatorname{Im}\lambda \neq 0$) be bounded implies the additional condition of Theorem 6 that concerns only the outer components of the operator-valued functions Θ_κ and Θ'_κ ($\kappa = iJ$).

2. For the boundedness of the wave operators it suffices to require the boundedness of the norms $\sup_{t\in\mathbf{R}}\|\exp(iL^{(\kappa)}t)\| < \infty$; therefore the complete wave operators for the pair of selfadjoint operators $(A + V, A)$ ($\kappa = J, 0$) exist under the sole condition $V \in \sigma_1$. The latter forms the content of the well-known Rosenblum-Kato theorem [8]. In this case it is obvious that $N_e^{(\kappa)}|_{\kappa=J} = H_a(A + V)$ as for the operator A (see above).

3. It was shown that the condition $V \in \sigma_1$ can be replaced by the condition $(L - \lambda_0)^{-1} - (A - \lambda_0)^{-1} \in \sigma_1$ ($\operatorname{Im}\lambda_0 \neq 0$). The proof of Theorems 5 and 6 remains the same. The results of this section are thus applicable to the Schrödinger operator with a decreasing, in general, complex-valued potential without invoking the Birman-Kato invariance principle.

4. We note that the wave operators in the case where L is a dissipative operator with an absolutely continuous spectrum and $(\operatorname{Im} L) \in \sigma_1$ have been studied by Sahnovič [4] in the framework of the triangular model of M. S. Livšic for nonselfadjoint operators. In [4] the possibility of carrying over the Rosenblum-Kato theorem to dissipative operators was proved. The existence of wave operators for $V \in \sigma_1$ (without the dissipative assumption) for operators similar to selfadjoint operators with an absolutely continuous spectrum was proved by Sahnovič [5], [6].

§6. The case of smooth perturbations

The existence of wave operators can be obtained also without the condition $V \in \sigma_1$ in a manner similar to the way in which this is done in the theory of smooth perturbations for the Friedrichs model. In particular, there is the following theorem.

THEOREM 7. *Let* Im $\kappa = 0$ *or* J, *and suppose that*

$$\sup_{k\in\mathbf{R}} \|(\Theta_\kappa(k)^{-1})\| < \infty, \quad \sup_{k\in\mathbf{R}} \|(\Theta'_\kappa(k))^{-1}\| < \infty, \quad \sup_{\operatorname{Im}\lambda>0} \|(I + S(\lambda))\| < \infty.$$

Then $\tilde{N}_e^{(\kappa)} = N_e^{(\kappa)}$, $\tilde{N}_e^0 = H_a(A) = H$, *and there exist the complete wave operators* $W_\pm(A, L^{(\kappa)})$ $(W_\pm(L^{(\kappa)}, A))$ *given by formulas* (10)–(13) *on the subspaces* $N_e^{(\kappa)}(H)$.

PROOF. Under our conditions the wave operators are well defined by formulas (10)–(13) on the linear manifolds $\tilde{N}_e^{(\kappa)}$ $(\tilde{N}_e^0)$. This follows easily from the fact that the operator-valued functions

$$(I + (1 - i\kappa)/2(S(k) - I)),\ (I + (1 + i\kappa)/2(S^*(k) - I)), \qquad k \in R,$$

and $(I + S^*(\bar{\lambda}))$, Im $\lambda < 0$, are also boundedly invertible (for the case $\kappa = iJ$ see [2], Russian p. 94). According to Lemma 4, the wave operators are bounded. It suffices to verify that $\tilde{N}_e^{(\kappa)} = N_e^{(\kappa)}$ and $\tilde{N}_e^0 = H$. The closedness of $\tilde{N}_e^{(\kappa)}$ follows from the identities (see §2)

$$\alpha(L^{-\text{II}} - \lambda)^{-1} = \Theta_\kappa(\lambda)\,\alpha(L^{(\kappa)} - \lambda)^{-1}, \quad \alpha(L^{\text{I}} - \lambda)^{-1} = \Theta'_\kappa(\lambda)\,\alpha(L^{(\kappa)} - \lambda)^{-1}$$

and Theorem 1, which makes it possible in the case of smooth vectors $u \in \tilde{N}_e^{(\kappa)}$ to pass to the limit in these identities, $\lambda = k + i0$ $(\lambda = k - i0)$. Because of the bounded invertibility of $\Theta_\kappa(k)$ and $\Theta'_\kappa(k)$ on the real axis, this readily implies the equivalence of the norms $\|g + S^*g\| + \|S\tilde{g} + g\|$ and

$$\left\{\|\alpha(L^{(\kappa)} - \lambda)^{-1}u\|_{H_2^+(E)} + \|\alpha(L^{(\kappa)} - \lambda)^{-1}u\|_{H_2^-(E)}\right\}$$

for smooth vectors $u = (\tilde{g}, g)$, and hence that $\tilde{N}_e^{(\kappa)} = N_e^{(\kappa)} \equiv \overline{\tilde{N}_e^{(\kappa)}}$. The proof of the second equality $\tilde{N}_e^0 = H$ proceeds by the same scheme with the use of the identities

$$\alpha(L^{-\text{II}} - \lambda)^{-1} = (1/2)(I + S(\lambda))\,\alpha(A - \lambda)^{-1}, \quad \alpha(L^{\text{II}} - \lambda)^{-1}$$
$$= (1/2)(I + S^*(\bar{\lambda}))\,\alpha(A - \lambda)^{-1}.$$

The invertibility of $(I + S(\lambda))$ and hence also of $(I + S^*(\bar{\lambda}))$, Im $\lambda < 0$, leads to the equivalence of the norms $(\|\tilde{g} + S^*g\|)$ and

$$\left\{\|\alpha(A - \lambda)^{-1}u\|_{H_2^+(E)} + \|\alpha(A - \lambda)^{-1}u\|_{H_2^-(E)}\right\}$$

already in the entire space H.

REMARK. With regard to Theorem 7 we mention the papers [6], [15] and [18]. A simple source for the invertibility of the operator-valued functions

$$(I + (1 - i\kappa)/2\,(S(\lambda) - I)), \quad (I + (1 + i\kappa)/2\,(S^*(\bar{\lambda}) - I)), \quad (I + S(\lambda)),$$

leading to the existence of the wave operators for the pair $(L^{(\kappa)}, A)$ is the inequality

$$\sup_{\operatorname{Im}\lambda\neq 0} \|(|V|)^{1/2}(A - \lambda)^{-1}(|V|)^{1/2}\| < (\|\kappa\|)^{-1}.$$

The corresponding assertion is a special case of Kato's theorem [15] on "relatively smooth perturbations". We note that the smoothness of $(|V|)^{1/2}$ relative to A in this case is a consequence of this inequality (see Theorem 7) and therefore does not constitute an added restriction.

§7. The scattering operator

Formulas (10)–(13) make it possible to consider the action of the scattering operator $S_0 \equiv W_+^{-1}(L^{(\kappa)}, A)W_-(L^{(\kappa)}, A)$ in the model representation. A direct computation leads to the formula (where $P_K(\tilde{g}, g) \in \tilde{N}_e^0$)

$$S_0 P_K\begin{pmatrix}\tilde{g}\\ g\end{pmatrix}$$

$$= P_K\begin{bmatrix} -\left(I + \frac{1 - i\kappa}{2}(S - I)\right)^{-1}\left(I + \frac{1 + i\kappa}{2}(S^* - I)\right)g \\ (I + S^*)^{-1}(I + S)\left(I + \frac{1 - i\kappa}{2}(S - I)\right)^{-1}\left(I + \frac{1 + i\kappa}{3}(S^* - I)\right)g \end{bmatrix}.$$

It is possible to attribute a meaning to this expression if it is required that $V \in \sigma_1$ as in §5. We restrict our considerations to the case of selfadjoint operators $\kappa = J$. For $P_K(\tilde{g}, g) \in \tilde{N}_e^0$, i.e. $(\tilde{g} + S^*g) + (S\tilde{g} + g) = 0$, there is the identity

$$\left\| P_K\begin{pmatrix}\tilde{g}\\ g\end{pmatrix}\right\|_{\mathcal{K}}^2 = (\tilde{g} + S^*g, \tilde{g} - P_+(\tilde{g} + S^*g)) + (S\tilde{g} + g, g - P_-(S\tilde{g} + g))$$
$$= -\|P_+(\tilde{g} + S^*g)\|^2 - \|P_-(S\tilde{g} + g)\|^2 + (\tilde{g} + S^*g, \tilde{g}) + (S\tilde{g} + g, g)$$
$$= -(S\tilde{g} + g, S\tilde{g} + g) + (\tilde{g} + S^*g, \tilde{g}) + (S\tilde{g} + g, g)$$
$$= (\tilde{g}, g) - (S\tilde{g}, S\tilde{g}) = ((I - S^*S)\tilde{g}, \tilde{g}),$$

which makes it possible to establish an isometry between $\tilde{N}_e^0$ and some subset $\{\tilde{g}\}$ belonging to the space of E-valued functions on the real axis with norm $[((I - S^*S)\tilde{g}, \tilde{g})]^{1/2}$. Under our condition $V \in \sigma_1$ it is easily proved that this isometry extends to an isometric mapping of N_e^0 onto the entire Hilbert space indicated above. Thus, the scattering operator S_0 for the pair of selfadjoint operators $(A + V, A)$ is unitarily equivalent to the operator of multiplication by the function([8]) of k

$$\left(I + \frac{1 - iJ}{2}(S - I)\right)^{-1}\left(I + \frac{1 + iJ}{2}(S^* - I)\right)(I + S^*)^{-1}(I + S),$$

which is defined for almost all k because of the presence of a scalar multiple for the operator-valued functions with the inverses required by this formula. We formulate the results of these considerations separately.

THEOREM 8. *The scattering operator for the selfadjoint operators $A + V$ and A ($V \in \sigma_1$) is unitarily equivalent to multiplication by the operator-valued function*

$$\left(I + \frac{1 - iJ}{2}(S - I)\right)^{-1}\left(I + \frac{1 + iJ}{2}(S^* - I)\right)(I + S^*)^{-1}(I + S)$$

(where $J \equiv$ sign V and S is the characteristic function of the dissipative operator $A + i|V|$) in the Hilbert space of E-valued ($E \equiv \overline{R(V)}$) vector-valued functions on the axis R which are square-summable with the operator "weight" $(I - S^*(k)S(k))$.

The theorem was proved above. We note only that in the absence of a completely nonselfadjoint operator $A + i|V|$ it is necessary to add orthogonally to the

([8]) It is not hard to see that the given operator-valued function is the scattering matrix for the pair of operators $(A + V, A)$. Indeed, in this Hilbert space the absolutely continuous part of the operator A is the operator of multiplication by k.

scattering operator described in Theorem 8 the identity operator on the corresponding subspace

$$H \ominus \bigvee_{(\operatorname{Im}\lambda \neq 0)} (A-\lambda)^{-1}(VH).$$

For Theorem 8 the condition $V \in \sigma_1$ can be replaced by $\sqrt{|V|}\,(A-\lambda_0)^{-1} \in \sigma_2$ or by a condition of relative smoothness (see §6). It is evident from the formulation of the theorem that the scattering operator admits a factorization as the product of four operator-valued functions of which two admit analytic continuation to the upper half plane $\operatorname{Im}\lambda > 0$ and the remaining two to the lower half plane.

In conclusion the author thanks B. S. Pavlov for his attention to the work.

BIBLIOGRAPHY

1. Béla Sz.-Nagy and Ciprian Foiaş, *Analyse harmonique des opérateurs de l'espace de Hilbert*, Masson, Paris, and Akad. Kiadó, Budapest, 1967; rev. English transl. North-Holland, Amsterdam and Amer. Elsevier, New York, 1970.

2. S. N. Naboko, *Absolutely continuous spectrum of a nondissipative operator, and a functional model*. I, Zap. Naučn. Sem. Leningrad. Otdel. Mat. Inst. Steklov. (LOMI) **65** (1976), 90–102; English transl., to appear in J. Soviet Math.

3. ______, *On the spectral analysis of nonselfadjoint operators*, Dokl. Akad. Nauk SSSR **232** (1977), 36–39; English transl. in Soviet Math. Dokl. **18** (1977).

4. L. A. Sahnovič, *Dissipative operators with absolutely continuous spectrum*, Trudy Moskov. Mat. Obšč. **19** (1968), 211–270; English transl. in Trans. Moscow Math. Soc. **19** (1968).

5. ______, *Operators which are similar to unitary operators with absolutely continuous spectrum*, Funkcional. Anal. i Priložen. **2** (1968), no. 1, 51–63; English transl. in Functional Anal. Appl. **2** (1968).

6. ______, *Nonunitary operators with absolutely continuous spectrum on the unit circle*, Dokl. Akad. Nauk SSSR **181** (1968), 558–561; English transl. in Soviet Math. Dokl. **9** (1968).

7. S. N. Naboko, *Wave operators for nonselfadjoint operators and the functional model*, Zap. Naučn. Sem. Leningrad. Otdel. Mat. Inst. Steklov. (LOMI) **69** (1977), 129–135; English transl. in J. Soviet Math. **10** (1978), no. 2.

8. Tosio Kato, *Perturbation theory for linear operators*, Springer-Verlag, 1966.

9. M. S. Brodskiĭ, *Triangular and Jordan representations of linear operators*, "Nauka", Moscow, 1969; English transl., Amer. Math. Soc., Providence, R. I., 1971.

10. B. S. Pavlov, *Selfadjoint dilation of the dissipative Schrödinger operator and its resolution in terms of eigenfunctions*, Mat. Sb. **102 (144)** (1977), 511–536; English transl. in Math. USSR Sb. **31** (1977).

11. Kenneth Hoffman, *Banach spaces of analytic functions*, Prentice-Hall, Englewood Cliffs, N. J., 1962.

12. B. S. Pavlov, *On separation conditions for the spectral components of a dissipative operator*, Izv. Akad. Nauk SSSR Ser. Mat. **39** (1975), 123–148; English transl. in Math. USSR Izv. **9** (1975).

13. Peter D. Lax and Ralph S. Phillips, *Scattering theory*, Academic Press, 1967.

14. E. C. Titchmarsh, *Introduction to the theory of Fourier integrals*, 2nd ed., Clarendon Press, Oxford, 1948.

15. Tosio Kato, *Wave operators and similarity for some non-selfadjoint operators*, Math. Ann. **162** (1965/66), 258–279.

16. S. N. Naboko, *Absolutely continuous spectrum of a nondissipative operator and a functional model*. II, Zap. Naučn. Sem. Leningrad. Otdel. Mat. Inst. Steklov. (LOMI) **73** (1977), 118–135; English transl., to appear in J. Soviet Math.

17. I. C. Gohberg and M. G. Kreĭn, *Introduction to the theory of linear nonselfadjoint operators in Hilbert space*, "Nauka", Moscow, 1965; English transl., Amer. Math. Soc., Providence, R. I., 1969.

18. B. Simon, *Scattering theory and quadratic forms: on a theorem of Schechter*, Comm. Math. Phys. **53** (1977), 151–153.

Translated by J. R. SCHULENBERGER

Труды Матем. Инст. Стеклов.
147(1980)

Proceedings of the Steklov
Institute of Mathematics
1981, Issue 2

ON UNIQUE SOLVABILITY OF BOUNDARY VALUE PROBLEMS FOR THE STOKES SYSTEM OF EQUATIONS IN DOMAINS WITH NONCOMPACT BOUNDARIES

UDC 517.946

K. I. PILECKAS

ABSTRACT. The first boundary value problem is considered for the stationary Stokes system in domains with several cone-shaped "arms" extending to infinity. It is shown that for unique solvability of this problem one can specify, besides the boundary conditions, limit values of the pressure at infinity in all except one of the "arms".

Bibliography: 5 titles.

§1. Introduction

As was shown in [1]–[4], it is necessary to specify certain functionals of the solution when formulating boundary value problems for the Navier-Stokes system of equations in certain domains with noncompact boundaries, this specification being in addition to the usual boundary conditions. If the domain $\Omega \subset R^3$ has more than one "arm" at infinity, i.e. if outside of the ball $K_R = \{x\colon |x| \leqslant R$, with $R \geqslant R_0$ arbitrary$\}$ the domain splits up into m connected components D_i, then as such functionals we can take velocity flows across sections of the $(m-1)$th "arm". For linearized problems we propose in the present paper a second and equivalent choice of supplementary conditions in which there appear values of the pressure in each of the D_i as $|x| \to \infty$. This result was established in [1] for a domain of special type.

We now give the definitions of our functional spaces.

$C_0^\infty(\Omega)$ is the set of infinitely differentiable vector-valued functions with compact supports in the domain Ω.

$\dot{J}^\infty(\Omega)$ is the set of solenoidal vectors from $C_0^\infty(\Omega)$.

$\mathring{D}(\Omega)$ and $\mathring{W}_2^1(\Omega)$ are the closures of $C_0^\infty(\Omega)$ in the norms generated, respectively, by the inner products

$$[\mathbf{u}, \mathbf{v}] = \int_\Omega \mathbf{u}_x \cdot \mathbf{v}_x dx = \int_\Omega \sum_{i,k=1}^3 u_{ix_k} v_{ix_k} dx, \tag{1}$$

$$(\mathbf{u}, \mathbf{v})_{2,\Omega}^{(1)} = \int_\Omega (\mathbf{u}\cdot\mathbf{v} + \mathbf{u}_x \cdot \mathbf{v}_x)\, dx; \tag{2}$$

$\hat{H}(\Omega)$ and $\hat{\mathring{J}}{}_2^1(\Omega)$ are subspaces of solenoidal vectors from $\mathring{D}(\Omega)$ and $\mathring{W}_2^1(\Omega)$.

$H(\Omega) \subset \hat{H}(\Omega)$ and $\mathring{J}_2^1(\Omega) \subset \hat{\mathring{J}}{}_2^1(\Omega)$ are the closures of $\dot{J}^\infty(\Omega)$ in the norms (1) and (2).

1980 *Mathematics Subject Classification*. Primary 35Q10, 35A05, 35D05; Secondary 46E40.

We assume that Ω belongs to the class of domains considered in [2], i.e. that the following conditions hold:

a) $\Omega \setminus K_{R_0} = \bigcup_{i=1}^m \mathfrak{D}_i$ for some $R_0 > 0$, where $\mathfrak{D}_i$ contains the infinite cone C_i, specified in some system of coordinates by the equation $\sqrt{(x_1^i)^2 + (x_2^i)^2} \equiv |x^{i'}| < \alpha^i x_3^i$.

b) The domain $\Omega_{R_1} \equiv \Omega \cap K_{R_1}$, $R_1 > R_0$, is such that

$$H(\Omega_{R_1}) = \hat{H}(\Omega_{R_1}).$$

c) $\mathfrak{D}_i \cap K_{R_1}$ belongs to the class K introduced in [2] (to this class there belong, in particular, domains with Lipschitz boundaries).

It was shown in [2] that under these conditions

$$\dim \hat{H}(\Omega)/H(\Omega) = \dim \hat{J}_2'(\Omega)/J_2'(\Omega) = m - 1.$$

§2. Stationary problem

In this section we consider the linear stationary Stokes problem

$$\begin{cases} \nu\Delta \mathbf{u} = \nabla p + \mathbf{f}, \quad \operatorname{div} \mathbf{u} = 0, \\ \mathbf{u}|_{\partial\Omega} = 0, \quad \mathbf{u}(x) \xrightarrow[|x|\to\infty]{} 0 \end{cases} \tag{3}$$

subject to the further conditions

$$p_1 - p_m = q_1, \ldots, p_{m-1} - p_m = q_{m-1}, \tag{4}$$

where the p_i are the limit values of the function $p(x)$, $x \in C_i$, as $|x| \to \infty$, and the q_i are specified real numbers. Relative to $\mathbf{f}(x)$ we assume that $\int_\Omega \mathbf{f} \cdot \mathbf{\Phi}\, dx$ is a bounded linear functional over $\mathbf{\Phi} \in H(\Omega)$ and that, in addition, $\mathbf{f}(x) \in L_2(\Omega)$.

By a generalized solution of the problem (3) we mean a vector-valued function $\mathbf{u}(x)$ from $\hat{H}(\Omega)$ satisfying the integral identity

$$\nu \int_\Omega \mathbf{u}_x \cdot \mathbf{\Phi}_x dx = \int_\Omega \mathbf{f} \cdot \mathbf{\Phi} dx \tag{5}$$

for arbitrary $\mathbf{\Phi} \in H(\Omega)$.

We show, subject to these restrictions on $\mathbf{f}(x)$, that the pressure $p(x)$, corresponding to an arbitrary generalized solution of the problem (3), tends toward a constant as $|x| \to \infty$, $x \in C_i$. For this purpose we require the following two lemmas.

LEMMA 1. *If the function $\varphi(x)$ is harmonic in R^3, and if its Dirichlet integral $\int_{R^3}\{\nabla\varphi(x)\}^2\, dx$ is finite, then* $\varphi(x) =$ const.

PROOF. Let $\zeta(x)$ be a function from $C_0^\infty(R^3)$, which depends only on $|x| = r$, is equal to 1 for $0 \leqslant r \leqslant 1$ and to zero for $r \geqslant 2$, and is monotonically decreasing for $1 < r < 2$. Then the function $\zeta_R(x) = \zeta(x/R)$ is equal to zero for $|x| \geqslant 2R$, is equal to 1 for $|x| \leqslant R$, and obeys the estimates

$$0 \leqslant \zeta_R(x) \leqslant 1, \quad |\nabla\zeta_R(x)| \leqslant \frac{c}{R},$$

where the constant c is independent of R.

It is obvious that the function $\varphi_{x_i} = \partial\varphi(x)/\partial x_i$ is harmonic in R^3. Therefore

$$0 = -\int_{R^3} \zeta_R^2 \sum_{i=1}^{3} \varphi_{x_i} \Delta \varphi_{x_i}^2 dx = \sum_{i,k=1}^{3} \int_{R^3} \zeta_R^2 \varphi_{x_i x_k}^2 dx + \sum_{i,k=1}^{3} \int_{R^3} 2\zeta_R \frac{\partial}{\partial x_k} \zeta_R(x) \varphi_{x_i} \varphi_{x_i x_k} dx.$$

It follows from this that

$$\int_{R^3} \zeta_R^2 \sum_{i,k=1}^{3} \varphi_{x_i x_k}^2 dx \leqslant 2 \left(\int_{R^3} \zeta_R^2 \sum_{i,k=1}^{3} \varphi_{x_i x_k}^2 dx \right)^{1/2} \left(\int_{R^3} |\nabla \zeta_R|^2 |\nabla \varphi|^2 dx \right)^{1/2}$$

$$\leqslant \frac{2c}{R} \left(\int_{R^3} \zeta_R^2 \sum_{i,k=1}^{3} \varphi_{x_i x_k}^2 dx \right)^{1/2} \left(\int_{|x| \leqslant 2R} |\nabla \varphi|^2 dx \right)^{1/2},$$

i.e.

$$\int_{R^3} \zeta_R^2 \sum_{i,k=1}^{3} \varphi_{x_i x_k}^2 dx \leqslant \frac{4c^2}{R^2} \int_{|x| \leqslant 2R} \{\nabla \varphi(x)\}^2 dx. \tag{6}$$

Letting $R \to \infty$ in (6), and taking into account the finiteness of the integral $\int_{R^3} \{\nabla\varphi(x)\}^2\, dx$, we find that $\varphi_{x_i x_k} = 0$, $i, k = 1, 2, 3$, and since $\partial\varphi/\partial x_i \in L_2(R^3)$, it follows that $\varphi(x) = \text{const}$. This proves the lemma.

LEMMA 2. *Let the function $q(x)$ be defined in the domain*

$$C_{R_0} = \{x : |x'| < \alpha x_3,\ \alpha > 0,\ |x| > R_0 > 0\}$$

and let $\int_{C_{R_0}} \{\nabla q(x)\}^2\, dx < \infty$. Then there exists a constant q_0 such that

$$\int_{C_{R_0}} \frac{(q(x) - q_0)^2}{|x|^2} dx < \infty.$$

PROOF. We extend the function $q(x)$ so that the resulting function will have a finite Dirichlet integral over all of R^3.

Upon passing to a spherical system of coordinates the domain C_{R_0} goes over into the domain

$$\tilde{C}_{R_0} = \left\{ (r, \varphi, \psi) : r > R_0,\ 0 \leqslant \varphi \leqslant 2\pi,\ 0 < \psi < \psi_0, \right.$$

$$\left. \text{where } \psi_0 = \arccos \frac{1}{\sqrt{\alpha^2 + 1}} < \frac{\pi}{2} \right\}$$

and

$$\int_{C_{R_0}} \{\nabla q(x)\}^2 dx$$

$$= \iiint_{\tilde{C}_{R_0}} \left\{ \left(\frac{\partial \tilde{q}}{\partial r} \right)^2 + \frac{1}{r^2} \left(\frac{\partial \tilde{q}}{\partial \psi} \right)^2 + \frac{1}{r^2 \sin^2 \psi} \left(\frac{\partial \tilde{q}}{\partial \varphi} \right)^2 \right\} r^2 \sin \psi\, dr\, d\varphi\, d\psi < \infty,$$

where $\tilde{q}(r, \varphi, \psi) = q(x(r, \varphi, \psi))$. Let

$$\tilde{q}_1(r, \varphi, \psi) = \begin{cases} \tilde{q}(r, \varphi, \psi), & \text{if } (r, \varphi, \psi) \in \tilde{C}_{R_0}, \\ \tilde{q}\left(r, \varphi, \dfrac{\psi_0}{\left(\frac{\pi}{2} - \psi_0\right)}\left(\dfrac{\pi}{2} - \psi\right)\right), & \text{if } (r, \varphi, \psi) \\ \in \left\{(r, \varphi, \psi) : r > R_0,\ 0 \leqslant \varphi \leqslant 2\pi,\ \psi_0 \leqslant \psi \leqslant \frac{\pi}{2}\right\}, \end{cases}$$

and

$$\tilde{q}_2(r, \varphi, \psi) = \begin{cases} \tilde{q}_1(r, \varphi, \psi), & \text{if } (r, \varphi, \psi) \in \left\{(r, \varphi, \psi) : r > R_0,\ 0 \leqslant \varphi \leqslant 2\pi,\ 0 \leqslant \psi \leqslant \frac{\pi}{2}\right\}, \\ \tilde{q}_1(r, \varphi, \pi - \psi), & \text{if } (r, \varphi, \psi) \in \left\{(r, \varphi, \psi) : r > R_0,\ 0 \leqslant \varphi \leqslant 2\pi,\ \frac{\pi}{2} \leqslant \psi \leqslant \pi\right\}. \end{cases}$$

Reverting to a cartesian coordinate system, we have

$$\int\limits_{\{x : |x| > R_0\}} \{\nabla q_2(x)\}^2\, dx = \iiint\limits_{\substack{r > R_0 \\ 0 \leqslant \varphi \leqslant 2\pi \\ 0 \leqslant \psi \leqslant \pi}} \left\{\left(\frac{\partial \tilde{q}_2}{\partial r}\right)^2 + \frac{1}{r^2}\left(\frac{\partial \tilde{q}_2}{\partial \psi}\right)^2 + \frac{1}{r^2 \sin^2 \psi}\left(\frac{\partial \tilde{q}_2}{\partial \varphi}\right)^2\right\} r^2 \sin \psi\, dr\, d\varphi\, d\psi < \infty.$$

We define a function $q_3(x)$ as follows:

$$q_3(x) = \begin{cases} q_2(x), & \text{if } |x| \geqslant R_0, \\ q_2\left(\dfrac{R_0^2}{|x|^2} x\right) \equiv q_2(y), & \text{if } |x| \leqslant R_0. \end{cases}$$

Since

$$\frac{\partial q_2(y)}{\partial x_l} = \sum_{k=1}^{3} \frac{\partial q_2(y)}{\partial y_k} \frac{\partial y_k}{\partial x_l} = \sum_{k=1}^{3} \frac{\partial q_2(y)}{\partial y_k} \left(\frac{R_0^2}{|x|^2}\delta_{kl} - \frac{2R_0^2 x_k x_l}{|x|^4}\right), \quad l = 1, 2, 3,$$

and

$$\int\limits_{R^3} \{\nabla_x q_3(x)\}^2\, dx = \int\limits_{\{x : |x| \geqslant R_0\}} \{\nabla_x q_2(x)\}^2\, dx + \int\limits_{\{x : |x| \leqslant R_0\}} \{\nabla_x q_2(y)\}^2\, dx,$$

it follows that

$$\begin{aligned} \int\limits_{R^3} \{\nabla_x q_3(x)\}^2\, dx &\leqslant \int\limits_{\{x : |x| \geqslant R_0\}} \{\nabla_x q_2(x)\}^2\, dx + c_1 \int\limits_{\{x : |x| \leqslant R_0\}} \{\nabla_y q_2(y)\}^2 \frac{R_0^4}{|x|^4}\, dx \\ &\leqslant \int\limits_{\{x : |x| \geqslant R_0\}} \{\nabla_x q_2(x)\}^2\, dx + c_2 \int\limits_{\{y : |y| \geqslant R_0\}} \{\nabla_y q_2(y)\}^2 \frac{R_0^2}{|y|^2}\, dy \\ &\leqslant c_3 \int\limits_{\{x : |x| \geqslant R_0\}} \{\nabla_x q_2(x)\}^2\, dx < \infty. \end{aligned}$$

Thus, $q_3(x)$ furnishes the necessary extension of $q(x)$.

We consider now the bounded linear functional

$$f(\Phi) = \int\limits_{R^3} \nabla q_3 \cdot \nabla \Phi\, dx,$$

prescribed for functions $\Phi(x)$ from $\mathring{D}(R^3)$. By Riesz's theorem we can put $f(\Phi)$ in the form $\int_{R^3}\nabla b\cdot\nabla\Phi\,dx$, where $b(x)$ is an element from $\mathring{D}(R^3)$. Then

$$\int_{R^3}\nabla(q_3-b)\cdot\nabla\Phi dx=0\quad\text{for all}\quad \Phi\in\mathring{D}(R^3).$$

Consequently, $\varphi(x) = q_3(x) - b(x)$ is a harmonic function in R^3, and

$$\int_{R^3}\{\nabla\varphi(x)\}^2dx<\omega.$$

From this we find, according to Lemma 1, that $q_3(x) - b(x) = q_0 = \text{const}$. Since for an arbitrary function $b(x)\in\mathring{D}(R^3)$ the estimate

$$\int_{R^3}\frac{b^2(x)}{|x|^2}dx\leqslant 4\int_{R^3}\{\nabla b(x)\}^2dx,$$

holds (see [5], Chapter 1, §1), it follows that

$$\int_{R^3-}\frac{(q_3(x)-q_0)^2}{|x|^2}dx=\int_{R^3}\frac{b^2(x)}{|x|^2}dx\leqslant 4\int_{R^3}\{\nabla b(x)\}^2dx<\infty. \tag{7}$$

From (7) we then have

$$\int_{C_{R_0}}\frac{(q(x)-q_0)^2}{|x|^2}dx<\infty.$$

This completes the proof of the lemma.

We return now to a consideration of problem (3) in a domain Ω satisfying conditions a)–c).

THEOREM 1. *Let the domains $C^{i,1}\subset C^i$, $i = 1,\dots,m$, be of the form*

$$C^{i,1}=\{x^i:|x^{i\prime}|\leqslant\alpha_1^ix_3^i,\ \alpha_1^i<\alpha^i\}$$

in the coordinate systems x^i, and let

$$C^{i,1}_{R_0+\delta}=C^{i,1}\setminus\bar{K}_{R_0+\delta},\ \delta>0.$$

If $u(x)$ is a generalized solution of the problem (3), *and if*

$$\mathbf{f}(x)\in L_2\left(\bigcup_{i=1}^{m}C^{i}_{R_0}\right),$$

then

$$\int_{C^{i,1}_{R_0+\delta}}|\mathbf{u}_{x_kx_l}(x)|^2dx<\infty\quad\forall i,\ i=1,\dots,m.$$

PROOF. The following estimate is known to hold for a solution of the problem (3) (see [5], Chapter 2, §1)

$$\|\mathbf{u}_{x_kx_l}\|_{2,\Omega''}\leqslant c_\zeta\|\mathbf{u}_x\|_{2,\Omega'\setminus\Omega''}+c_\zeta\|\mathbf{f}\|_{2,\Omega'}+c_1\|(\Delta\zeta)\mathbf{u}\|_{2,\Omega'}. \tag{8}$$

Here Ω'' is any strictly interior subdomain of Ω', and Ω' is any strictly interior subdomain of Ω; $\zeta(x)$ is a twice continuously differentiable function which is equal to one in Ω'', does not exceed one on $\Omega' \setminus \Omega''$, and is equal to zero outside of Ω'; and c_ζ depends only on $\max_{x\in\Omega'\setminus\Omega''}|\nabla\zeta(x)|$.

For $\delta \leqslant R$ and $R \geqslant R_0$, the closure of the domain $C^{i,1}_{R_0+\delta,R} \equiv C^{i,1}_{R_0+\delta} \cap K_R$ is contained in the domain $C^i_{R_0,2R} \equiv C^i_{R_0} \cap K_{2R}$, and $\overline{C}^i_{R_0,2R} \subset \Omega$ for all $i = 1, \ldots, m$. Let

$$\zeta_1^i(x^i) = \begin{cases} 1, & \text{if} \quad x^i \in C^{i,1}, \\ \dfrac{(a^i)^2 (x_3^i)^2 - |x^{i\prime}|^2}{[(a^i)^2 - (a_1^i)^2](x_3^i)^2}, & \text{if} \quad x^i \in C^i \setminus C^{i,1}, \\ 0, & \text{if} \quad x^i \overline{\in} C^i, \end{cases}$$

and

$$\zeta_{2,R}^i(x^i) = \begin{cases} 0, & \text{if} \quad |x^i| \leqslant R_0, \\ (|x^i| - R_0)/\delta, & \text{if} \quad R_0 \leqslant |x^i| \leqslant R_0 + \delta, \\ 1, & \text{if} \quad R_0 + \delta \leqslant |x^i| \leqslant R, \\ R^{-3}(|x^i| - 2R)^2(2|x^i| - R), & \text{if} \quad R \leqslant |x^i| \leqslant 2R, \\ 0, & \text{if} \quad |x^i| \geqslant 2R. \end{cases}$$

We now take the function $\zeta_R^i(x) = \zeta_1^i(x^i)\zeta_{2,R}^i(x^i)$. It is equal to one in $C^{i,1}_{R_0+\delta,R}$, decreases monotonically in $C^i_{R_0,2R} \setminus C^{i,1}_{R_0+\delta,R}$, and is equal to zero outside of $C^i_{R_0,2R}$. Therefore, by virtue of (8),

$$\|\mathbf{u}_{x_k x_l}\|_{2,C^{i,1}_{R_0+\delta,R}} \leqslant c_{\zeta_R^i}\|\mathbf{u}_x\|_{2,C^i_{R_0,2R}\setminus C^{i,1}_{R_0+\delta,R}} + c_{\zeta_R^i}\|\mathbf{f}\|_{2,C^i_{R_0,2R}} + c_1\|(\Delta\zeta_R^i)\mathbf{u}\|_{2,C^i_{R_0,2R}}. \tag{9}$$

It is not difficult to see that

$$|\nabla\zeta_R^i|(x^i)| \leqslant \frac{c}{|x^i|} \leqslant \frac{c}{R_0}, \quad |\Delta\zeta_R^i(x^i)| \leqslant \frac{c}{|x^i|^2}.$$

Then

$$\|(\Delta\zeta_R^i)\mathbf{u}\|^2_{2,C^i_{R_0,2R}} = \int_{C^i_{R_0,2R}} (\Delta\zeta_R^i)^2\mathbf{u}^2 dx^i \leqslant \frac{c}{R_0^2}\int_{C^i_{R_0,2R}} \frac{\mathbf{u}^2}{|x^i|^2}dx^i$$
$$\leqslant \frac{4c}{R_0^2}\int_\Omega \mathbf{u}_x^2 dx < \infty \quad \forall R > R_0.$$

Letting $R \to \infty$ in the inequality (9), and noting that

$$\mathbf{f}(x) \in L_2\Big(\bigcup_{i=1}^m C^i_{R_0}\Big),$$

and that $\max|\nabla\zeta_R^i(x^i)|$ and $\|(\Delta\zeta_R^i)\mathbf{u}\|_{2,C^i_{R_0,2R}}$ are bounded by constants independent of R, we obtain

$$\|\mathbf{u}_{x_k x_l}\|_{2,C^{i,1}_{R_0+\delta}} < \infty, \quad i = 1, \ldots, m. \tag{10}$$

THEOREM 2. *Let* $\mathbf{u}(x)$ *be a generalized solution of the problem* (3), *let* $p(x)$ *be the corresponding pressure, and let*

$$\mathbf{f}(x) \in L_2\Big(\bigcup_{i=1}^m C^i_{R_0}\Big).$$

Then there exist constants p_i *such that*

$$\int_{C^{i,1}_{R+\delta}} \frac{(p(x)-p_i)^2}{|x|^2}\,dx<\infty \quad \textit{for } \forall i,\ i=1,\ldots,m. \tag{11}$$

PROOF. By virtue of (3) and Theorem 1 we have

$$\int_{C^{i,1}_{R_0+\delta}} \{\nabla p(x)\}^2\,dx<\infty,$$

whence, according to Lemma 2, constants p_i exist satisfying (11).

We consider the homogeneous problem for the Stokes system of equations:

$$\begin{cases} \nu\Delta\mathbf{u}=\nabla p, & \operatorname{div}\mathbf{u}=0,\\ \mathbf{u}|_{\partial\Omega}=0, & \mathbf{u}(x)\underset{|x|\to\infty}{\longrightarrow}0. \end{cases} \tag{3'}$$

THEOREM 3. *If all the* $q_i=p_i-p_m=0$, $i=1,\ldots,m-1$, *where the* p_i *are the constants from* (11), *then the only solution of problem* (3′) *is* $\mathbf{u}(x)=0$.

For an arbitrary set of numbers $\{q_i\}_{i=1}^{m-1}$, *specified beforehand, there exists a unique solution* $\mathbf{u}(x)$ *of problem* (3′), *with corresponding pressure* $p(x)$, *for which* (11) *holds.*

PROOF. Let $q_i=0$, $i=1,\ldots,m-1$. With no loss of generality we can assume that

$$p_1=\ldots=p_m=0. \tag{12}$$

From the results obtained in [2] it follows that we can represent an arbitrary vector-valued function $\mathbf{u}(x)\in\hat{H}(\Omega)$ in the form

$$\mathbf{u}(x)=\mathbf{a}(x)+\mathbf{v}(x), \tag{13}$$

where $\mathbf{v}(x)\in H(\Omega)$; $\mathbf{a}(x)$ and $\mathbf{u}(x)$ belong to the quotient class $\hat{H}(\Omega)/H(\Omega)$, and, for $|x|$ sufficiently large,

$$|\mathbf{a}(x)|\leqslant\frac{c}{|x|^2}. \tag{14}$$

Since the generalized solution of (3′) belongs to $\hat{H}(\Omega)$, it has the representation (13) with some vector-valued function $\mathbf{a}^{(1)}(x)$; this function can be chosen so as to be equal to zero close to the boundary of the cones $C^{i,1}$.

We multiply both sides of the equation $\nu\Delta u=\nabla p$ scalarly by $\mathbf{a}^{(1)}(x)$ and then integrate over Ω_R with $R>R_0+\delta$. Then

$$\nu\int_{\Omega_R}\Delta\mathbf{u}\cdot\mathbf{a}^{(1)}dx=\int_{\Omega_R}\nabla p\cdot\mathbf{a}^{(1)}dx. \tag{15}$$

Integrating (15) by parts, and noting that div $a^{(1)}=0$, we obtain

$$-\nu\int_{\Omega_R}\mathbf{u}_x\cdot\mathbf{a}^{(1)}_x dx+\nu\int_{S_R}\mathbf{a}^{(1)}\cdot\frac{\partial\mathbf{u}}{\partial n}dS=\int_{S_R}p\mathbf{a}^{(1)}\cdot\mathbf{n}dS, \tag{16}$$

where

$$S_R = \partial K_R \cap \left(\bigcup_{i=1}^{m} C^{i,1}_{R_0+\delta} \right).$$

From (14) it follows that $\mathbf{a}^{(1)} \in L_2(\Omega)$. Therefore, there exists a sequence of numbers $\{R_l\} \to \infty$ for which

$$\nu \int\limits_{S_{R_l}} \mathbf{a}^{(1)} \cdot \frac{\partial \mathbf{u}}{\partial n} dS \underset{l\to\infty}{\longrightarrow} 0.$$

For sufficiently large R we have

$$\left| \int\limits_{S_R} p\mathbf{a}^{(1)} \cdot \mathbf{n} dS \right| \leqslant c \int\limits_{S_R} \frac{|p(x)|}{|x|^2} dS \leqslant c \left(\int\limits_{S_R} \frac{p^2(x)}{|x|^2} dS \right)^{1/2} \left(\int\limits_{S_R} \frac{dS}{|x|^2} \right)^{1/2}.$$

By virtue of (11) and (12) we can select from $\{R_l\}$ a subsequence $\{R_{l_k}\}$ such that

$$\int\limits_{S_{R_{l_k}}} p^2(x)/|x|^2 dS \underset{l_k\to\infty}{\longrightarrow} 0.$$

The expression $\int_{S_R} dS/|x|^2$ stays bounded as $R \to \infty$. Therefore, if in (16) we let R become infinite along the sequence $\{R_{l_k}\}$, we obtain

$$\nu \int\limits_{\Omega} \mathbf{u}_x \cdot \mathbf{a}^{(1)}_x dx = 0.$$

From this equation and the definition of a generalized solution of (3′) it follows that

$$\nu \int\limits_{\Omega} \mathbf{u}_x^2 dx = \nu \int\limits_{\Omega} \mathbf{u}_x \cdot \mathbf{a}^{(1)}_x dx + \nu \int\limits_{\Omega} \mathbf{u}_x \cdot \mathbf{v}_x dx = 0,$$

i.e. $\mathbf{u}(x) = 0$.

We now prove the second assertion of the theorem. We consider an arbitrary $(m-1)$-dimensional vector $\mathbf{F} = (F_1, \ldots, F_{m-1})$. Then the vector

$$\mathbf{F}^* = (F_1^*, \ldots, F_m^*) = \left(F_1, \ldots, F_{m-1}, -\sum_{i=1}^{m-1} F_i \right)$$

will satisfy the condition $\Sigma_1^m F_i^* = 0$. It follows from the results obtained in [1] and [2] that a unique solution of (3′) exists which satisfies the equations

$$\int\limits_{S_k^i} \mathbf{u} \cdot \mathbf{n} dS = F_i^*, \quad i = 1, \ldots, m.$$

By Theorem 2 we can find, in accordance with this solution, numbers p_i, $i = 1, \ldots, m$, such that

$$\int\limits_{C^{i,1}_{R_0+\delta}} \frac{(p(x) - p_i)^2}{|x|^2} dx < \infty \quad \forall i, \quad i = 1, \ldots, m.$$

We take $\mathbf{Q} = (q_1, \dots, q_{m-1}) = (p_1 - p_m, \dots, p_{m-1} - p_m)$, and we denote by A an operator which makes correspond to each vector $\mathbf{F} = (F, \dots, F_{m-1})$ a vector $\mathbf{Q}$ according to the scheme presented above. From the linearity of the problem (3′) it follows that A is a linear operator in R^{m-1}. We consider the equation

$$\mathbf{AF} = \mathbf{Q}.$$

From what has been proved, the only possible solution of the equation $\mathbf{AF} = 0$ is $\mathbf{F} = 0$; consequently (17) is uniquely solvable for arbitrary $\mathbf{Q}$. Therefore, for an arbitrary set of numbers $\{q_i\}_1^{m-1}$ satisfying the conditions of the theorem, there exists a unique solution of (3′). This completes the proof of the theorem.

We now consider the problem (3), (4).

THEOREM 4. *If* $\mathbf{f}(x) \in L_2(\bigcup_{i=1}^m C_{R_0}^i)$, *and if* $\int_\Omega \mathbf{f} \cdot \mathbf{\Phi}\, dx$ *is a bounded linear functional over* $\mathbf{\Phi} \in H(\Omega)$, *then for an arbitrary set of real numbers* $\{q_i\}_1^{m-1}$ *there exists a unique solution of the problem* (3), (4).

PROOF. The uniqueness of the solution of the problem (3), (4) is a consequence of its linearity and Theorem 3. We prove the existence of the solution. Let $\mathbf{v}(x)$ be a generalized solution of the Stokes system belonging to $H(\Omega)$ (its existence was proved in [5]). Let $h_1, \dots, h_m$ be "limiting" values of the pressure $q(x)$ corresponding to $\mathbf{v}(x)$. By virtue of Theorem 3, for the set of numbers $s_i = q_i + (h_i - h_m)$, $i = 1, \dots, m - 1$, there exists a unique generalized solution $\mathbf{w}(x) \in \hat{H}(\Omega)$ of the problem (3′), (4). Then $\mathbf{u}(x) = \mathbf{v}(x) + \mathbf{w}(x)$ is a generalized solution of the problem (3), (4). This completes the proof of the theorem.

REMARK. In case the spaces $\mathring{J}_2'(\Omega)$ and $J_2'(\Omega)$ do not coincide, analogous results can be obtained for the problem

$$\begin{cases} \nu\Delta\mathbf{u} + \mathbf{u} = \nabla p + \mathbf{f}, \quad \operatorname{div}\mathbf{u} = 0, \\ \mathbf{u}|_{\partial\Omega} = 0, \quad \mathbf{u}(x) \underset{|x|\to\infty}{\to} 0. \end{cases} \tag{3″}$$

Its generalized solution is a vector-valued function $\mathbf{u}(x) \in \mathring{J}_2'(\Omega)$ which satisfies the integral identity

$$\nu \int_\Omega \mathbf{u}_x \cdot \mathbf{\Phi}_x dx + \int_\Omega \mathbf{u} \cdot \mathbf{\Phi} dx = \int_\Omega \mathbf{f} \cdot \mathbf{\Phi} dx,$$

for all $\mathbf{\Phi}(x) \in \mathring{J}_2'(\Omega)$.

In conclusion, the author expresses his deep gratitude to V. A. Solonnikov for his guidance in the work.

BIBLIOGRAPHY

1. John G. Heywood, *On uniqueness questions in the theory of viscous flow*, Acta Math. **136** (1976), 61–102.

2. O. A. Ladyženskaja and V. A. Solonnikov, *On some problems of vector analysis and generalized formulations of boundary problems for the Navier-Stokes equations*, Zap. Naučn. Sem. Leningrad. Otdel. Mat. Inst. Steklov. (LOMI) **59** (1976), 81–116; English transl. in J. Soviet Math. **10** (1978), no. 2.

3. ______, *On the solvability of boundary value and initial-boundary value problems for the Navier-Stokes equations in regions with noncompact boundaries*, Vestnik Leningrad. Univ. **1977**, no. 13 (Ser. Mat. Meh. Astr. vyp. 3), 39–48; English transl. in Vestnik Leningrad Univ. Math. **10** (1982).

4. V. A. Solonnikov and K. I. Pileckas, *On some spaces of solenoidal vectors and the solvability of a boundary-value problem for the system of Navier-Stokes equations in regions with noncompact boundaries*, Zap. Naučn. Sem. Leningrad. Otdel. Mat. Inst. Steklov. (LOMI) 73 (1977), 136–151; English transl., to appear in J. Soviet Math.

5. O. A. Ladyženskaja, *Mathematical problems in the dynamics of a viscous incompressible fluid*, 2nd rev. aug. ed., "Nauka", Moscow, 1970; English transl. of 1st ed., *The mathematical theory of viscous incompressible flow*, Gordon and Breach, New York, 1963; rev. 1969.

Translated by J. F. HEYDA

Труды Матем. Инст. Стеклов.
147(1980)

Proceedings of the Steklov
Institute of Mathematics
1981, Issue 2

SMALL PERTURBATIONS OF AN INTEGRABLE DYNAMICAL SYSTEM WITH AN INTEGRAL INVARIANT

UDC 517.9

N. V. SVANIDZE

ABSTRACT. In this paper one proves a theorem concerning the existence of a smooth family of tori invariant under a nonlinear transformation T in $\mathbb{R}^n$ when T is a small perturbation of a translation map. The new aspects relating to the results of Kolmogorov, Arnol'd and Moser are that one establishes the smooth dependence of the tori with respect to the rotation numbers considered as parameters and proves the smallness of the Lebesgue measure of the complement of the set covered by this family of tori.

Bibliography: 6 titles.

In this paper we consider a nonlinear transformation (1.1) for small values of the parameter γ. Several problems of mathematical physics and celestial mechanics reduce to the study of such a transformation. In previous papers by Kolmogorov, [1], Arnol'd [2] and Moser [3] it was established under certain conditions that the transformation (1.1) has invariant tori which fill up a set of positive measure in the phase space, when γ is small enough. Below we present a proof of the existence of a smooth family of tori invariant under the transformation (1.1). In doing so, we are able to establish two new results relating to the work of Kolmogorov, Arnol'd and Moser cited above. First we show that the constructed tori depend smoothly on the rotation numbers considered as parameters, and, second, we prove that this family of tori fills up a set $\Pi(I)$ whose complement has Lebesgue measure $O(\gamma)$ (or, in the notation of [4], $O(\sqrt{\varepsilon})$) in the phase space. Our theorem generalizes to higher dimensions the results obtained in [4] for the two-dimensional case. At the same time, the majority of the methods and constructions used in [4] and [5] can be directly extended to the higher-dimensional case.

§1. Formulation of the theorem

We consider a transformation $T\colon (\mathbb{R}^n/\mathbb{Z}^n) \times I \to (\mathbb{R}^n/\mathbb{Z}^n) \times \mathbb{R}^n$ of the form

$$(x, y) \mapsto (x + y + \gamma f(x, y) \pmod 1,\ y + \gamma^2 g(x, y)), \tag{1.1}$$

where $\mathbb{R}^n$ is n-dimensional real space, $\mathbb{Z}^n$ is the n-dimensional lattice of integers, $I = \{y \in \mathbb{R}^n \mid a \leqslant y \leqslant b\}$ is a parallelipiped in $\mathbb{R}^n$ and γ is a small parameter. We will use the notation $\Pi = (\mathbb{R}^n/\mathbb{Z}^n) \times \mathbb{R}^n$ and $\Pi(I) = (\mathbb{R}^n/\mathbb{Z}^n) \times I$.

It is not hard to see that for $\gamma = 0$ the transformation (1.1) has the invariant tori $y = \text{const}$, whose union is the whole space $\Pi(I)$. For $\gamma \neq 0$ small enough and under additional assumptions, the transformation (1.1) also has invariant tori, which fill up a nowhere dense set of positive Lebesgue measure (see [1], [2] and [3]).

1980 *Mathematics Subject Classification*. Primary 58F30; Secondary 70H99, 70F15.

Let us introduce an object which will be called a family of invariant tori (see [5]). Let $\Pi(I \cap E) = (\mathbb{R}^n/\mathbb{Z}^n) \times (I \cap E)$, where E is a subset of $\mathbb{R}^n$, and let T_0: $\Pi(I \cap E) \to \Pi$ be the transformation defined by $(x, y) \mapsto (x + y, y)$.

We say that the map Ξ: $\Pi(I) \to \Pi$ defines a C^s-family of tori invariant under T and having rotation numbers from the set E if the following conditions hold:

1) Ξ is a diffeomorphism of class C^s of $\Pi(I)$ onto $\Xi(\Pi(I))$.

2) $\Xi(\Pi(I \cap E))$ is invariant under the transformation T, i.e. $T\Xi(\Pi(I \cap E)) = \Xi(\Pi(I \cap E))$.

3) $\Xi^{-1}T\Xi|\Pi(I \cap E) = T_0$.

We call the set $\Xi(\Pi(I \cap E))$ the *family of invariant tori* (defined by Ξ). In $\Pi(I \cap E)$ every torus y = const is invariant under T_0. Moreover, the restriction of T_0 to such a torus is the translation by y. The condition 3) says that, for $y \in I \cap E$, the image of every torus y = const is a smooth torus in $\mathbb{R}^n/\mathbb{Z}^n$, invariant under T. The corresponding value of y is uniquely defined mod 1, and we call it the *rotation number* of the given invariant torus.

Here we prove the existence of a smooth family of invariant tori with rotation numbers from the set

$$E = \bigcap_k I_{k, N_k, M_{k-1}}, \tag{1.2}$$

where the set $I_{k,N_k,M_{k-1}}$ will be defined later by (3.5) and (3.6).

We assume that the functions f and g which appear in (1.1) satisfy the following condition.

CONDITION 1. The functions f and g are of class C^1, and are estimated by

$$\sum_{i=1}^{n} |f_i| + |g_i| + \sum_{|\rho_1|=1} |D_x^{\rho_1} f_i| + \sum_{1 \leqslant |\rho_2| \leqslant 2(\sigma+2)} |D_x^{\rho_2} g_i| + \sum_{|\rho_2|=1} |D_y^{\rho_2} g_i| < \delta_0, \tag{1.3}$$

$$\sum_{i=1}^{n} |D_x^{\rho_1} D_y^{\rho_2} f_i| + |D_x^{\rho_1} D_y^{\rho_2} g_i| \leqslant c_0, \quad |\rho_1| + |\rho_2| \leqslant l, \tag{1.4}$$

Here δ_0 and c_0 are positive constants; D_x and D_y denote the derivation operations; for the vectors $\rho_i = \{\rho_{1,i}, \ldots, \rho_{n,i}\} \subset \mathbb{N}^n$ $(i = 1, 2)$ we set $|\rho_i| = \sum_{k=1}^n \rho_{k,i}$; σ is a positive constant bigger than $n + 1$; and l is a natural number which for the purpose of our proof must satisfy the inequality $l > l_0(s, \sigma)$ where

$$l_0(s, \sigma) = \frac{(2 - (3\sigma + 2n + 4)/((\sigma + 2)(s + 1) + 2\sigma + 2n + 1))^3 (\sigma + 2)(s + 5)}{1 - (3\sigma + 2n + 4)/((\sigma + 2)(s + 1) + 2\sigma + 2n + 1)}, \tag{1.5}$$

In this formula, $s \geqslant 3$ is the number of continuous derivatives which we will require for the future family of tori invariant under T and having rotation numbers from E (1.2).

In general, a transformation T defined by (1.1) need not have any invariant tori. To see this, one can take $f = 0$ and $g = 1$. Therefore, to exclude such a "break-down", we introduce a further condition.

CONDITION 2. We assume that the transformation (1.1) has an integral invariant–a differential 1-form

$$\omega^1 = \sum_{i=1}^{n} X_i(x, y)\, dx_i + Y_i(x, y)\, dy_i + dZ(x, y), \tag{1.6}$$

with coefficients X_i, Y_i and Z of class C^3 satisfying

$$|D_x^{\rho_1} D_y^{\rho_2} X_i| + \gamma |D_x^{\rho_1} D_y^{\rho_2} Y_i| \leqslant A\gamma^{3+1-|\rho_2|} \tag{1.7}$$

for $1 \leqslant |\rho_1| + |\rho_2| \leqslant 3$, $i = 1, \dots, n$, and

$$|J_{[X]}(y)| \geqslant B^n \gamma^{\beta n} \tag{1.8}$$

with positive constants A and B and real number β. In (1.8) we used the notation

$$J_{[X]}(y) = \det([D_{y_j} X_i])_{i,j=1}^n, \quad \text{where } [f(x, y)] = \int_0^1 f(x, y)\, dx.$$

Condition 2 introduced above is the analogue of condition (α) imposed on the nonlinear transformation in [3] and [4].

In what follows, a family of invariant tori of class C^s with rotation numbers from the set E, as defined by (1.2), will be called an *L-family of invariant tori* if the following conditions hold:

a) It is defined by an application $\Xi\colon \Pi(I') \to \Pi(I)$ of the form

$$(\xi, \eta) \mapsto (\xi + U(\xi, \eta) \pmod 1,\ \eta + V(\xi, \eta)), \tag{1.9}$$

where the functions U and V satisfy the estimate

$$\sum_{i=1}^n |D_\xi^{\rho_1} D_\eta^{\rho_2} U_i| + \gamma^{-1} |D_\xi^{\rho_1} D_\eta^{\rho_2} V_i| \leqslant c_1 \gamma^{-|\rho_2|} \delta_0^{1-p(\rho_1, \rho_2)},$$
$$0 \leqslant |\rho_1| + |\rho_2| \leqslant s, \tag{1.10}$$

with

$$p(\rho_1, \rho_2) = \frac{(2 - (3\sigma + 2n + 4)/((\sigma + 2)(s + 1) + 2\sigma + 2n + 1))(|\rho_1| + |\rho_2|) + 2\sigma + n}{(2 - (3\sigma + 2n + 4)/((\sigma + 2)(s + 1) + 2\sigma + 2n + 1)) \times (\sigma + 2)(s + 1) + 2\sigma + 2n + 1},$$

and $I' = \{y \in \mathbb{R}^n | a' \leqslant y \leqslant b'\}$ is a parallelepiped in $\mathbb{R}^n$, $I' \subset I$, and

$$a_i < a_i' \leqslant a_i + c_2\gamma, \quad b_i - c_2\gamma \leqslant b_i' < b_i. \tag{1.11}$$

b) The Lebesgue measure of $I' \cap E$ satisfies (see §6)

$$\operatorname{mes}(I' \cap E) \geqslant (1 - c_3\gamma(1 + \delta_0^q)) \operatorname{mes} I, \tag{1.12}$$

with a positive constant q.

In (1.10)–(1.12), the c_i ($i = 1, 2, 3$) are positive constants.

In the formulation of the following theorem, $s \geqslant 3$ is the number of continuous derivatives we require for the family of invariant tori which will be constructed; n is the dimension of the space $\mathbb{R}^n$, σ is an arbitrary constant satisfying $\sigma > n + 1$, and c_0, A and B are the positive constants introduced in (1.4), (1.7) and (1.8) respectively.

THEOREM. *There exist positive constants δ_0 and c_i ($i = 1, 2, 3$), depending only on s, n, σ, c_0, A and B, such that if the transformation* (1.1) *satisfies Conditions* 1 *and* 2, *then it has an L-family of invariant tori.*

§2. The localizing multiplier, the smoothing operator and the finite difference equation

In this section we set forth some constructions (see [4]–[6]) which are used later to define an iterative process converging to the solution we desire. All the positive constants appearing below, denoted by c_i ($i = 1, \dots, 21$) and "const", depend only on s, n, σ, c_0, A, B, a and b.

The localizing multiplier. Let N be a positive integer. By definition, the set E_N consists of all the vectors $y \in I$ such that for every $m_0 \in \mathbb{Z}$ and $m \in \mathbb{Z}^n$, $0 < |m| \leqslant N$, the inequality

$$|my - m_0| \geqslant \gamma |m|^{-\sigma} \tag{2.1}$$

holds. Here $\sigma > n + 1$, $0 < \gamma < 1$, $|m| = \sum_1^n m_i$ and $my = \sum_1^n m_i y_i$ (see Figure 1).

FIGURE 1

We introduce two further sets, E'_N and E''_N, defined exactly as E_N, except that we replace γ in (2.1) by $\gamma/2$ and $\gamma/4$ respectively. Clearly

$$E''_N \supset E'_N \supset E_N. \tag{2.2}$$

We construct a smooth function $\theta_{NM}(y)$ equal to 1 for $y \in E_N$ and 0 for $y \notin E'_N$, and we call it a localizing* multiplier.

Let us take $l^* = 2s + 8$ and choose a function $\varphi\colon \mathbb{R}^1 \to [0, 1]$ with the following properties:

1) $\varphi(y) = 1$ for $|y| \leqslant 1/2$, and $\varphi(y) = 0$ for $|y| \geqslant 1$.

2) $\varphi(y) \leqslant cl^*(1 - |y|)^{l^*-1}/2^n(6n)^{l^*}$ for $3/4 \leqslant |y| \leqslant 1$, where $c = \int_{\mathbb{R}^1} \varphi(y)\,dy$.

(2.3)

Now we define the localizing multiplier $\theta_{NM}(\eta)$ with $M > 0$, through the formula

$$\theta_{NM}(\eta) = \left(\frac{M}{c}\right)^n \int_{E_{N+\varepsilon}} \prod_{i=1}^n \varphi(M(\eta_i - y_i))\,dy. \tag{2.4}$$

Editor's note. The Russian term is срезающий

Here ε is a positive constant depending on γ and N, which satisfies

$$\varepsilon < \frac{1}{4}\gamma N^{-\sigma-1}, \tag{2.5}$$

and $E_N + \varepsilon$ is the ε-neighborhood of the set E_N.

LEMMA 1. 1) $\theta_{NM} \in C^\infty$, $0 \leqslant \theta_{NM}(\eta) \leqslant 1$.

2) $|D_\eta^\rho \theta_{NM}| \leqslant c_4 M^{|\rho|}$ *for* $0 \leqslant |\rho| \leqslant l$.

3) $\theta_{NM}(\eta) = 1$ *for* $\eta \in E_N$ *and* $\theta_{NM}(\eta) = 0$ *for* $\eta \notin E'_N$, *if we take*

$$M > \sqrt{n}\,\varepsilon^{-1}. \tag{2.6}$$

4) *For* $M_+ \geqslant 24nM$, $N_+ \geqslant N$ *and* $|\Delta| \leqslant 2\sqrt{n}\, M_+^{-1}$, *we have, for* $\eta \in \operatorname{supp} \theta_{N_+M_+}$

$$1 - \theta_{NM}(\eta + \Delta) \leqslant \left(\frac{M}{M_+}\right)^{l^*}. \tag{2.7}$$

In (2.7) $\Delta \in \mathbb{R}^n$ *is a vector and* $|\Delta|$ *is the Euclidean norm of* Δ *in* $\mathbb{R}^n$.

PROOF. The parameter ε appearing in the definition of θ_{NM} is constrained by the inclusions $E_N \subset E_N + \varepsilon \subset E_N + 2\varepsilon \subset E'_N$. To realize them it is enough to assume (2.5), and now the first three properties of θ_{NM} are evident. Since $E_{N_+} \subset E_N$ for $N_+ \geqslant N$, the function $1 - \theta_{NM}(\eta + \Delta) \neq 0$ for $\eta + \Delta \in E_N$, i.e. it is enough to consider

$$\eta \in \overline{(E_{N_+} + 2\varepsilon_+ + |\Delta|)} \setminus \overline{(E_N - |\Delta|)}.$$

We take $|\Delta| < \varepsilon_+$ and consider the cube

$$I_M(\eta + \Delta) = \{y \in \mathbb{R}^n \,|\, |\eta_i + \Delta_i - y_i| \leqslant 1/M\}.$$

It is clear that for $\eta \in \operatorname{supp} \theta_{N_+M_+}$ the maximum distance between the center $\eta + \Delta$ of the cube $I_M(\eta + \Delta)$ and that part of the boundary ∂E_N lying in $I_M(\eta + \Delta)$ is not bigger than $3\varepsilon_+$. However, $I_M(\eta + \Delta)$ can leave $E_N + \varepsilon$, and the point of $I_M(\eta + \Delta)$ at maximum distance from the part of the boundary of $E_N + \varepsilon$ lying in $I_M(\eta + \Delta)$ is one of the vertices of the cube; this distance is also not bigger than $3\varepsilon_+$ (see Figure 2).

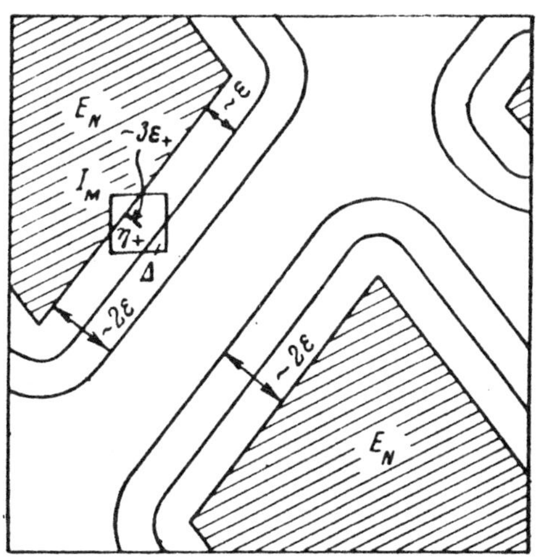

FIGURE 2

As a consequence

$$1-\theta_{NM}(\eta+\Delta)=\left(\frac{M}{c}\right)^n \int_{I_M\setminus(E_N+\varepsilon)} \prod_{i=1}^{n} \varphi(M(\eta_i+\Delta_i-y_i))\,dy. \tag{2.8}$$

Let us make a translation and move the origin of the system of coordinates to the point $\eta+\Delta$. In the new coordinates we can represent the domain of integration $I_M \setminus (E_N+\varepsilon)$ as a union of no more than 2^n domains Ω_k with disjoint interiors. Moreover, for every Ω_k there is an index j_0 such that $y_{j_0}^{\sim}$-coordinate of all the points in Ω_k satisfies

$$\frac{1}{M}-3\varepsilon_+\sqrt{n}\leqslant y_{j_0}^{\sim}\leqslant\frac{1}{M}, \tag{2.9}$$

or

$$-\frac{1}{M}\leqslant y_{j_0}^{\sim}\leqslant-\frac{1}{M}+3\varepsilon_+\sqrt{n}. \tag{2.10}$$

Therefore we have

$$I_M\setminus(E_N+\varepsilon)=\bigcup_{k=1}^{p}\Omega_k,\quad p\leqslant 2^n. \tag{2.11}$$

Let us choose one of the sets Ω_k, and let $y_{j_0}^{\sim}$ be that coordinate which takes values, for example, in the interval (2.9). If we assume that ε_+ satisfies

$$3\varepsilon_+\sqrt{n}\leqslant\frac{1}{4M}, \tag{2.12}$$

then from (2.12) and (2.3) we obtain the following chain of inequalities:

$$\begin{aligned}
&\left(\frac{M}{c}\right)^n\int_{\Omega_k}\prod_{i=1}^{n}\varphi(My_i^{\sim})\,dy^{\sim}\\
&\quad\leqslant\left(\frac{M}{c}\right)^n\int_{1/M-3\varepsilon_+\sqrt{n}\leqslant y_{j_0}^{\sim}\leqslant 1/M}\varphi(My_{j_0}^{\sim})\,dy_{j_0}^{\sim}\int_{\mathbb{R}^{n-1}}\prod_{\substack{i=1\\ i\neq j_0}}^{n}\varphi(My_i^{\sim})\,dy_i^{\sim}\\
&\quad\leqslant\frac{l^*M}{2^n(6n)^{l^*}}\int_{1/M-3\varepsilon_+\sqrt{n}}^{1/M}(1-My_{j_0}^{\sim})^{l^*-1}\,dy_{j_0}^{\sim}=\left(\frac{\varepsilon_+}{2\sqrt{n}}\right)^{l^*}M^{l^*}\leqslant 2^{-n}\left(\frac{M}{M_+}\right)^{l^*}.
\end{aligned} \tag{2.13}$$

In a similar way one can estimate the integrals over domains Ω_k which satisfy (2.10). Finally, summing the estimates of type (2.13) over all the Ω_k, we obtain (2.7).

The smoothing operator. We construct a smoothing operator $\mathbb{T}_{NM}$ which transforms continuous functions defined on $\Pi(I^{\sim})$ into C^∞-functions defined on $\Pi(I^{\sim}-1/M)$, where $I^{\sim}=\{y\in\mathbb{R}^n|\ a^{\sim}<y<b^{\sim}\}$ and $I^{\sim}\subset I$. Let us assume that $b_i^{\sim}-a_i^{\sim}>2M^{-1}$. The operator $\mathbb{T}_{NM}$ acts on a function $h(x,y)$ according to the formula

$$(\mathbb{T}_{NM}h)(x,y)=\iint_{\Pi(I^{\sim})}\prod_{i=1}^{n}X_N(x_i-\xi_i)\,Y_M(y_i-\eta_i)\,h(\xi,\eta)\,d\xi\,d\eta. \tag{2.14}$$

Here $X_N(x)$ is a C^∞-function, periodic with period 1, and $Y_M(y)$ is a C^∞-function equal to zero for $|y|\geqslant M^{-1}$. These functions were constructed in [5] and [6], where the following lemma is also proven.

LEMMA 2. *If N, $M > c_5$, where c_5 is a sufficiently large positive constant, then there exist functions $X_N(x)$ and $Y_M(y)$ such that the operator $\mathbb{T}_{NM}$ defined by* (2.14) *has the following properties*:

1) *The coefficients of the Fourier expansion of $(\mathbb{T}_{NM}h)(x,y)$ relative to the functions* $\exp(i2\pi mx)$ *will be zero for $|m| > N$.*

2) *If $h(x,y) = 0$ for every $(x,y) \in \mathbb{R}^n \times \Gamma(d)$, where $\Gamma(d)$ is a strip of width d in $\mathbb{R}^n$, with $d > 2\sqrt{n}M^{-1}$, then $(\mathbb{T}_{NM}h)(x,y) = 0$ for every $(x,y) \in \mathbb{R}^n \times \Gamma(d - 2\sqrt{n}M^{-1})$.*

3) *There is a positive constant c_6 such that*

$$|D_x^{\rho_1}D_y^{\rho_2}\mathbb{T}_{NM}h| \leqslant c_6 N^{|\rho_1|+n}M^{|\rho_2|}\max|h|, \quad 0 \leqslant |\rho_1| + |\rho_2| \leqslant l+1,$$

and for $l' \leqslant l$

$$|h - \mathbb{T}_{NM}h| \leqslant c_6 \max_{\substack{|\rho_1|+|\rho_2|=l' \\ x,\,y}} N^{-|\rho_1|}M^{-|\rho_2|}|D_x^{\rho_1}D_y^{\rho_2}h|.$$

The finite difference equation. Let us consider the equation (see [5])

$$w(x+y,\ y) - w(x,\ y) = h(x,\ y), \tag{2.15}$$

where $h(x,y)$ is a given continuous function defined on $\Pi(I^{\sim})$, with $I^{\sim} \subset I$. We must find a function w which is also defined on $\Pi(I^{\sim})$ and satisfies (2.15).

LEMMA 3. *Assume that the function $h(x,y)$ has the following properties*:

1) $[h](y) = 0$.

2) $h(x,y) = 0$ *for* $y \notin E_N''$.

3) *The coefficients of the Fourier expansion of $h(x,y)$ relative to the functions* $\exp(i2\pi mx)$ *will be zero for $|m| > N$.*

4) *The function $h(x,y)$ is of class C^{l+1}.*

Then the equation (2.15) *has a unique solution w of class C^{l+1}, defined on $\Pi(I^{\sim})$ and satisfying $[w](y) = 0$ and*

$$|D_x^{\rho_1}D_y^{\rho_2}w| \leqslant c_7 \max_{\substack{x,\,y \\ 0\leqslant p\leqslant \rho_2}} \gamma^{-|p|-1}|D_x^{\rho_1+p}D_y^{\rho_2-p}h|\,N^{\sigma(|p|+1)} \tag{2.16}$$

for $0 \leqslant |\rho_1| + |\rho_2| \leqslant l+1$.

PROOF. If the components of the vector y are incommensurable, then the homogeneous equation $u(x+y) - u(x) = 0$ has a unique integrable solution with mean value equal to zero, namely the function identically equal to zero. This gives the uniqueness of the solution of (2.15). To prove existence and the estimate (2.16), we analyze the explicit expression of the solution

$$w(x,\ y) = \sum_{0<|m|\leqslant N} \frac{h_m(y)}{\exp(i2\pi my) - 1}\exp(i2\pi mx), \tag{2.17}$$

where

$$h_m(y) = \int_0^1 e^{-i2\pi mx}h(x,\ y)\,dx. \tag{2.18}$$

Since $h_m(y) = 0$ for $y \in E''_N$, $w(x, y)$ is also zero for such values of y. If $y \in E''_N$, then the small denominators $d_m(y) = \exp(i2\pi my) - 1$ are estimated by

$$|d_m(y)| \geqslant 4|my - m_0| \geqslant \gamma |m|^{-\sigma}. \tag{2.19}$$

The inequality (2.19) shows that $d_m^{-1}(y)$ is a smooth bounded function in a neighborhood of the set where $h_m(y) \neq 0$. Therefore $w(x, y)$ is of class C^{l+1}. It remains to prove (2.16). We compute

$$D_x^{\rho_1} D_y^{\rho_2} w(x, y) = \sum_{0<|m|\leqslant N} \sum_{0\leqslant k\leqslant \rho_2} \binom{\rho_2}{k} D_y^{\rho_2-k} h_m(y) \times D_y^k (d_m^{-1}(y)) (i2\pi m)^{\rho_1} \exp(i2\pi mx). \tag{2.20}$$

Using induction, it is not hard to prove that

$$D_y^k (d_m^{-1}(y)) = (i2\pi m)^k \frac{P_{|k|}(\exp(i2\pi my))}{d_m(y)^{|k|+1}}, \tag{2.21}$$

where $P_{|k|}$ is a polynomial of degree k. Moreover,

$$|D_y^{\rho_2-k} h_m(y) (i2\pi m)^{\rho_1+k}| \leqslant \max_{x,y} |D_x^{\rho_1+k} D_y^{\rho_2-k} h|. \tag{2.22}$$

It is clear now that (2.20)–(2.22) and (2.19) imply (2.16). A direct substitution into (2.15) proves that the function $w(x, y)$ defined by (2.17) is a solution of (2.15).

§3. The description of the iterative process

We introduce three sequences of positive numbers: N_k, M_k and δ_k, the first two of them increasing and the last one decreasing. The relations between these sequences are as follows (see [2], [4], and [5]):

$$M_k = \gamma^{-1} N_k^{\nu}, \quad N_{k+1} = N_k^{\varkappa}, \quad \delta_k = N_k^{-\mu}. \tag{3.1}$$

The number κ from the open interval $(1 + 1/(\sigma + 1), \kappa_0)$, where

$$\varkappa_0 = 2 - \frac{3\sigma + 2n + 4}{(\sigma + 2)(s + 1) + 2\sigma + 2n + 1}, \tag{3.2}$$

will be chosen and fixed in §8.

Also, in (3.1)

$$\nu = \varkappa(\sigma + 2) \quad \text{and} \quad \mu = \varkappa(\nu(s + 1) + 2\sigma + 2n + 1). \tag{3.3}$$

For the moment, the only free parameter is N_0. In any case, it must satisfy $N_0 > c_5$, with a constant c_5 such as to ensure that the operator $\mathbb{T}_{NM}$ makes sense for $N \geqslant N_0$.

We also introduce two sequences of sets:

1) a sequence of parallelepipeds $I_k \subset \mathbb{R}^n$, $k = 1, 2, \ldots$, defined by

$$I_k = I_{k-1} - d_{k-1}, \tag{3.4}$$

with initial parallelepipeds $I_0 = I = [a, b]$;

2) a sequence $I_{k,N_k,M_{k-1}}$ defined by

$$I_{k,N_k,M_{k-1}} = \operatorname{supp} \theta_{N_k, M_{k-1}} \cap (I_{k-1,N_{k-1},M_{k-2}} - d_{k-1}), \tag{3.5}$$

where $k = 1, 2, \ldots$ and $I_{0,N_0,M_{-1}} = I_0$.

In (3.4) and (3.5) we put

$$d_{k-1} = \sqrt{n} M_k^{-1} + 3\gamma\beta_{k-1} + \gamma^2\delta_{k-1}, \tag{3.6}$$

with $\beta_{k-1} = 2nc_7 N_k^{2\sigma+n}\delta_{k-1}$.

The map Ξ: $\Pi(I') \to \Pi(I)$ whose existence is stated in the theorem is then obtained as the limit $\Xi = \lim_{m\to\infty} \Xi_m$ of a sequence of maps Ξ_m: $\Pi(I_m) \to \Pi(I)$ of class C^l ($m = 1, 2, \ldots$). Every map Ξ_m has the form

$$\Xi_m = R_0 \circ R_1 \circ \ldots \circ R_{m-1}, \tag{3.7}$$

where in turn R_k: $\Pi(I_{k+1}) \to \Pi(I_k)$ is a map of class C^l ($k = 0, 1, \ldots$) given by

$$(\xi, \eta) \mapsto (\xi + u_k(\xi, \eta) (\mathrm{mod}\, 1), \ \eta + v_k(\xi, \eta)). \tag{3.8}$$

The set $I' \subset \bigcap_k I_k$ will be defined in §5.

At every step of the iterative process we introduce a transformation $T_{k+1} = R_k^{-1} \circ T_k \circ R_k$ ($k = 0, 1, \ldots$) which is a diffeomorphism of $\Pi(I_{k+1}, N_{k+1}, M_k)$ into a certain part of $\Pi(I_{k,N_k,M_{k-1}})$. To define it we ask for the following diagram to be commutative:

$$\begin{array}{ccc}
\Pi\left(I_{k,N_k,M_{k-1}} - \dfrac{\sqrt{n}}{M_{k+1}} - 2\gamma\beta_k\right) & \xleftarrow{T_k} & \Pi\left(I_{k,N_k,M_{k-1}} - \dfrac{\sqrt{n}}{M_{k+1}} - 2\gamma\beta_k - \gamma^2\delta_k\right) \\
\Big\downarrow R_k^{-1} & & \Big\uparrow R_k \\
\Pi\left(I_{k,N_k,M_{k-1}} - \dfrac{\sqrt{n}}{M_{k+1}}\right) & \xleftarrow{T_{k+1} = R_k^{-1} T_k R_k} & \Pi\left(I_{k,N_k,M_{k+1}} - \dfrac{\sqrt{n}}{M_{k+1}} - 3\gamma\beta_k - \gamma^2\delta_k\right).
\end{array} \tag{3.9}$$

In coordinates, the transformation T_k has the form

$$\begin{aligned}
&(x, y) \mapsto (x_1, y_1), \\
&x_1 = x + y + \gamma f_k(x, y) (\mathrm{mod}\, 1), \\
&y_1 = y + \gamma^2 g_k(x, y).
\end{aligned} \tag{3.10}$$

Necessary conditions for the commutativity of the diagram are

$$I_{k+1} \cap \operatorname{supp} \mathbb{T}_{N_{k+1}, M_{k+1}} \theta_{N_{k+1}, M_k} \subset I_{k+1, N_{k+1}, M_k}, \tag{3.11}$$

$$\operatorname{supp} \mathbb{T}_{N_{k+1}, M_{k+1}} \theta_{N_{k+1}, M_k} \subset E''_{N_{k+1}}, \tag{3.12}$$

But in order to satisfy them it is enough to ensure the validity of the inequalities

$$\frac{\gamma}{4N_{k+1}^{\sigma+1}} + \frac{\sqrt{n}}{M_k} + \frac{\sqrt{n}}{M_{k+1}} < \frac{2\sqrt{n}}{M_{k-1}} - d_k, \tag{3.13}$$

$$\frac{\gamma}{4N_{k+1}^{\sigma+1}} + \frac{\sqrt{n}}{M_k} + \frac{\sqrt{n}}{M_{k+1}} < \frac{3}{4}\frac{\gamma}{N_{k+1}^{\sigma+1}}. \tag{3.14}$$

If we compare with the definition (3.1) of the parameters N_k and M_k, then (3.14) reduces to the inequality

$$N_k^{\nu-\varkappa(\sigma+1)} > 4\sqrt{n};$$

Since $\nu = \kappa(\sigma + 2)$, we can realize this inequality by choosing the number N_0 in a right way. One can follow the same argument in order to obtain (3.13), using (3.2), (3.3) and choosing the parameter.

We define the functions u_k and v_k appearing in (3.8) as solutions of the following system which are smooth and periodic with period 1 in ξ:

$$u_k(\xi+\eta, \eta) - u_k(\xi, \eta) = \gamma \hat{f}_k(\xi, \eta) + v_k(\xi, \eta),$$
$$v_k(\xi+\eta, \eta) - v_k(\xi, \eta) = \gamma^2(\hat{g}_k(\xi, \eta) - [\hat{g}_k](\eta)), \tag{3.15}$$

Here

$$\hat{f}_k = \mathbb{T}_{N_{k+1}, M_{k+1}} \theta_{N_{k+1}, M_k} f_k^{\sim},$$
$$\hat{g}_k = \mathbb{T}_{N_{k+1}, M_{k+1}} \theta_{N_{k+1}, M_k} g_k^{\sim}, \tag{3.16}$$

and the functions $f_k^{\sim} = f_k$ and $g_k^{\sim} = g_k$ for $y \in I_{k,N_k,M_{k-1}}$, and $f_k^{\sim} = 0$, $g_k^{\sim} = 0$ for $y \in I_k \setminus I_{k,N_k,M_{k-1}}$.

At every step of the process, the following inequalities will be satisfied: for $\eta \in \operatorname{supp} \theta_{N_{k+1},M_k}$

$$\sum_{i=1}^{n} |f_{i,k}(\xi, \eta)| + |g_{i,k}(\xi, \eta)| \leqslant \delta_k, \tag{3.17}$$

$$\sum_{i=1}^{n} |D_\xi^{\rho_1} D_\eta^{\rho_2} N_k f_{i,k}(\xi, \eta)| + |D_\xi^{\rho_1} D_\eta^{\rho_2} N_k^{\nu} g_{i,k}(\xi, \eta)| \leqslant N_k^{|\rho_1|} M_k^{|\rho_2|},$$
$$k = 0, 1, \ldots, \quad 1 \leqslant |\rho_1| + |\rho_2| \leqslant l. \tag{3.18}$$

Here we have set $f_k = \{f_{i,k}(\xi, \eta)\}_{i=1}^n$ and $g_k = \{g_{i,k}(\xi, \eta)\}_{i=1}^n$.

In §§4 and 5, assuming that the inequalities (3.17) and (3.18) are valid at every step of the iterative process, we show that the described sequence maps of Ξ_m converges to a map Ξ of class C^s possessing all the necessary properties.

In §§7 and 8 we prove that for N_0 large enough ($N_0 > N_0^{\sim}(s, n, \sigma, c_0, A, B)$), the estimates (3.17) and (3.18) are valid at every step of the iterative process.

§4. Estimates for the functions u_k and v_k

For N_0 large enough and assuming that $[u_k] = 0$, the functions u_k and v_k are uniquely defined by (3.15). Indeed, the right-hand side of the second of the equations (3.15) is defined in the strip $I_k - \sqrt{n}\, M_{k+1}^{-1}$ and satisfies condition 1) of Lemma 3. The conclusions 4) of Lemma 1 and 2) of Lemma 2, supplemented by (3.12), show that the right-hand side of the second of the equations (3.15) also satisfies condition 2) of Lemma 3 for $N = N_{k+1}$. Statement 1) of Lemma 2 guarantees that condition 3) of Lemma 3 is also satisfied for $N = N_{k+1}$. The smoothness of the right-hand side in (3.15) (condition 4) of Lemma 3) is also a consequence of Lemma 2. If we apply Lemma 3 to the second of the equations (3.15), we deduce that it has a unique solution $v_k^{\sim}$ of class C^{l+1} satisfying $[v_k^{\sim}] = 0$ and the estimate

$$D_\xi^{\rho_1} D_\eta^{\rho_2} v_{i,k}^{\sim}| \leqslant c_7 \max_{\substack{0 \leqslant p \leqslant \rho_2 \\ \xi, \eta}} \gamma^{1-|p|} |D_\xi^{\rho_1+p} D_\eta^{\rho_2-p}(\hat{g}_k - [\hat{g}_k])|$$
$$\leqslant 2c_6 c_7 \gamma N_{k+1}^{|\rho_1|+\sigma+n} M_{k+1}^{|\rho_2|} \delta_k,$$
$$i = 1, \ldots, n, \quad 0 \leqslant |\rho_1| + |\rho_2| \leqslant l+1. \tag{4.1}$$

To solve the first of the equations (3.15), it is necessary to have $[v_k] = -\gamma[\hat{f}_k]$. To this end we take $v_k = v_k^{\sim} - \gamma[\hat{f}_k]$. From (4.1) and statement 3) of Lemma 2 we obtain the following estimate for v_k:

$$|D_\xi^{\rho_1} D_\eta^{\rho_2} v_{i,k}| \leqslant c_8 \gamma N_{k+1}^{|\rho_1|+\tau+n} M_{k+1}^{|\rho_2|} \delta_k,$$
$$i=1, \dots, n, \quad 0 \leqslant |\rho_1| + |\rho_2| \leqslant l+1. \tag{4.2}$$

The same kind of argument used above convinces us that, assuming (3.12), the right-hand side of the first of the equations (3.15) satisfies all the conditions of Lemma 3. Therefore this equation has a (unique) solution of class C^{l+1} satisfying $[u_k] = 0$ and the estimate

$$|D_\xi^{\rho_1} D_\eta^{\rho_2} u_{i,k}| \leqslant c_8 N_{k+1}^{|\rho_1|+2\tau+n} M_{k+1}^{|\rho_2|} \delta_k,$$
$$i=1, \dots, n, \quad 0 \leqslant |\rho_1| + |\rho_2| \leqslant l+1. \tag{4.3}$$

§5. Proof of the convergence of the iterative process and estimates of the functions $U(\xi, \eta)$ and $V(\xi, \eta)$

Let us introduce the set $I^{\sim} = \bigcap_k I_k$ and prove that the sequence $\Xi_k|\Pi(I^{\sim})$ converges together with all the derivatives of order $\leqslant s$ to a transformation ξ: $\Pi(I^{\sim}) \to \Pi(I)$. The transformation Ξ_k defined by (3.7) has the form

$$(\xi, \eta) \mapsto (\xi + U_k(\xi, \eta) \pmod 1, \ \eta + V_k(\xi, \eta)), \tag{5.1}$$

where the functions U_k and V_k are defined on $\Pi(I_k)$ and satisfy the recursion relations

$$U_{k+1}(\xi, \eta) - U_k(\xi + u_k(\xi, \eta), \eta + v_k(\xi, \eta)) = u_k(\xi, \eta),$$
$$V_{k+1}(\xi, \eta) - V_k(\xi + u_k(\xi, \eta), \eta + v_k(\xi, \eta)) = v_k(\xi, \eta),$$
$$k = 0, 1, \dots \tag{5.2}$$

By definition, $U_0 = 0$ and $V_0 = 0$.

From (4.2) and (4.3) we immediately obtain the uniform convergence $U_k(\xi, \eta) \to U(\xi, \eta)$ and $V_k(\xi, \eta) \to V(\xi, \eta)$ in the strip $I^{\sim}$. The inequality

$$\sum_{k=0}^{\infty} N_0^{-\alpha \varkappa^k} \leqslant N_0^{-\alpha} + \int_0^{\infty} N_0^{-\alpha \varkappa^k} dk \leqslant N_0^{-\alpha} + \frac{1}{\alpha \log \varkappa \log N_0} N_0^{-\alpha} \quad \text{for } \alpha > 0 \tag{5.3}$$

shows that the limit functions satisfy

$$\sum_{i=1}^{n} |U_i(\xi, \eta)| + \gamma^{-1} |V_i(\xi, \eta)| \leqslant \text{const}\, N_0^{-\varkappa(\nu(s+1)+1)}. \tag{5.4}$$

Let us estimate the derivatives of U and V. For simplicity we denote by W_k (w_k) the pair (U_k, V_k) (respectively (u_k, v_k)). We introduce the seminorm

$$\|W\|_\tau = \sum_{i=1}^{n} \max_{\substack{1 \leqslant |\rho_1| + |\rho_2| \leqslant \tau \\ \xi, \eta}} |D_\xi^{\rho_1} D_\eta^{\rho_2} U_i(\xi, \eta)| \gamma^{|\rho_2|} + \max_{\substack{1 \leqslant |\rho_1| + |\rho_2| \leqslant \tau \\ \xi, \eta}} |D_\xi^{\rho_1} D_\eta^{\rho_2} V_i(\xi, \eta)| \gamma^{|\rho_2|-1}.$$

If we use (4.2), (4.3) and the equalities

$$\delta_k = N_0^{-\varkappa^{k+1}(\nu(s+1)+2\sigma+2n+1)}, \quad N_{k+1} = N_0^{\varkappa^{k+1}}, \quad M_{k+1} = \gamma^{-1} N_0^{\varkappa^{k+1}\nu}, \tag{5.5}$$

it is not hard to deduce a recursive inequality from (5.2), valid for $\tau \leqslant s + 1$:

$$\| W_{k+1} \|_\tau \leqslant 2c_8 N_0^{-\varkappa^k \gamma_\tau} + (1 + c_9 N_0^{-\varkappa^k \gamma_\tau}) \| W_k \|_\tau, \tag{5.6}$$

where

$$\gamma_\tau = \varkappa\nu\left(s + 1 - \tau + \frac{n+1}{\nu}\right). \tag{5.7}$$

The first term in the right-hand side of (5.6) is obtained by differentiating the right members of (5.2), and the second by differentiating U_k and V_k.

Since the following inequalities, for $\tau \leqslant s + 1$, are immediate consequences of (5.6), (5.3):

$$\| W_k \|_\tau \leqslant \left[\prod_{i=0}^{k-1} (1 + c_9 N_0^{-\varkappa^i \gamma_\tau}) \right] 2c_8 N_0^{-\gamma_\tau} + \sum_{i=0}^{k-1} \left[\prod_{j=i+1}^{k-1} (1 + c_9 N_0^{-\varkappa^j \gamma_\tau}) \right] 2c_8 N_0^{-\varkappa^i \gamma_\tau}, \tag{5.8}$$

$$\left| \prod_{i=0}^{\infty} (1 + c_9 N_0^{-\varkappa^i \gamma_\tau}) - 1 \right| \leqslant c_{10} N_0^{-\gamma_\tau}, \tag{5.9}$$

we obtain $\| W_k \|_\tau \leqslant c_1 N_0^{-\gamma_\tau}$. This last inequality shows that all the derivatives of order $\leqslant s + 1$ of U_m and V_m have in turn derivatives of order $\leqslant s$, and

$$\| W \|_\tau \leqslant c_1 N_0^{-\gamma_\tau} \quad \text{for } 1 \leqslant \tau \leqslant s. \tag{5.10}$$

To conclude this section, we remark that the set I' appearing in the formulation of the theorem can be taken as

$$I' = I^{\sim} - 3\gamma\beta^{\sim} - \gamma^2\delta_0, \tag{5.11}$$

where $\beta^{\sim} = \text{const}\, N^{-\varkappa(\nu(s+1)+1)}$.

§6. The estimate of the "width" of the invariant set

Let us show that for the parallelepiped I' defined by (5.11) the estimate (1.11) is valid. For the parallelepipeds $I_k = [a_k, b_k]$ constructed above (see (3.4)), the numbers $a_{i,k}$ and $b_{i,k}$ $(k = 0, 1, \dots)$ form two monotonic sequences $a_{i,0} = a_i$, $a_{i,k+1} > a_{i,k}$ and $b_{i,0} = b_i$, $b_{i,k+1} < b_{i,k}$ $(i = 1, \dots, n)$. Let us denote the limits by $a_i^{\sim} = \lim_{k\to\infty} a_{i,k}$ and $b_i^{\sim} = \lim_{k\to\infty} b_{i,k}$ and estimate the difference $a_i^{\sim} - a_i$ (the evaluation of $b_i^{\sim} - b_i$ is similar). From (3.4) it is clear that $a_i^{\sim} - a_i \leqslant \Sigma_0^\infty d_k$, where d_k is defined by (3.6). As a consequence of (5.5) the series $\Sigma_0^\infty d_k$ converges, and by (5.3)

$$a_i^{\sim} - a_i \leqslant \gamma \,\text{const}\, N_0^{-\varkappa^2(s+2)}. \tag{6.1}$$

Finally, using the definition (5.11) of the set I', we arrive at the necessary estimate

$$\sum_{i=1}^{n} |a_i' - a_i| + |b_i' - b_i| \leqslant c_2\gamma. \tag{6.2}$$

We can pass now to the estimate of the "width" of the invariant set, and for this we find a lower bound for the Lebesgue measure of the set $E = \bigcap_k I_{k,N_k,M_{k-1}}$, where $I_{k,N_k,M_{k-1}}$ is defined by (3.5). Instead of $I_{k,N_k,M_{k-1}}$ we consider the new sets

$$I_{k,N_k} = E_{N_k} \cap (I_{k-1,N_{k-1}} - d_{k-1}), \quad k = 1, 2, \ldots, \tag{6.3}$$

where the numbers d_k were introduced above (see (3.6)).

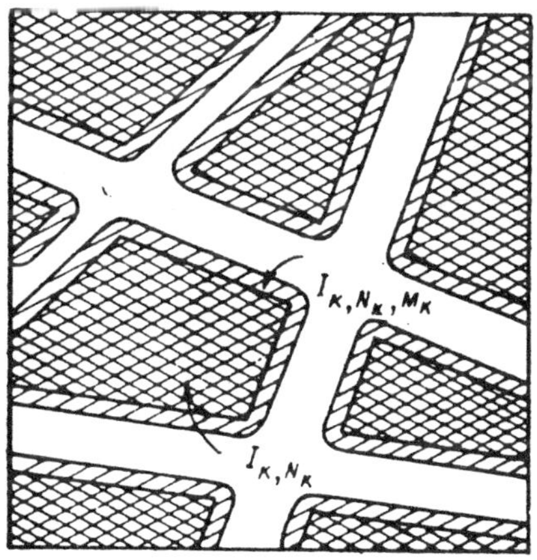

FIGURE 3

It is clear that $I_{k,N_k} \subset I_{k,N_k,M_{k-1}}$ (see Figure 3) and

$$\operatorname{mes} \bigcap_k I_{k,N_k,M_{k-1}} \geqslant \operatorname{mes} \bigcap_k I_{k,N_k}. \tag{6.4}$$

Therefore it is enough to find a lower bound for the Lebesgue measure of $\bigcap_k I_{k,N_k}$. To do this, we need to prove some auxiliary lemmas (see [2]).

LEMMA 4. *Let Γ_{m_0m} be the strip defined in $\mathbb{R}^n$ by $|my - m_0| < \gamma|m|^{-\sigma}$. Then*

$$\operatorname{mes}(I \cap \Gamma_{m_0m}) \leqslant \frac{2}{\min_k (b_k - a_k)} \gamma |m|^{-\sigma} \operatorname{mes} I. \tag{6.5}$$

PROOF. Let us consider the hyperplane $\Gamma^{\sim}$ defined by $my - m_0 = 0$, with normal vector N. We let $\|m\| = (\Sigma_1^n m_i^2)^{1/2}$. Since $\operatorname{mes}(I \cap \Gamma^{\sim}) = \int |N|\, dy$ and $|N| \leqslant \|m\|$, we obtain the inequality

$$\operatorname{mes}(I \cap \Gamma^{\sim}) \leqslant (\|m\|/\min(b_k - a_k)) \operatorname{mes} I.$$

If we observe that the width of Γ_{m_0m} is $2\gamma/(|m|^\sigma \|m\|)$, we obtain the required estimate (6.5).

LEMMA 5. *The parallelepiped $I = [a, b]$ is a domain of type D (see [2]), i.e. for every $0 < d_1 < d_2 \leqslant 0.5 \min_i(b_i - a_i)$ the inequality*

$$\operatorname{mes}[(I - d_1) \setminus (I - d_2)] \leqslant D(d_2 - d_1) \operatorname{mes} I \tag{6.6}$$

is satisfied, with $D = 2^n \cdot n/\min_i(b_i - a_i)$.

PROOF. Let us consider the function $P_n(d) = \text{mes}(I - d)$ and write it in the form $\Sigma_0^n p_r d^r$. An easy induction shows that the the coefficients p_r satisfy

$$|p_r| \leqslant \frac{2^r\binom{n}{r}}{\left[\min\limits_i (b_i - a_i)\right]^r} \text{mes } I.$$

We use it to obtain the following estimate of the derivative of $P_n(d)$:

$$|P_n'(d)| \leqslant \sum_{r=1}^{n} |p_r| r d^{r-1} \leqslant D \text{ mes } I. \tag{6.7}$$

To obtain (6.6) it is now enough to use the Lagrange identity

$$|P_n(d_1) - P_n(d_2)| \leqslant \max |P_n'| (d_2 - d_1)$$

and (6.7).

Following the terminology of the paper [2], we say that $I^\sim$ has in I the type N ($I^\sim \subset I$) if $I^\sim = (I - d) \setminus (\bigcup_j \Gamma_j)$, where $d > 0$ and $\bigcup_j \Gamma_j$ is a union of no more than N strips. It is clear that for $d_2 > d_1 > 0$

$$(I^\sim - d_1) \setminus (I^\sim - d_2) \subset [(I - d - d_1) \setminus (I - d - d_2)] \\ \cup \left\{\left[\left(\bigcup_j \Gamma_j + d_2\right) \setminus \left(\bigcup_j \Gamma_j + d_1\right)\right] \cap I\right\},$$

Hence from Lemmas 4 and 5 we obtain

$$\text{mes}\,[(I^\sim - d_1) \setminus (I^\sim - d_2)] \leqslant D(d_2 - d_1) \text{ mes } I \\ + 2(d_2 - d_1) ND \text{ mes } I = D(d_2 - d_1)(1 + 2N) \text{ mes } I. \tag{6.8}$$

LEMMA 6 (see [2]).

$$\text{mes}\,(I_{k-1, N_{k-1}} \setminus I_{k, N_k}) \leqslant LD[\gamma_1 \Delta_k + d_{k-1} N_{k-1}^{n+1}] \text{ mes } I, \tag{6.9}$$

where

$$L = 2^{n+3}(|b| + 2), \quad |b| = \left(\sum_i b_i^2\right)^{1/2}, \quad \Delta_k = \sum_{N_{k-1} < r \leqslant N_k} r^{n-2}. \tag{6.10}$$

PROOF. We have

$$I_{k-1, N_{k-1}} \setminus I_{k, N_k} = (I_{k-1, N_{k-1}} \setminus E_k) \cup [I_{k-1, N_{k-1}} \setminus (I_{k-1, N_{k-1}} - d_{k-1})],$$

so let us first estimate the Lebesgue measure of the set

$$I_{k-1, N_{k-1}} \setminus (I_{k-1, N_{k-1}} - d_{k-1}).$$

a. The set $I_{k-1, N_{k-1}}$ has in I the type

$$\sum_{1 \leqslant r \leqslant N_{k-1}} 2^n r^{n-1} [2r(|b| + 1) + 1] \leqslant 2^n N_{k-1}^n [2N_{k-1}(|b| + 1) + 1], \tag{6.11}$$

where $|b|$ is the distance from the origin of coordinates to the furthest vertex of the parallelepiped $I = [a, b]$.

Indeed, for fixed m_0 and $|m| = r$, the number of strips Γ_{m_0m} is equal to $2^n r^{n-1}$. But m_0 takes no more than $2r(|b| + 1) + 1$ values; hence (6.11). Taking (6.8) into account, we obtain

$$\operatorname{mes}[I_{k-1,N_{k-1}} \setminus (I_{k-1,N_{k-1}} - d_{k-1})] \leqslant Dd_{k-1}\Big[1 + 2^{n+1} \sum_{1\leqslant r\leqslant N_{k-1}} r^{n-1}(2r(|b|+1)+1)\Big] \operatorname{mes} I$$
$$\leqslant Dd_{k-1} 2^{n+3} N_{k-1}^{n+1}(|b|+2) \operatorname{mes} I. \quad (6.12)$$

b. Let us evaluate the Lebesgue measure of $I_{k-1,N_{k-1}} \setminus E_{N_k}$. Since for fixed m_0 and $|m| = r$ the number of distinct strips Γ_{m_0m} is $2^n r^{n-1}$, Lemma 4 gives

$$\sum_{|m|=r} \operatorname{mes}(I \cap \Gamma_{m_0 m}) \leqslant \frac{2}{n} D\gamma r^{-\sigma} \operatorname{mes} I$$

and

$$\operatorname{mes}(I_{k-1,N_{k-1}} \setminus E_{N_k}) \leqslant \sum_{N_{k-1}<r\leqslant N_k} \sum_{\substack{|m|=r\\ |m_0|\leqslant r(|b|+1)}} \operatorname{mes}(I \cap \Gamma_{m_0 m})$$
$$\leqslant \frac{4}{n}\gamma D(|b|+2)\Delta_k \operatorname{mes} I. \quad (6.13)$$

Combining (6.12) and (6.13), we obtain the desired inequality (6.9).

Going back to an estimate for $\operatorname{mes}(\bigcap_k I_{k,N_k,M_{k-1}})$, we use Lemma 6 to infer that

$$\operatorname{mes}(I \setminus I_{k,N_k,M_{k-1}}) \leqslant \operatorname{mes}(I \setminus I_{k,N_k}) \leqslant \operatorname{mes}(I \setminus I_{1,N_1}) + \sum_{i=2}^{k} \operatorname{mes}(I_{i-1,N_{i-1}} \setminus I_{i,N_i})$$
$$\leqslant LD \sum_{i=1}^{k} [\gamma\Delta_i + d_{i-1}N_{i-1}^{n+1}] \operatorname{mes} I. \quad (6.14)$$

The definition (3.6) of the numbers d_{i-1}, the equalities (5.5) and the inequality (5.3) easily imply the convergence of $\sum_1^\infty d_{i-1}N_{i-1}^{i+1}$ and the bound

$$\sum_{i=1}^{\infty} d_{i-1}N_{i-1}^{n+1} \leqslant \text{const}\, \gamma N_0^{-\varkappa^2\nu+n+1}. \quad (6.15)$$

If $\sigma > n + 1$, the series $\sum_1^\infty \Delta_i$ also converges. Therefore (6.14) and (6.15) yield the inequality

$$\operatorname{mes}\Big[I \setminus \Big(\bigcap_k I_{k,N_k,M_{k-1}}\Big)\Big] \leqslant \Big[\Big(LD\sum_{k=1}^{\infty}\Delta_k\Big)\gamma + \gamma LD \,\text{const}\, N_0^{-\varkappa^2\nu+n+1}\Big] \operatorname{mes} I. \quad (6.16)$$

Finally, combining (6.16) and Lemma 5, we obtain the necessary estimate of the "width" of the invariant set:

$$\operatorname{mes}(I' \cap E) \geqslant [1 - c_3\gamma(1 + N_0^{-\varkappa^2\nu+n+1})] \operatorname{mes} I. \quad (6.17)$$

§7. The proof of the estimates (3.18)

All that remains is to prove (3.17) and (3.18), i.e. to show that N_0 can be chosen large enough to guarantee that these estimates are automatically fulfilled. To this aim we choose κ in a special way and impose a lower bound on l.

As we already observed, (3.17) and (3.18) are valid for $k = 0$. We prove them in general by induction, i.e. we assume that they are valid for $k = m - 1$ and prove that they are valid also for $k = m$. To simplify, we introduce some new notation which will be used in this and the following section:

$$f_{m-1}=f,\ g_{m-1}=g,\ f_m=\varphi,\ g_m=\psi,\ u_{m-1}=u,\ v_{m-1}=v,$$
$$\hat f_{m-1}=\hat f,\ \hat g_{m-1}=\hat g,\ N_{m-1}=N_-,\ M_{m-1}=M_-,\ \delta_{m-1}=\delta_-,$$
$$N_m=N,\ M_m=M,\ \delta_m=\delta,\ N_{m+1}=N_+.$$

From (3.8), (3.10) and (3.15) we obtain the functional equations which φ and ψ must satisfy:

$$\varphi=\gamma^{-1}(u_1-\tilde u_1)+f(\xi+u,\ \eta+v)-\hat f,$$
$$\psi=\gamma^{-2}(v_1-\tilde v_1)+g(\xi+u,\ \eta+v)-\hat g+[\hat g]. \tag{7.1}$$

Here we also use a shorter notation:

$$u_1(\xi,\ \eta)=u(\xi+\eta,\ \eta),\ \tilde u_1(\xi,\ \eta)=u(\xi+\eta+\gamma\varphi(\xi,\ \eta),\ \eta+\gamma^2\psi(\xi,\ \eta))$$

and similarly for v_1 and $\tilde v_1$. We let $x = \xi + u$ and $y = \eta + v$, and rescale the parameters: $\bar\xi = N\xi$, $\bar\eta = M\eta$, $\bar x = Nx$, $\bar y = My$, $\bar f = Nf$, $\bar g = N^\nu g$, $\bar\varphi = N\varphi$, $\bar\psi = N^\nu\psi$, $\bar u = Nu$, $\bar v = Mv$, $\hat{\bar f} = N\hat f$ and $\hat{\bar g} = N^\nu\hat g$. One can now rewrite (7.1) in the form

$$\bar\varphi(\bar\xi,\ \bar\eta)=\gamma^{-1}\Big(\bar u\Big(\bar\xi+\frac{N}{M}\bar\eta,\ \bar\eta\Big)-\bar u\Big(\bar\xi+\frac{N}{M}\bar\eta+\gamma\bar\varphi,\ \bar\eta+\gamma^2\bar\psi\Big)\Big)$$
$$+\bar f(\bar x,\ \bar y)=\hat{\bar f}(\bar\xi,\ \bar\eta),$$
$$\bar\psi(\bar\xi,\ \bar\eta)=\gamma^{-2}\Big(\bar v\Big(\bar\xi+\frac{N}{M}\bar\eta,\ \bar\eta\Big)-\bar v\Big(\bar\xi+\frac{N}{M}\bar\eta+\gamma\bar\varphi,\ \bar\eta+\gamma^2\bar\psi\Big)\Big)$$
$$+\bar g(\bar x,\ \bar y)-\hat{\bar g}(\bar\xi,\ \bar\eta)+[\hat{\bar g}](\bar\eta). \tag{7.2}$$

From (4.2) and (4.3) we obtain

$$|D_{\bar\xi}^{\rho_1}D_{\bar\eta}^{\rho_2}\bar v_i|\leqslant c_8\gamma N^{\sigma+n}M\delta_-\ll 1,$$
$$|D_{\bar\xi}^{\rho_1}D_{\bar\eta}^{\rho_2}\bar u_i|\leqslant c_8 N^{2\sigma+n+1}\delta_-\ll 1,$$
$$i=1,\ \ldots,\ n,\quad 0\leqslant|\rho_1|+|\rho_2|\leqslant l+1. \tag{7.3}$$

LEMMA 7. *If* $\eta\in\operatorname{supp}\theta_{N_+,M}$, *then*

$$\sum_{i=1}^{n}|D_{\bar\xi}^{\rho_1}D_{\bar\eta}^{\rho_2}(\bar f_i(\bar\xi+\bar u,\ \bar\eta+\bar v)-\hat{\bar f}_i(\bar\xi,\ \bar\eta))|$$
$$+|D_{\bar\xi}^{\rho_1}D_{\bar\eta}^{\rho_2}(\bar g_i(\bar\xi+\bar u,\ \bar\eta+\bar v)-\hat{\bar g}_i(\bar\xi,\ \bar\eta))|\ll 1,$$
$$0\leqslant|\rho_1|+|\rho_2|\leqslant l. \tag{7.4}$$

In (7.3) and (7.4) the notation $a \ll b$ has the following meaning: for any $\alpha > 0$ and for N_0 large enough, $a < \alpha b$, the lower bound on N_0 being independent of m.

PROOF. Since $\eta \in \operatorname{supp}\theta_{N_+M}$ and since we must have (3.11), it follows that $y \in \operatorname{supp}\theta_{NM_-}$. Consequently to prove the lemma we can use (3.17) and (3.10) for $k = m - 1$.

Let us compute the derivatives of the composite function

$$D_{\xi}^{\rho_1}D_{\bar\eta}^{\rho_2}f_i(\xi+\bar v,\ \bar\eta+\bar v)=\sum_{1\leqslant|s_1|+|s_2|\leqslant|\rho_1|+|\rho_2|} D_{\bar x}^{s_1}D_{\bar y}^{s_2}f_i(\bar x,\ \bar y)$$

$$\times\sum_{\substack{\{\alpha_{i_1k_1},\ldots,\alpha_{i_nk_n},\\ \beta_{l_1m_1},\ldots,\beta_{l_nm_n}\}}} C_{\{\alpha_{i_1k_1},\ldots,\beta_{l_nm_n}\}}\prod_{\substack{i,\ldots,i_n,k_1,\ldots,k_n\\ l_1,\ldots,l_n,m_1,\ldots,m_n\\ 1\leqslant j\leqslant n}}\{D_{\xi}^{i_j}D_{\bar\eta}^{k_j}(\xi_j+\bar u_j(\xi,\ \bar\eta))\}^{\alpha_{i_jk_j}}$$

$$\times\{D_{\xi}^{l_j}D_{\bar\eta}^{m_j}(\bar\eta_j+\bar v_j(\xi,\ \bar\eta))\}^{\beta_{l_jm_j}}. \quad (7.5)$$

In (7.5) we have to sum over all the sets $\{\alpha_{i_1k_1},\ldots,\alpha_{i_nk_n},\beta_{l_1m_1},\ldots,\beta_{l_nm_n}\}$ of natural numbers which satisfy the equalities

$$\sum_{i_j,k_j}\alpha_{i_jk_j}=s_{j,1},\quad \sum_{l_j,m_j}\beta_{l_jm_j}=s_{j,2},$$

$$\sum_{r=1}^{n}\sum_{i_r,k_r} i_{j,r}\alpha_{i_rk_r}+\sum_{l_r,m_r} l_{j,r}\beta_{l_rm_r}=\rho_{j,1},$$

$$\sum_{r=1}^{n}\sum_{i_r,k_r} k_{j,r}\alpha_{i_rk_r}+\sum_{l_r,m_r} m_{j,r}\beta_{l_rm_r}=\rho_{j,2},$$

$$1\leqslant j\leqslant n. \quad (7.6)$$

Let us estimate a typical term of the sum (7.5), using (3.18) for $k = m - 1$. If $(i_j, k_j) \neq ((1)_j, 0)$ and $(l_j, m_j) \neq (0, (1)_j)$ (here $(1)_j$ is the unit vector with j-component equal to 1 and all the others equal to zero), then

$$|D_{\xi}^{i_j}D_{\bar\eta}^{k_j}(\xi_j+\bar u_j(\xi,\ \bar\eta))|=|D_{\xi}^{i_j}D_{\bar\eta}^{k_j}u_j|\leqslant c_8N_-^{-\mu+\varkappa(2\sigma+n+1)},$$

$$|D_{\xi}^{l_j}D_{\bar\eta}^{m_j}(\bar\eta_j+\bar v_j(\xi,\ \bar\eta))|=|D_{\xi}^{l_j}D_{\bar\eta}^{m_j}\bar v_j|\leqslant c_8N_-^{-\mu+\varkappa(\sigma+n+\nu)}. \quad (7.7)$$

The special factors with $(i_j, k_j) = ((1)_j, 0)$ and $(l_j, m_j) = (0, (1)_j)$ of the product (7.5) are estimated by

$$|D_{\xi_j}(\xi_j+\bar u_j)|\leqslant 1+|D_{\xi_j}\bar u_j|\leqslant 1+c_8N_-^{-\mu+\varkappa(2\sigma+n+1)}\leqslant 2,$$

$$|D_{\bar\eta_j}(\bar\eta_j+\bar v_j)|\leqslant 1+|D_{\bar\eta_j}\bar v_j|\leqslant 1+c_8N_-^{-\mu+\varkappa(\sigma+n+\nu)}\leqslant 2. \quad (7.8)$$

The estimate of the absolute value of a typical term in (7.5) is

$$D_{\bar x}^{s_1}D_{\bar y}^{s_2}f_i(\bar x,\bar y)\prod_{\substack{i_1,\ldots,i_n,k_1,\ldots,k_n\\ l_1,\ldots,l_n,m_1,\ldots,m_n\\ 1\leqslant j\leqslant n}}\{D_{\xi}^{i_j}D_{\bar\eta}^{k_j}(\xi_j+\bar u_j(\xi,\bar\eta))\}^{\alpha_{i_jk_j}}\{D_{\xi}^{l_j}D_{\bar\eta}^{m_j}(\bar\eta_j+\bar v_j(\xi,\bar\eta))\}^{\beta_{l_jm_j}}$$

$$\leqslant \mathrm{const}\left(\frac{N_-}{N}\right)^{|s_1|-1}\left(\frac{M_-}{M}\right)^{|s_2|}\prod_{\substack{1\leqslant j\leqslant n\\ i_1,\ldots,i_n,k_1,\ldots,k_n\\ l_1,\ldots,l_n,m_1,\ldots,m_n\\ (i_j,k_j)\neq((1)_j,0),\\ (l_j,m_j)\neq(0,(1)_j)}} N_-^{-\alpha_{i_jk_j}(\mu-\varkappa(2\sigma+n+1))-\beta_{l_jm_j}(\mu-\varkappa(\sigma+n+\nu))}$$

$$=\mathrm{const}\ N_-^{\gamma f(s_1,s_2,\alpha_{(1)_j0},\beta_{0(1)_j})}, \quad (7.9)$$

where

$$\gamma_f(s_1, s_2, \alpha_{(1)_j0}, \beta_{0(1)_j}) = -[\mu - \varkappa(2\sigma + n) - 1]\left[|s_1| - \sum_{j=1}^{n} \alpha_{(1)_j0}\right]$$

$$-[\mu - \varkappa(\sigma + n) - \nu]\left[|s_2| - \sum_{j=1}^{n} \beta_{0(1)_j}\right] - (\varkappa - 1)\left[\sum_{j=1}^{n} \alpha_{(1)_j0} - 1\right]$$

$$-(\varkappa - 1)\nu \sum_{j=1}^{n} \beta_{0(1)_j}. \quad (7.10)$$

We can estimate $D_\xi^{\rho_1} D_\eta^{\rho_2} \bar{g}(\xi + \bar{u}, \bar{\eta} + \bar{v})$ in the same way, and for a typical term in the sum analogous to (7.5) we obtain an estimate of the same kind as (7.9); the only difference is that in the right-hand side the number $\gamma_f(s_1, s_2, \alpha_{(1)_j0}, \beta_{0(1)_j})$ is replaced by

$$\gamma_g(s_1, s_2, \alpha_{(1)_j0}, \beta_{0(1)_j}) = \gamma_f(s_1, s_2, \alpha_{(1)_j0}, \beta_{0(1)_j}) + (\nu - 1)(\varkappa - 1) > \gamma_f. \quad (7.11)$$

If the numbers γ_f and γ_g are negative, the corresponding terms are small. From (7.6) we have

$$|s_1| - \sum_{j=1}^{n} \alpha_{(1)_j0} \geqslant 0 \quad \text{and} \quad |s_2| - \sum_{j=1}^{n} \beta_{0(1)_j} \geqslant 0.$$

If at least one of these differences is positive, (7.11) and (7.10) imply

$$\gamma_f < \gamma_g \leqslant -\mu + \varkappa(\sigma + n) + \nu + (\varkappa - 1)(\nu - 1) = -\varkappa\nu s - \nu + 1 < -\varkappa\nu s < 0.$$

If $\sum_{j=1}^{n} \alpha_{(1)_j0} = |s_1|$ and $\sum_{j=1}^{n} \beta_{0(1)_j} = |s_2|$ simultaneously, then we deduce from (7.6) that all the remaining numbers $\alpha_{l_j k_j}$ and $\beta_{l_j m_j}$ are equal to zero, $s_1 = \rho_1$, $s_2 = \rho_2$, and the term we want to evaluate reduces to

$$D_{\bar{x}}^{\rho_1} D_{\bar{y}}^{\rho_2} \bar{f}_i \ (\text{or } \bar{g}_i) \prod_{j=1}^{n} (1 + D_{\xi_j} \bar{u}_j)^{\rho_{j,1}} (1 + D_{\eta_j} \bar{v}_j)^{\rho_{j,2}}. \quad (7.12)$$

Let us use Newton's binomial formula and represent the right-hand side of (7.12) as the sum of terms of the form

$$D_{\bar{x}}^{\rho_1} D_{\bar{y}}^{\rho_2} \bar{f}_i \ (\text{or } \bar{g}_i) \prod_{j=1}^{n} (D_{\xi_j} \bar{u}_j)^{k_{j,1}} (D_{\eta_j} \bar{v}_j)^{k_{j,2}}. \quad (7.13)$$

If $|k_1| + |k_2| > 0$, we have for (7.13) the upper bound

$$\text{const}\left(\frac{N_-}{N}\right)^{|\rho_1|}\left(\frac{M_-}{M}\right)^{|\rho_2|} N_-^{\varkappa(\sigma+n+\nu)-\mu} \ll 1.$$

Finally, we must estimate the terms with $|k_1| = 0$ and $|k_2| = 0$, i.e. $D_{\bar{x}}^{\rho_1} D_{\bar{y}}^{\rho_2} \bar{f}_i(\bar{x}, \bar{y})$ and respectively $D_{\bar{x}}^{\rho_1} D_{\bar{y}}^{\rho_2} \bar{g}_i(\bar{x}, \bar{y})$. To do this, it is convenient to subtract from these terms $D_{\bar{\xi}}^{\rho_1} D_{\bar{\eta}}^{\rho_2} \bar{\bar{f}}_i(\bar{\xi}, \bar{\eta})$ (respectively, $D_{\bar{\xi}}^{\rho_1} D_{\bar{\eta}}^{\rho_2} \bar{\bar{g}}_i(\bar{\xi}, \bar{\eta})$). But before we do this, let us estimate separately the derivatives of the functions $\bar{\bar{f}}_i$ and $\bar{\bar{g}}_i$, using Lemma 2:

$$|D_{\xi}^{\rho_1} D_{\eta}^{\rho_2} \bar{\bar{f}}_i(\xi, \eta)| + |D_{\xi}^{\rho_1} D_{\eta}^{\rho_2} \bar{\bar{g}}_i(\xi, \eta)|$$
$$= N^{-|\rho_1|} M^{-|\rho_2|} \{|D_{\xi}^{\rho_1} D_{\eta}^{\rho_2} \bar{f}_i(\xi, \eta)| N + |D_{\xi}^{\rho_1} D_{\eta}^{\rho_2} \bar{g}_i(\xi, \eta)| N^{\nu}\}$$
$$\leqslant c_6 \{N^{n+1} + N^{n+\nu}\} \delta_- \leqslant 2c_6 N_-^{-\mu+\varkappa(\nu+n)} \ll 1. \quad (7.14)$$

We see that these quantities are always small.

Now we estimate the difference

$$D_{\bar{x}}^{\rho_1} D_{\bar{y}}^{\rho_2} \bar{f}_i(\bar{x}, \bar{y}) - D_{\bar{\xi}}^{\rho_1} D_{\bar{\eta}}^{\rho_2} \bar{\hat{f}}_i(\bar{\xi}, \bar{\eta}).$$

We remark that

$$\gamma_f(\rho_1, \rho_2, \rho_1, \rho_2) = -(\varkappa - 1)(|\rho_1| - 1 + |\rho_2| \nu)$$

and if $|\rho_1| = 1$ and $|\rho_2| = 0$ this quantity is not negative. Therefore it is enough to estimate the difference $D_{\bar{x}_j} \bar{f}_i(\bar{x}, \bar{y}) - D_{\bar{\xi}_j} \bar{\hat{f}}_i(\bar{\xi}, \bar{\eta})$. The remaining cases are then covered by the results obtained above. Let us represent the previous difference in the form

$$\begin{aligned} D_{\bar{x}_j} \bar{f}_i(\bar{x}, \bar{y}) - D_{\bar{\xi}_j} \bar{\hat{f}}_i(\bar{\xi}, \bar{\eta}) &= D_{x_j} f_i(x, y) - D_{\xi_j} \hat{f}_i(\xi, \eta) \\ &= D_{x_j} f_i(x, y) - D_{\xi_j} f_i(\xi, \eta) + (1 - \theta_{N, M_-}(\eta)) D_{\xi_j} f_i(\xi, \eta) \\ &\quad + D_{\xi_j} (\mathrm{Id} - \mathbb{T}_{NM}) \theta_{NM_-}(\eta) f_i(\xi, \eta). \end{aligned} \tag{7.15}$$

We use the inequalities

$$\begin{aligned} |\xi_i - x_i| &\leqslant c_8 N^{2\sigma + n} \delta_-, \\ |\eta_i - y_i| &\leqslant c_8 \gamma N^{\sigma + n} \delta_-, \end{aligned} \tag{7.16}$$

resulting from (4.2) and (4.3), and obtain

$$\begin{aligned} |D_{x_j} f_i(x, y) - D_{\xi_j} f_i(\xi, \eta)| &\leqslant n \max |D^2_{\xi_j \xi_k} f_i(\xi, \eta)| \max |x_k - \xi_k| \\ &\quad + n \max |D_{\xi_j} D_{\eta_k} f_i(\xi, \eta)| \max |y_k - \eta_k| \leqslant n N_- c_8 N^{2\sigma + n} \delta_- \\ &\quad + n M_- c_8 \gamma N^{\sigma + n} \delta_- \leqslant \mathrm{const}\, N_-^{-\mu + \varkappa(\nu + \sigma + n)} \ll 1. \end{aligned} \tag{7.17}$$

The inequality (7.16) shows that $|\eta_i - y_i| < M^{-1}$ Hence we can apply conclusion 4) of Lemma 1 to the expression $1 - \theta_{NM_-}(\eta)$. Thus from Lemma 1 and (3.18) for $k = m - 1$ we obtain

$$|(1 - \theta_{NM_-}(\eta)) D_{\xi_j} f_i(\xi, \eta)| \leqslant \left(\frac{M_-}{M}\right)^{l^*} = N_-^{-(\varkappa - 1) l^* \nu} \ll 1. \tag{7.18}$$

Moreover, from Lemma 2, 3), for $l' = l - 1$, and (3.18) for $k = m - 1$, we obtain

$$\begin{aligned} |D_{\xi_j} (\mathrm{Id} - \mathbb{T}_{NM}) \theta_{NM_-}(\eta) f_i(\xi, \eta)| &= |(\mathrm{Id} - \mathbb{T}_{NM}) \theta_{NM_-}(\eta) D_{\xi_j} f_i(\xi, \eta)| \\ \leqslant \mathrm{const} \max_{|\rho_1| + |\rho_2| = l - 1} N^{-|\rho_1|} M^{-|\rho_2|} N_-^{|\rho_1|} M_-^{|\rho_2|} &\leqslant \mathrm{const}\, N_-^{-(\varkappa - 1)(l - 1)} \ll 1. \end{aligned} \tag{7.19}$$

Therefore the right-hand side in (7.15) is small. Let us find now the analogous estimates for the function $\bar{g}(\bar{x}, \bar{y})$. The quantity

$$\gamma_g(\rho_1, \rho_2, \rho_1, \rho_2) = -(\varkappa - 1)[|\rho_1| + \nu(|\rho_2| - 1)]$$

is negative, except in two cases: a) $0 < |\rho_1| < \nu$, $|\rho_2| = 0$, and b) $|\rho_1| = 0$, $|\rho_2| = 1$. It is enough to consider only these two cases.

For a) we represent the difference $D_{\bar{x}}^{\rho_1}\bar{g}_i(\bar{x},\bar{y}) - D_{\bar{\xi}}^{\rho_1}\hat{\bar{g}}_i(\bar{\xi},\bar{\eta})$ as before, i.e. in the form

$$
\begin{aligned}
N_-^{\nu-|\rho_1|}(D_x^{\rho_1}g_i(x,y) - D_\xi^{\rho_1}\hat{g}_i(\xi,\eta)) = N_-^{\nu-|\rho_1|}(D_x^{\rho_1}g_i(x,y) \\
- D_\xi^{\rho_1}g_i(\xi,\eta)) + N_-^{\nu-|\rho_1|}(1-\theta_{NM_-}(\eta))\, D_\xi^{\rho_1}g_i(\xi,\eta) \\
+ N_-^{\nu-|\rho_1|}D_\xi^{\rho_1}(\mathrm{Id} - T_{NM})\,\theta_{NM_-}(\eta)\, g_i(\xi,\eta). \quad (7.20)
\end{aligned}
$$

To estimate the first term in the right-hand side of (7.20) one proceeds the same way as for (7.17). The second term is estimated by

$$
\begin{aligned}
|N_-^{\nu-|\rho_1|}(1-\theta_{NM_-}(\eta))\, D_\xi^{\rho_1}g_i(\xi,\eta)| \leqslant \left(\frac{N_-}{N}\right)^{\rho_1-\nu}\left(\frac{M_-}{M}\right)^{l^*} \\
\leqslant N_-^{-(\varkappa-1)\nu(l^*-1)} \ll 1. \quad (7.21)
\end{aligned}
$$

and the third by

$$
|N_-^{\nu-|\rho_1|}D_\xi^{\rho_1}(\mathrm{Id} - T_{NM})\,\theta_{NM_-}(\eta)\, g_i(\xi,\eta)| \leqslant \mathrm{const}\, N_-^{-(\varkappa-1)(l-\nu)} \ll 1.
$$

In case b),

$$
\begin{aligned}
D_{\bar{y}_j}\bar{g}_i(\bar{x},\bar{y}) - D_{\bar{\eta}_j}\hat{\bar{g}}_i(\xi,\eta) = \gamma(D_{y_j}g_i(x,y) - D_{\eta_j}\hat{g}_i(\xi,\eta)) \\
= \gamma(D_{y_j}g_i(x,y) - D_{\eta_j}g_i(\xi,\eta)) + \gamma D_{\eta_j}(1-\theta_{NM_-}(\eta))\, g_i(\xi,\eta) \\
+ \gamma D_{\eta_j}(\mathrm{Id} - T_{NM})\,\theta_{NM_-}(\eta)\, g_i(\xi,\eta). \quad (7.22)
\end{aligned}
$$

In order to obtain the subsequent estimates, we use (3.18) for $k = m - 1$. For the first term in (7.22), (7.16) gives

$$
\begin{aligned}
|\gamma(D_{y_j}g_i(x,y) - D_{\eta_j}g_i(\xi,\eta))| \leqslant n\gamma \max|D_{x_k}D_{y_j}g_i| \max|x_k - \xi_k| \\
+ n\gamma \max|D^2_{y_jy_k}g_i| \max|y_k - \eta_k| \leqslant 2nc_8 N_-^{-\mu+\varkappa(\sigma+n)+\nu} \ll 1. \quad (7.23)
\end{aligned}
$$

The second term in the right-hand side of (7.22) is estimated by Lemma 1:

$$
\begin{aligned}
|\gamma D_{\eta_j}(1-\theta_{NM_-}(\eta))\, g_i(\xi,\eta)| \leqslant \gamma\left(\frac{M_-}{M}\right)^{l^*} N_-^{\nu}M_- \\
+ \gamma c_4 M_-\delta_- \leqslant N_-^{-(\varkappa-1)l^*\nu} + c_4 N_-^{\nu-\mu} \ll 1. \quad (7.24)
\end{aligned}
$$

and the third (using Lemma 3) by

$$
\begin{aligned}
|\gamma D_{\eta_j}(\mathrm{Id} - T_{NM})\,\theta_{NM_-}(\eta)\, g_i(\xi,\eta)| = |\gamma(\mathrm{Id} - T_{NM})\, D_{\eta_j}\theta_{NM_-}(\eta)\, g_i(\xi,\eta)| \\
\leqslant c_6 \max_{|\rho_1|+|\rho_2|=l-1} N^{-|\rho_1|}M^{-|\rho_2|}\, |D_\eta^{\rho_2+(1)j}\theta_{NM_-}(\eta)\, D_\xi^{\rho_1}g_i(\xi,\eta)| \\
\leqslant \mathrm{const}\, N_-^{-(\varkappa-1)(l-1)} \ll 1.
\end{aligned}
$$

The following lemma of local character is evident.

LEMMA 8. *For every positive integer l there is a positive number $\alpha(l,n)$ with the following property. Let $\varphi(\xi,\eta)$ and $\psi(\xi,\eta)$ be vector-valued functions of class C^l which satisfy the equations*

$$
\varphi(\xi,\eta) = \Phi(\varphi(\xi,\eta), \psi(\xi,\eta), \xi, \eta),
$$

$$
\psi(\xi,\eta) = \Psi(\varphi(\xi,\eta), \psi(\xi,\eta), \xi, \eta),
$$

where $\Phi(\varphi, \psi, \xi, \eta)$ and $\Psi(\varphi, \psi, \xi, \eta)$ are functions of class C^l. If all the derivatives of Φ_i and Ψ_i with respect to all the variables are bounded ($\leqslant$) in absolute value by $\alpha(l, n)$, then all the derivatives of order $\leqslant l$ of the functions φ_i and ψ_i are bounded ($\leqslant$) in absolute value by $(2n)^{-1}$.

The estimates (7.3) and (7.4) show that for N_0 large enough one can apply the result of the Lemma 8 to (7.2). Going back to the initial scale of parameters, we obtain (3.18) for $k = m$.

§8. The proof of the estimate (3.17)

The equations (7.1) imply

$$|\varphi_i| + |\psi_i| \leqslant \sum_{p=1}^{n} (\max|D_{\xi_p} u_i| + \gamma^{-1} \max|D_{\xi_p} v_i|)|\varphi_p| + (\gamma \max|D_{\eta_p} u_i| + \max|D_{\eta_p} v_i|)|\psi_p| + |f_i(\xi + u, \eta + v) - \hat{f}_i(\xi, \eta)| + |g_i(\xi + u, \eta + v) - \hat{g}_i(\xi, \eta)| + |[\hat{g}_i](\eta)|. \tag{8.1}$$

Taking into account (4.2) and (4.3), where N_0 is large enough, we require that

$$\max|D_{\xi_p} u_i| + \gamma^{-1} \max|D_{\xi_p} v_i| + \gamma \max|D_{\eta_p} u_i| + \max|D_{\eta_p} v_i| \leqslant 4c_8 N_-^{\varkappa(\nu+2\sigma+n)-\mu} \leqslant \frac{1}{2n}. \tag{8.2}$$

Now we can rewrite (8.1) in the form

$$\sum_{i=1}^{n} |\varphi_i| + |\psi_i| \leqslant 2 \sum_{i=1}^{n} |f_i(\xi + u, \eta + v) - \hat{f}_i(\xi, \eta)| + |g_i(\xi + u, \eta + v) - \hat{g}_i(\xi, \eta)| + |[\hat{g}_i](\eta)|. \tag{8.3}$$

Let us estimate the difference

$$|f_i(\xi + u, \eta + v) - \hat{f}_i(\xi, \eta)|.$$

For $g_i(\xi + u, \eta + v) - \hat{g}_i(\xi, \eta)$ one can use the same argument.

We represent this difference as a sum of three terms:

$$f_i(x, y) - \hat{f}_i(\xi, \eta) = \mathrm{I}_{f_i} + \mathrm{II}_{f_i} + \mathrm{III}_{f_i},$$

where

$$\mathrm{I}_{f_i} = (1 - \theta_{NM_-}(y)) f_i(x, y),$$
$$\mathrm{II}_{f_i} = (\mathrm{Id} - \mathbb{T}_{NM})(\theta_{NM_-} f_i)(x, y),$$
$$\mathrm{III}_{f_i} = (\mathbb{T}_{NM} \theta_{NM_-} f_i)(x, y) - (\mathbb{T}_{NM} \theta_{NM_-} f_i)(\xi, \eta).$$

We have to evaluate I_{f_i}, II_{f_i} and III_{f_i} for $\eta \in \operatorname{supp} \theta_{N_+M}$. But as we remarked earlier, in this case (3.11) guarantees that $y \in \operatorname{supp} \theta_{NM_-}$; hence one can use (3.17) for $k = m - 1$, and (3.18). Now (7.16) shows that $|\eta_i - y_i| \leqslant M^{-1}$, and therefore we can apply conclusion 4) of Lemma 1 to the expression $1 - \theta_{NM_-}(y)$, and obtain

$$|\mathrm{I}_{f_i}| \leqslant \left(\frac{M_-}{M}\right)^{l^*} \frac{\delta_-}{\delta} \delta \leqslant N_-^{(\varkappa-1)(\mu-\nu l^*)} \delta. \tag{8.4}$$

Conclusion 3) of Lemma 2 gives us

$$|\mathrm{II}_{f_i}| \leqslant c_6 \max_{|\rho_1|+|\rho_2|=l} N^{-|\rho_1|} M^{-|\rho_2|} |D_x^{\rho_1} D_y^{\rho_2} 0_{NM} f_i| \leqslant \mathrm{const}\, N_-^{-(\varkappa-1)l-1+\mu\varkappa}\delta. \quad (8.5)$$

The same conclusion 3) of Lemma 2 plus (7.16) give

$$|\mathrm{III}_{f_i}| \leqslant \sum_{p=1}^{n} \max|D_{\xi_p}\hat{f}_i||x_p-\xi_p| + \max|D_{\eta_p}\hat{f}_i||y_p-\eta_p|$$

$$\leqslant \sum_{p=1}^{n} c_6 c_8 N^{n+1}\delta_- N^{2\sigma+n}\delta_- + c_6 c_8 N^n M\delta_- \gamma N^{\sigma+n}\delta_- \leqslant 2nc_6c_8 N_-^{-(2-\varkappa)\mu+\varkappa(2n+\nu+\sigma)}\delta. \quad (8.6)$$

Now if we combine (8.4) – (8.6) we obtain

$$\sum_{i=1}^{n} |f_i(\xi+u,\ \eta+v) - \hat{f}_i(\xi,\ \eta)| \leqslant G_1(N_-)\,\delta, \quad (8.7)$$

where

$$G_1(N_-) = c_{11} N_-^{(\varkappa-1)(\mu-\nu l^*)} + c_{12} N_-^{-(\varkappa-1)l-1+\mu\varkappa} + c_{13} N_-^{-(2-\varkappa)\mu+\varkappa(2n+\nu+\sigma)}. \quad (8.8)$$

The conditions we impose on the parameters μ, κ, ν, l and l^* are such that

$$G_1(N_-) \leqslant \frac{1}{6}. \quad (8.9)$$

Lastly, we need to evaluate $\Sigma_1^n \|[\hat{g}_i]\|$. For this we make the following observation: the transformation T_m, expressed in the coordinates $(x_m, y_m) \equiv (\xi, \eta)$ by

$$\xi_1 = \xi + \eta + \gamma\varphi(\xi,\ \eta)\ (\mathrm{mod}\ 1),$$
$$\eta_1 = \eta + \gamma^2\psi(\xi,\ \eta),$$

preserves the differential 1-form $\omega^{1\sim}$. In other words,

$$\oint_\Gamma \omega^{1\sim} = \oint_{T_m\Gamma} \omega^{1\sim} = \oint_\Gamma T_m^*\omega^{1\sim} \quad (8.10)$$

for every closed contour Γ. The following lemma describes some properties of the 1-form $\omega^{1\sim}$.

LEMMA 9. *Suppose that the* 1-*form* ω^1 *defined by* (1.6) *satisfies* (1.7) *and* (1.8). *Then in the coordinates* $(x_m, y_m) = (\xi, \eta)$, *the form* $\omega^{1\sim}$ *has the expression*

$$\omega^{1\sim} = \sum_{i=1}^{n} X_i^\sim(\xi,\ \eta)\,d\xi_i + Y_i^\sim(\xi,\ \eta)\,d\eta_i + dZ^\sim(\xi,\ \eta), \quad (8.11)$$

The functions $X_i^\sim$, $Y_i^\sim$ *and* $Z^\sim$ *are of class* C^2, *and their derivatives satisfy*

$$|D_\xi^{\rho_1} D_\eta^{\rho_2} X_i^\sim| + \gamma|D_\xi^{\rho_1} D_\eta^{\rho_2} Y_i^\sim| \leqslant \frac{3}{2} A\gamma^{\beta+1-|\rho_2|}, \quad (8.12)$$

$$i=1,\ \ldots,\ n, \quad 1 \leqslant |\rho_1| + |\rho_2| \leqslant 2,$$

$$|J_{[X^\sim]}(y)| \geqslant \frac{1}{2} B^n \gamma^{\beta_n}. \quad (8.13)$$

We do not prove Lemma 9 here. It is simple, but the proof is very cumbersome and recalls the deduction of the estimate (3.18). We only mention that, at every step of the iterative process, evaluating terms of the type $\Sigma_1^n \|[\hat{g}_p]\|$ involves only the derivatives of the functions $X_i^\sim$ and $Y_i^\sim$ and not the functions themselves (see (8.18)). It is therefore natural to require the possibility of estimating the derivatives of $X_i^\sim$ and $Y_i^\sim$ only in terms of the derivatives of the coefficients X_i and Y_i of the initial 1-form ω^1. Moreover, we want the estimates obtained this way, after the transformation of coordinates $x = \xi + u$, $y = \eta + v$, to reproduce themselves at every step. It happens that these two requirements completely determine the form of the estimates which must be satisfied by the derivatives of the functions X_i, Y_i, $X_i^\sim$ and $Y_i^\sim$. More exactly, these estimates coincide with (1.7), (1.8), (8.12) and (8.13).

The ranges of the parameters κ, ν and μ are restricted by the condition

$$\varkappa(4\nu+2\sigma+2n)-\mu<0 \tag{8.14}$$

which appears in the proof of Lemma 9.

In (8.10) we take, successively, the basic cycles Γ_i of the tori defined by the equations $\xi_j = \xi_j^0$, $j \neq i$, and $\eta_k = \eta_k^0$, as the closed contour Γ. Choosing Γ_i in this way and writing (8.10) in coordinates, we have

$$\begin{aligned}\int_0^1 \{X_i^\sim(\xi,\eta)-X_i^\sim(\xi+\eta+\gamma\varphi,\ \eta+\gamma^2\psi)\\ -\sum_{j=1}^n(X_j^\sim(\xi+\eta+\gamma\varphi,\ \eta+\gamma^2\psi)\gamma D_{\xi_i}\varphi_j\\ +Y_j^\sim(\xi+\eta+\gamma\varphi,\ \eta+\gamma^2\psi)\gamma^2 D_{\xi_i}\psi_j)\}\,d\xi_i=0.\end{aligned} \tag{8.15}$$

We expand the second term in the left member of (8.15) in a Taylor series, and we integrate the third and fourth terms by parts:

$$\begin{aligned}X_i^\sim(\xi+\eta+\gamma\varphi,\ \eta+\gamma^2\psi)=X_i^\sim(\xi+\eta,\ \eta)\\ +\sum_{p=1}^n(D_{\xi_p}X_i^\sim)(\xi+\eta,\ \eta)\gamma\varphi_p+(D_{\eta_p}X_i^\sim)(\xi+\eta,\ \eta)\gamma^2\psi_p+H_i(\xi,\ \eta).\end{aligned} \tag{8.16}$$

The function $H_i(\xi, \eta)$ is estimated by

$$\begin{aligned}|H_i(\xi,\eta)|\leqslant\frac{1}{2}\sum_{|p_1|+|p_2|=2}\frac{2}{p_1!\,p_2!}\max|D_x^{p_1}D_y^{p_2}X_i^\sim|\gamma^{|p_1|+2|p_2|}|\varphi^{p_1}\psi^{p_2}|\\ \leqslant A\gamma^{\beta+3}\left(\sum_{p=1}^n|\varphi_p|+|\psi_p|\right)^2.\end{aligned} \tag{8.17}$$

Here one has to use (8.12).

We now transform the new expression of (8.15) in the following way: in the terms containing ψ_i to the first power we replace ψ by $\psi = \psi^\sim + [\hat{g}]$; we isolate the terms containing $[\hat{g}]$ and integrate all the expressions with respect to all the variables ξ_j. We thus obtain (8.15) in the form

$$\sum_{p=1}^n[\hat{g}_p](\eta)[D_{\eta_p}X_i^\sim](\eta)=\gamma^{-1}b_i(\eta),\quad 1\leqslant i\leqslant n, \tag{8.18}$$

where

$$b_i(\eta) = -\sum_{p=1}^{n} [(D_{\xi_p} X_i^{\sim})_1 \varphi_p + (D_{\eta_p} X_i^{\sim})_1 \gamma \psi_p^{\sim}] + [H_i]$$
$$+ \sum_{j,p=1}^{n} [\varphi_i \{\widetilde{(D_{\xi_p} X_j^{\sim})_1} (\delta_{ip} + \gamma D_{\xi_i} \varphi_p) + \widetilde{(D_{\eta_p} X_j^{\sim})_1} \gamma^2 D_{\xi_i} \psi_p\}$$
$$+ \gamma \psi_j^{\sim} \{\widetilde{(D_{\xi_p} Y_j^{\sim})_1} (\delta_{ip} + \gamma D_{\xi_i} \varphi_p) + \widetilde{(D_{\eta_p} Y_j^{\sim})_1} \gamma^2 D_{\xi_i} \psi_p\}]. \quad (8.19)$$

In (8.19) we use again the simpler notation

$$f_1(\xi, \eta) = f(\xi + \eta, \eta), \quad \tilde{f}_1(\xi, \eta) = f(\xi + \eta + \gamma\varphi(\xi, \eta), \eta + \gamma^2 \psi(\xi, \eta))$$

and $[f](\eta) = \int_0^1 f(\xi, \eta)\, d\xi$. If we use now (3.18) for $k = m$ and (8.12), we obtain

$$|b_i(\eta)| \leqslant \text{const}\, \gamma^{\beta+1} A \sum_{p=1}^{n} |\varphi_p| + |\psi_p^{\sim}|$$
$$+ \text{const}\, \gamma^{\beta+1} A \left(\sum_{p=1}^{n} |\varphi_p| + |\psi_p| \right)^2. \quad (8.20)$$

From (8.3) and Lemmas 1 and 2 we can directly deduce the estimate

$$\sum_{p=1}^{n} |\varphi_p| + |\psi_p| \leqslant c_{14} N_-^{n-\mu}. \quad (8.21)$$

Now let us estimate $\Sigma_1^n |\varphi_p| + |\psi_p^{\sim}|$. In order to do this, we replace ψ by $\psi^{\sim} + [\hat{g}]$ in (7.1) and repeat the argument which has been used to infer (8.3) and in particular (8.2). It is easy then to obtain

$$\sum_{i=1}^{n} |\varphi_i| + |\psi_i^{\sim}| \leqslant G_1(N_-)\delta + G_2(N_-) \sum_{p=1}^{n} \|\hat{g}_p\|, \quad (8.22)$$

where $G_1(N_-)$ is defined by (8.8) and

$$G_2(N_-) = c_{15} N_-^{\varkappa(\nu+2\sigma+n)-\mu}. \quad (8.23)$$

If we substitute (8.21) and (8.22) into (8.20), we obtain

$$|b_i(\eta)| \leqslant A\gamma^{\beta+1} (G_3(N_-)\delta + G_2(N_-) \sum_{p=1}^{n} \|[\hat{g}_p]\|), \quad (8.24)$$

with $G_3(N_-) = G_1(N_-) + c_{16} N_-^{2n-(2-\kappa)\mu}$.

Let us return to the system (8.18). It has the determinant $J_{[X^{\sim}]}$ estimated by (8.13). Cramer's rule gives

$$[\hat{g}_p] = \gamma^{-1} J_{[X^{\sim}]}^{-1} \Delta_p, \quad (8.25)$$

where Δ_p denotes the determinant obtained from $J_{[X^{\sim}]}$ when one replaces the pth column by the column made of the elements $\{b_i\}_1^n$ from (8.19). It is easy to obtain the following inequality from (8.25) and using (8.12), (8.13) and (8.24):

$$\sum_{p=1}^{n} \|[\hat{g}_p]\| \leqslant c_{17} N_-^{\varkappa(\nu+2\sigma+n)-\mu} \sum_{p=1}^{n} \|[\hat{g}_p]\| + \{c_{18} N_-^{(\varkappa-1)(\mu-\nu l^*)} + c_{19} N_-^{(\varkappa-1)l-1+\mu\varkappa}$$
$$+ c_{20} N_-^{(2-\varkappa)\mu+\varkappa(2n+\nu+\sigma)} + c_{21} N_-^{2n-(2-\varkappa)\mu}\}\delta. \quad (8.26)$$

We want to choose N_0 so that the coefficient of $\Sigma_1^n \|[\hat{g}_p]\|$ in the right-hand side of (8.26) is smaller than $1/2$ and the coefficient of δ is smaller than $1/12$. For this it is enough to impose the following inequalities upon the parameters l, l^*, κ, μ and ν:

$$\begin{gathered}(\varkappa-1)(\mu-\nu l^*)<0,\\ \mu\varkappa-(\varkappa-1)l-1<0,\\ \varkappa(\nu+\sigma+2n)-(2-\varkappa)\mu<0.\end{gathered}\tag{8.27}$$

Since the enumerated parameters satisfy (8.27), we have

$$\sum_{p=1}^{n}\|[\hat{g}_p]\|\leqslant\frac{1}{6}\delta.\tag{8.28}$$

Finally, we consider (8.3), taking into account (8.7), (8.9) and (8.28). In doing so, we conclude that (3.17) is valid for $k = m$.

BIBLIOGRAPHY

1. A. N. Kolmogorov, *On conservation of conditionally periodic motions for a small change in Hamilton's function*, Dokl. Akad. Nauk SSSR **98** (1954), 527–530. (Russian)

2. V. I. Arnol'd, *Small denominators and problems of stability of motion in classical and celestial mechanics*, Uspehi Mat. Nauk **18** (1963), no. 6(114), 91–192; English transl. in Russian Math. Surveys **18** (1963).

3. J. Moser, *On invariant curves of area preserving mappings of an annulus*, Nachr. Akad. Wiss. Göttingen Math.-Phys. Kl. II **1962**, 1–20.

4. V. F. Lazutkin, *Moser's theorem on invariant curves*, Questions in the Dynamical Theory of the Propagation of Seismic Waves, no. 14, "Nauka", Leningrad, 1974, pp. 109–120. (Russian) RŽ Mat. **1975** #1Б1088.

5. ______, *The Kolmogorov-Arnol'd-Moser theory and the eigenfunctions of the Laplace operator*, Doctoral Dissertation, Leningrad. Otdel. Mat. Inst. Steklov., Leningrad, 1977, pp. 134–177. (Russian)

6. ______, *The existence of caustics for a billiard problem in a convex domain*, Izv. Akad. Nauk SSSR Ser. Mat. **37** (1973), 186–216; English transl. in Math. USSR Izv. **7** (1973).

Translated by A. IACOB

Труды Матем. Инст. Стеклов.
147(1980)

Proceedings of the Steklov
Institute of Mathematics
1981, Issue 2

ESTIMATES FOR SOLUTIONS OF PARABOLIC INITIAL-BOUNDARY VALUE PROBLEMS IN WEIGHTED HÖLDER NORMS

UDC 517.946

V. A. SOLONNIKOV AND A. G. HAČATRJAN

ABSTRACT. Sharp estimates are obtained for solutions of parabolic initial-boundary value problems in spaces of functions whose derivatives for $t > 0$ are Hölder continuous, and at $t = 0$ may have algebraic singularities. A particular case of the estimates so obtained are estimates for the norms in C^s of solutions $u(x, t)$ of a homogeneous parabolic initial-boundary value problem in terms of the same norm of the function $u(x, 0)$.

Bibliography: 4 titles.

1. Introduction

We consider the initial-boundary value problem

$$\mathcal{L}\left(x, t, \frac{\partial}{\partial x}, \frac{\partial}{\partial t}\right)u(x, t) = f(x, t), \quad x \in \Omega \subset R^n, \quad t \in (0, T),$$

$$\mathcal{B}\left(x, t, \frac{\partial}{\partial x}, \frac{\partial}{\partial t}\right)u(x, t)|_{x\in\partial\Omega=S} = \Phi(x, t),$$

$$\mathcal{C}\left(\frac{\partial}{\partial t}\right)u|_{t=0} = \varphi(x) \tag{1}$$

under the following assumptions about the matrix differential operators $\mathfrak{L}$, $\mathfrak{B}$ and $\mathcal{C}$.

$\mathfrak{L}$ is $2b$-parabolic in the sense of Petrovskiĭ, the orders of its elements $L_{kj}(x, t, \partial/\partial x, \partial/\partial t)$ $(k, j = 1, \ldots, m)$ not surpassing $2br_j$, where b and r_j are positive integers. By the order of the operator L_{kj} we understand the degree of the polynomial $L_{kj}(x, t, \lambda\xi, \lambda^{2b}p)$ with respect to λ. It is assumed that the system $\mathfrak{L}u = f$ is solvable for the highest order derivatives of u with respect to t, i.e. that

$$L_{kj}(x, t, \xi, p) = \delta_{kj}p^{r_j} + A_{kj}(x, t, \xi, p),$$

where A_{kj} does not contain p^{r_j}.

$\mathfrak{B}$ is the operator associated with $\mathfrak{L}$ by the condition of complementarity [1]; the orders of its elements

$$B_{qj}\left(x, t, \frac{\partial}{\partial x}, \frac{\partial}{\partial t}\right) \left(q = 1, \ldots, b\sum_{j=1}^{m} r_j \equiv R, \quad j = 1, \ldots, m\right)$$

do not surpass $\sigma_q + 2br_j$, where $\sigma_q \leqslant -1$, and if $\sigma_q + 2br_j < 0$, then $B_{qj} = 0$.

1980 *Mathematics Subject Classification.* Primary 35K50, 35B45.

$\mathcal{C}(p) = \operatorname{diag}(1, \ldots, p^{r_1 - 1}, 1, \ldots, p^{r_m - 1})$, so that the initial conditions have the form

$$\left.\frac{\partial^{k-1} u_j}{\partial t^{k-1}}\right|_{t=0} = \varphi_j^{(k)}(x), \quad k = 1, \ldots, r_j, \quad j = 1, \ldots, m.$$

The domain Ω will be assumed bounded or unbounded; in the latter case the vector u will be supposed bounded at infinity.

In [1], the unique solvability of (1) in the spaces $C^{l,l/2b}(Q_T)$, $Q_T = \Omega = \times (0, T)$, was proved, and an exact estimate of the solution was obtained in the norms of these spaces. Here this result is extended to the somewhat wider class of spaces $C_s^{l,l/2b}(Q_T)$, whose elements may have derivatives with singularities at $t = 0$. Such a generalization allows us to prove the solvability of (1) with fewer conditions of compatibility than in [1], and in the case $f = 0$, $\Phi = 0$ to obtain an exact estimate for solutions in the norm of $C^s(\Omega)$ for any $t > 0$ in terms of the corresponding norms of the initial data. For $r_1 = \cdots = r_m = 1$, this was done by one of the authors [2] and even earlier in the one-dimensional case by Belonosov and Zelenjak [3].

We define the function spaces necessary for our purposes. By $C^s(\Omega)$ with arbitrary $s \geqslant 0$ we understand the space of functions $[s]$ times differentiable in Ω with finite norm

$$|u|_\Omega^{(s)} = \sum_{|\alpha|<s} |D^\alpha u|_\Omega + [u]_\Omega^{(s)},$$

where

$$|u|_\Omega = \sup_{x\in\Omega} |u(x)|, \quad [u]_\Omega^{(s)} = \sum_{|\alpha|=s} |D^\alpha u|_\Omega$$

for s an integer, and

$$[u]_\Omega^{(s)} = \sum_{|\alpha|=[s]} \sup_{x,\, y\in\Omega} |x-y|^{[s]-s} |D^\alpha u(x) - D^\alpha u(y)|$$

for s nonintegral. The space $C_s^{l,l/2b}(Q_T)$ with any nonintegral $l > 0$ and $s \in [0, l)$ is the space of functions defined in Q_T with finite norm

$$|u|_{s,Q_T}^{(l)} = \sup_{t<T} t^{(l-s)/2b} [u]_{Q'_t}^{(l)} + \sum_{s<2bk+|\alpha|<l} \sup_{t<T} t^{(2bk+|\alpha|-s)/2b} |D_t^k D_x^\alpha u(x, t)|_\Omega$$
$$+ [u]_{Q_T}^{(s)} + \sum_{2bk+|\alpha|\leqslant s} \sup_{t<T} |D_t^k D_x^\alpha u(x, t)|_\Omega, \qquad (2)$$

where

$$[u]_{Q_T}^{(s)} = \sup_{t<T} [u]_\Omega^{(s)} + \sum_{s-2b<2bk+|\alpha|<s} \sup_{x,\, t,\, \tau} \frac{|D_t^k D_x^\alpha u(x, t) - D_\tau^k D_x^\alpha u(x, \tau)|}{(t-\tau)^{(s-2bk-|\alpha|)/2b}},$$

and $Q'_t = \Omega \times (t/2, t)$. For $s = l$ this space coincides with $C^{l,l/2b}(Q_T)$. For $s < 0$ we define the norm in $C_s^{l,l/2b}(Q_T)$ by the same formula (2), but without the latter two terms. Furthermore, for $s \geqslant 0$ we define $\mathring{C}_s^{l,l/2b}(Q_T)$ as the space of functions in $C_s^{l,l/2b}(Q_T)$ satisfying zero initial conditions

$$D_t^k D_x^\alpha u(x, t)|_{t=0} = 0, \quad 2bk + |\alpha| \leqslant s.$$

For $s < 0$ we set

$$\mathring{C}_s^{l,\,l/2b}(Q_T) = C_s^{l,\,l/2b}(Q_T).$$

In the same way, we define the spaces $C_s^{l,l/2b}(S_T)$ and $\mathring{C}_s^{l,l/2b}(S_T)$ on the surface $S_T = S \times (0, T)$.

LEMMA. *In the space $\mathring{C}_s^{l,l/2b}(Q_T)$ the norm (2) is equivalent to the norm*

$$\|u\|_{s,Q_T}^{(l)} = \sup_{t<T} t^{l-s/2b}[u]_{Q_t'}^{(l)} + \sup_{t<T} t^{-s/2b}|u(x,\ t)|_{\Omega}.$$

PROOF. The estimate $|u|_{s,Q_T}^{(l)} \leqslant c\|u\|_{s,Q_T}^{(l)}$ follows from the known "interpolation" inequalities

$$|D_x^\alpha D_t^k u(x,\ t)|_\Omega \leqslant \varepsilon_1^{l-2bk-|\alpha|}[u]_{Q_t'}^{(l)} + c_1\varepsilon_1^{-2bk-|\alpha|}|u|_{Q_t'},$$

$$[u]_{Q_t'}^{(s)} \leqslant \varepsilon_2^{l-s}[u]_{Q_t'}^{(l)} + c_2\varepsilon_2^{-s}|u|_{Q_t'},$$

valid for $s < l$ and $2bk + |\alpha| < l$ with arbitrarily small ε_i, which one takes proportional to $t^{1/2b}$, and also from the inequality

$$[u]_{Q_T}^{(s)} \leqslant \sup_{t<T}[u]_{Q_t'}^{(s)} + c_3 \sum_{2bk+|\alpha|<s}\ \sup_{t<T} t^{\frac{2bk+|\alpha|-s}{2b}}|D_t^k D_x^\alpha u(x,\ t)|_\Omega.$$

The estimate $\|u\|_{s,Q_T}^{(l)} \leqslant c|u|_{s,Q_T}^{(l)}$ for $s \leqslant 0$ is clear, and for $s > 0$ it follows from the inequality

$$t^{-s/2b}|u(x,t)| \leqslant \frac{t^{-s/2b}}{(k-1)!}\int_0^t (t-\tau)^{k-1}|D_\tau^k u(x,\ \tau) - D_\tau^k u(x,\ 0)|_\Omega\, d\tau \leqslant c_4[u]_{Q_t'}^{(s)}$$

in the case of nonintegral $s/2b$ (here $k = [s/2b]$) or from

$$t^{-s/2b}|u(x,\ t)|_\Omega \leqslant c_5|D_\tau^k u(x,\ \tau)|_{Q_t}$$

in the case of integral $s/2b = k$. This proves the lemma.

In the present paper, the following theorems are established.

THEOREM 1. *Let $\mu \in (0, 1), -\mu < k < 2b$, and $S \in C^{2br_{\max}+\mu}$. Let the coefficients of the operators A_{ij} belong to $C^{\mu,\mu/2b}(Q_T)$, and let those of the operators B_{qj} to belong to $C^{\mu-\sigma_q,(\mu-\sigma_q)/2b}(Q_T)$. Then problem (1) has a unique solution*

$$u_j \in C_{2br_j-k}^{2br_j+\mu,\ r_j+\mu/2b}(Q_T)$$

for any $f_i \in C_{-k}^{\mu,\mu/2b}(Q_T)$, $\varphi_j^{(i)} \in C^{2b(r_j-i+1)-k}(\Omega)$ and $\Phi_q \in C_{-k-\sigma_q}^{\mu-\sigma_q,(\mu-\sigma_q)/2b}(S_T)$ satisfying the compatibility conditions[1]

$$\frac{\partial^{i_q}}{\partial t^{i_q}} \sum_{j=1}^m B_{qj}\left(x,\ t,\ \frac{\partial}{\partial x},\ \frac{\partial}{\partial t}\right)u_j\bigg|_{t=0} = \frac{\partial^{i_q}\Phi_q(x,\ t)}{\partial t^{i_q}}\bigg|_{t=0} \tag{3}$$

(1) The derivatives of u_j on the left side of (3) must be expressed in terms of the $\varphi_j^{(i)}$.

for all q and i_q such that $\sigma_q + 2bi_q \leqslant -k$. For all $t \leqslant T$, the following estimate holds:

$$\sum_{j=1}^{m} |u_j|_{2br_j-k, Q_t}^{(2br_j+\mu)} \leqslant c\left(\sum_{j=1}^{m} |f_j|_{-k, Q_t}^{(\mu)} + \sum_{q=1}^{R} |\Phi_q|_{-k-\sigma_q, S_t}^{(\mu-\sigma_q)} + \sum_{j=1}^{m}\sum_{i=1}^{r_j} |\varphi_j^{(i)}|_{\Omega}^{(2b(r_j-i+1)-k)}\right). \quad (4)$$

THEOREM 2. *Let $k = 2b$. Let the assumptions of Theorem 1 be fulfilled, and let the coefficients of the operators A_{ij} have bounded continuous derivatives with respect to the x_j. Let*

$$f_i(x, t) = \sum_{k=1}^{n} \frac{\partial f_{ik}(x, t)}{\partial x_k} + f_i'(x, t),$$

where

$$\sum_{i=1}^{m}\sum_{k=1}^{n} \sup_{t<T} t^{1-1/2b} |f_{ik}(x, t)|_{\Omega} + \sum_{i=1}^{m} \sup_{t<T} t^{1-\delta/2b} |f_i'|_{\Omega} < \infty, \quad \delta \in (0, 1].$$

Then problem (1) *has a unique solution in the space*

$$\prod_{j=1}^{m} C_{2br_j-2b}^{2br_j+\mu, r_j+\mu/2b}(Q_T)$$

and for all $t \leqslant T$

$$\sum_{j=1}^{m} |u_j|_{2br_j-2b, Q_t}^{(2br_j+\mu)} \leqslant c\left(\sum_{j=1}^{m} |f_j|_{-2b, Q_t}^{(\mu)} + \sum_{q=1}^{R} |\Phi_q|_{-2b-\sigma_q, S_t}^{(\mu-\sigma_q)} + \sum_{j=1}^{m}\sum_{i=1}^{r_j} |\varphi_j^{(i)}|_{\Omega}^{(2b(r_j-i+1)-k)} + \sum_{j=1}^{m}\sum_{k=1}^{n} \sup_{t<T} t^{1-1/2b} |f_{ik}(x, t)|_{\Omega} + \sum_{i=1}^{m} \sup_{t<T} t^{1-\delta/2b} |f_i'(x, t)|_{\Omega}\right). \quad (5)$$

From (4) and (5) it follows that the solution of (1) for $f = 0$ and $\Phi = 0$ satisfies the inequality

$$\sum_{j=1}^{m} |u_j(x, t)|_{\Omega}^{(2br_j-k)} \leqslant c \sum_{j=1}^{m}\sum_{i=1}^{r_j} |\varphi_j^{(i)}|_{\Omega}^{(2b(r_j-i+1)-k)},$$

which was established for $k \in [-\mu, 0)$ in [1].

§2. The case of zero initial data

We set $\varphi = 0$ and consider problem (1) as a problem for the determination of the vector $u = (u_1, \ldots, u_m)$, $u_j \in \mathring{C}_{2br_j-k}^{2br_j+\mu, r_j+\mu/2b}(Q_T)$, satisfying the relation

$$\mathcal{L}u = f, \quad \mathcal{B}u|_{x\in S} = \Phi \quad (6)$$

with given $f = (f_1, \ldots, f_m)$, $\Phi = (\Phi_1, \ldots, \Phi_R)$, $f_j \in \mathring{C}_{-k}^{\mu,\mu/2b}(Q_T)$ and $\Phi_q \in \mathring{C}_{-k-\sigma_q}^{\mu-\sigma_q,(\mu-\sigma_q)/2b}(S_T)$.

THEOREM 3. *Let* $-\mu \leqslant k < 2b$, *and let the conditions of Theorem* 1 *be fulfilled. Then for any* $f_j \in \mathring{C}^{\mu,\mu/2b}_{-k}(Q_T)$ *and* $\Phi_q \in \mathring{C}^{\mu-\sigma_q,(\mu-\sigma_q)/2b}_{-k-\sigma_q}(S_T)$ *the problem* (6) *has a unique solution* $u = (u_1, \ldots, u_m)$, $u_j \in \mathring{C}^{2br_j+\mu,r_j+\mu/2b}_{2br_j-k}(Q_T)$ *for which the estimate*

$$\sum_{j=1}^{m} \| u_j \|^{(2br_j+\mu)}_{2br_j-k, Q_t} \leqslant c \left(\sum_{j=1}^{m} \| f_j \|^{(\mu)}_{-k, Q_T} + \sum_{q=1}^{R} \| \Phi_q \|^{(\mu-\sigma_q)}_{-k-\sigma_q, S_t} \right) \tag{7}$$

is valid for all $t \leqslant T$. *If, however,* $k = 2b$ *and the conditions of Theorem* 2 *are fulfilled, then* (6) *has a unique solution* u *with* $u_j \in \mathring{C}^{2br_j+\mu,r_j+\mu/2b}_{2br_j-2b}(Q_T)$ *and the inequality*

$$\sum_{j=1}^{m} \| u_j \|^{(2br_j+\mu)}_{2br_j-k, Q_t} \leqslant c \left(\sum_{j=1}^{m} \| f_j \|^{(\mu)}_{k, Q_t} + \sum_{j=1}^{m} \sum_{p=1}^{n} \sup_{\tau<t} \tau^{1-1/2b} | f_{jp}(x, \tau) |_\Omega \right.$$
$$\left. + \sum_{j=1}^{m} \sup_{\tau<t} \tau^{1-\delta/2b} | f'_j(x, \tau) |_\Omega + \sum_{q=1}^{R} \| \Phi_q \|^{(\mu-\sigma_q)}_{k-\sigma_q, S_t} \right) \tag{7'}$$

holds.

PROOF. We first prove (7) and (7′). Let $\eta(r)$ be a monotone infinitely differentiable function equal to unity for $r \geqslant 1/2$ and zero for $r \leqslant 1/4$. We fix a number $t_1 \in (0, T)$. The vector $v(x, t) = u(x, t)\eta(t/t_1)$ satisfies

$$\mathfrak{L}\left(x, t, \frac{\partial}{\partial x}, \frac{\partial}{\partial t}\right) v(x, t) = f\eta\left(\frac{t}{t_1}\right) + [\mathfrak{L}, \eta] u = g(x, t),$$
$$\mathscr{B}\left(x, t, \frac{\partial}{\partial x}, \frac{\partial}{\partial t}\right) v(x, t)|_{x\in S} = \Phi\eta\left(\frac{t}{t_1}\right) + [\mathscr{B}, \eta] u|_{x\in S} = \Psi(x, t),$$

where $[\mathfrak{L}, \eta]u = \mathfrak{L}(u\eta) - \eta\mathfrak{L}u$ and vanishes for $t \leqslant t_1/4$. If $u_j \in \mathring{C}^{2br_j+\mu,r_j+\mu/2b}_{2br_j-k}(Q_T)$, then $v_j \in \mathring{C}^{2br_j+\mu,r_j+\mu/2b}(Q_T)$ and, as was shown in [1],

$$\sum_{j=1}^{m} [v_j]^{(2br_j+\mu)}_{Q_{t_1}} \leqslant c_1 \left(\sum_{j=1}^{m} [g_j]^{(\mu)}_{Q_{t_1}} + \sum_{q=1}^{R} [\psi_q]^{(\mu-\sigma_q)}_{S_{t_1}} \right). \tag{8}$$

Since the operators $[\mathfrak{L}, \eta]u$ and $[\mathfrak{B}, \eta]u$ contain, as compared with $\mathfrak{L}u$ and $\mathfrak{B}u$, only lower order terms, it is easy to estimate the right side of (8) in terms of

$$c_2 \left\{ \sum_{j=1}^{m} \left([f_j]^{(\mu)}_{Q''_{t_1}} + t_1^{-\mu/2b} | f_j |_{Q''_{t_1}} \right) + \sum_{q=1}^{R} \left([\Phi_q]^{(\mu-\sigma_q)}_{S''_{t_1}} + t_1^{-(\mu-\sigma_q)/2b} | \Phi_q |_{S''_{t_1}} \right) \right.$$
$$\left. + \sum_{j=1}^{m} \left(\varepsilon [u_j]^{(2br_j+\mu)}_{Q''_{t_1}} + c_\varepsilon t^{-r_j-\mu/2b} | u_j |_{Q''_{t_1}} \right) \right\},$$

where $Q''_{t_1} = \Omega \times (t_1/4, t_1)$ and $S''_{t_1} = S \times (t_1/4, t_1)$. After that is done, multiply both sides of (8) by $t^{(\mu+k)/2b}$ and take the maximum with respect to $t_1 \in (t_0/2, t_0)$ and $t_0 < T$. Then, taking ε sufficiently small, we obtain

$$\sum_{j=1}^{m} \| u_j \|^{(2br_j+\mu)}_{2br_j-k, Q_{t_0}} \leqslant c_3 \left\{ \sum_{j=1}^{m} \| f_j \|^{(\mu)}_{-k, Q_{t_0}} + \sum_{q=1}^{R} \| \Phi_q \|^{(\mu-\sigma_q)}_{-k-\sigma_q, S_{t_0}} \right.$$
$$\left. + \sum_{j=1}^{m} \sup_{t_1<t_0} t_1^{-r_j+k/2b} | u_j(x, t_1) |_\Omega \right\}. \tag{9}$$

Proceeding to the estimate of $|u_j(x, t)|$, we again fix $t_1 < t_0$. Let w be the solution of the problem

$$\mathcal{L} w = f\eta\left(\frac{t}{t_1}\right), \quad \mathcal{B} w|_{x\in S} = \Phi\eta\left(\frac{t}{t_1}\right).$$

Since $f_j\eta(t/t_1) \in \mathring{C}^{\mu,\mu/2b}(Q_T)$ and $\Phi_q\eta(t/t_1) \in \mathring{C}^{-\sigma_q+\mu,(\mu-\sigma_q)/2b}(S_T)$, it follows that $w_j \in \mathring{C}^{2br_j+\mu,r_j+\mu/2b}(Q_T)$ and

$$\sum_{j=1}^{m} t_1^{-r_j+k/2b}|w_j(x, t_1)| \leqslant t_1^{(\mu+k)/2b}\sum_{j=1}^{m}[w_j]_{Q_{t_1}}^{(2br_j+\mu)}$$
$$\leqslant c_4\left\{\sum_{j=1}^{m}\|f_j\|_{-k,Q_{t_1}}^{(\mu)} + \sum_{q=1}^{R}\|\Phi_q\|_{-k-\sigma_q,S_{t_1}}^{(\mu-\sigma_q)}\right\}.$$

The vector $v = u - w$ satisfies the relations

$$\mathcal{L} v = f\left[1-\eta\left(\frac{t}{t_1}\right)\right] = \tilde{f}, \quad \mathcal{B} v|_{x\in S} = \Phi\left[1-\eta\left(\frac{t}{t_1}\right)\right] = \tilde{\Phi}$$

and can be expressed as

$$v(x, t) = \int_0^t\int_\Omega \mathcal{G}(x, y, t, \tau)\tilde{f}(y, \tau)\,dy\,d\tau + \int_0^t\int_S \mathcal{P}(x, y, t, \tau)\tilde{\Phi}(y, \tau)\,dS\,d\tau, \quad (10)$$

where $\mathcal{G}$ and $\mathcal{P}$ are the Green's matrices constructed by Ivasišen [4]. Their elements G_{ij} and P_{iq} are subject to the inequalities

$$|G_{ij}(x, y, t, \tau)| \leqslant c_5(t-\tau)^{-n/2b+r_i-1}\exp\left(-c_6\frac{|x-y|^{2b/(2b-1)}}{(t-\tau)^{1/(2b-1)}}\right),$$

$$\left|\frac{\partial}{\partial y_k}G_{ij}(x, y, t, \tau)\right| \leqslant c_7(t-\tau)^{-(n+1)/2b+r_i-1}\exp\left(-c_8\frac{|x-y|^{2b/(2b-1)}}{(t-\tau)^{1/(2b-1)}}\right),$$

$$|P_{iq}(x, y, t, \tau)| \leqslant c_9(t-\tau)^{-(n-1)/2b+r_i-1+\sigma_q/2b}\exp\left(-c_{10}\frac{|x-y|^{2b/(2b-1)}}{(t-\tau)^{1/(2b-1)}}\right). \quad (11)$$

Formula (10) was established in [4] for

$$\tilde{f}_i \in C^{\mu,\mu/2b}(Q_T), \quad \tilde{\Phi}_q \in C^{\mu-\sigma_q,(\mu-\sigma_q)/2b}(S_T),$$

but it makes sense also for

$$\tilde{f}_i \in \mathring{C}_{-k}^{\mu,\mu/2b}(Q_T), \quad \tilde{\Phi}_q \in \mathring{C}_{-k-\sigma_q}^{\mu-\sigma_q,(\mu-\sigma_q)/2b}(S_T),$$

and may be proved in the following fashion. First one needs to write it for the vector $v\eta(t/\varepsilon)$, and then pass to the limit as $\varepsilon \to 0$. For $k = 2b$, in place of (10) we have

$$v_i(x, t) = -\sum_{j=1}^{m}\sum_{p=1}^{n}\int_0^t\int_\Omega \frac{\partial}{\partial y_p} G_{ij}(x, y, t, \tau)\tilde{f}_{jp}(y, \tau)\,dy\,d\tau$$
$$+\sum_{j=1}^{m}\sum_{p=1}^{n}\int_0^t\int_S G_{ij}(x, y, t, \tau)\tilde{f}_{jp}(y, \tau)\,n_p(y)\,dS\,d\tau$$
$$+\sum_{j=1}^{m}\int_0^t\int_\Omega G_{ij}(x, y, t, \tau)\tilde{f}'_j(y, \tau)\,dy\,d\tau$$
$$+\sum_{q=1}^{R}\int_0^t\int_S P_{iq}(x, y, t, \tau)\tilde{\Phi}_q(y, \tau)\,dS\,d\tau, \qquad (10')$$

where $\tilde{f}_{i_p} = f_{i_p}[1 - \eta(t/t_1)]$, $\tilde{f}'_j = f'_j[1 - \eta(t/t_1)]$, and the n_k are the components of the unit vector normal (outward) to S. This is proved in the same way as (10).

In (10) and (10′) we set $t = t_1$ and estimate $|v_i(x, t_1)|$ with the aid of (11). We obtain

$$\sum_{i=1}^{m} t_1^{-r_i+k/2b}|v_i(x, t_1)|_\Omega \leqslant c_{11}\Bigg(\sum_{i, j=1}^{m} t_1^{-r_i+k/2b}\int_0^{t_1/2}(t_1-\tau)^{r_i-1}|f_j(y, \tau)|_\Omega\,d\tau$$
$$+\sum_{i=1}^{m}\sum_{q=1}^{R}\int_0^{t_1/2}(t_1-\tau)^{\sigma_q/2b+r_i-1}|\Phi_q(y, \tau)|_S\,d\tau\Bigg)$$
$$\leqslant c_{12}\left(\sum_{j=1}^{m}\|f_j\|_{-k, Q_{t_1}}^{(\mu)}+\sum_{q=1}^{R}\|\Phi_q\|_{-k-\sigma_q, S_{t_1}}^{(\mu-\sigma_q)}\right)$$

for $k < 2b$ and

$$\sum_{i=1}^{m} t_1^{-r_i+k/2b}|v_i(x, t_1)|_\Omega \leqslant c_{13}\Bigg(\sum_{i=1}^{m}\sum_{p=1}^{n}\sup_{\tau<t_1}\tau^{1-1/2b}|f_{ip}(x, \tau)|_\Omega$$
$$+t_1^{\delta/2b}\sum_{i=1}^{m}\sup_{\tau<t_1}\tau^{1-\delta/2b}|f'_i|_\Omega+\sum_{q=1}^{R}\|\Phi_q\|_{-2b-\sigma_q, S_{t_1}}^{(\mu-\sigma_q)}\Bigg)$$

for $k = 2b$. Since t_1 was arbitrary, these estimates and (9) imply (7) and (7′).

To prove the solvability of (6), we replace f and Φ by $f\eta(t/\varepsilon)$ and $\Phi\eta(t/\varepsilon)$. As is shown in [1], the problem obtained in this fashion has a solution u_ε with $u_{\varepsilon j} \in \mathring{C}^{2br_j+\mu, r_j+\mu/2b}(Q_T)$ for which the estimate

$$\sum_{j=1}^{m}\|u_{\varepsilon j}\|_{2br_j-k, Q_T}^{(2br_j+\mu)} \leqslant c\left(\sum_{j=1}^{m}\left\|f_j\eta\left(\frac{t}{\varepsilon}\right)\right\|_{-k, Q_T}^{(\mu)}+\sum_{q=1}^{R}\left\|\Phi_q\eta\left(\frac{t}{\varepsilon}\right)\right\|_{-k-\sigma_q, S_t}^{(\mu-\sigma_q)}\right) \leqslant c'$$

($k \in [-\mu, 2b)$) was just now proved, c' not depending on ε. Such an inequality is valid also for $k = 2b$. Therefore there exists a sequence $\{u^{\varepsilon_n}\}$, $\varepsilon_n \to 0$ as $n \to \infty$, such that the functions $u_j^{\varepsilon_n}$ and their derivatives $D_x^\alpha D_t^l u_j^{\varepsilon_n}$, $2bl + |\alpha| \leqslant 2br_j$, converge uniformly on any compact $K \subset \Omega$ to a solution $u(x, t)$ of (6), and

$$u_j(x, t) \in C_{2br_j-k}^{2br_j+\mu, r_j+\mu/2b}(Q_T).$$

The theorem is proved.

§3. Proof of Theorems 1 and 2

We reduce (1) to a problem with zero initial data. We first assume that $k \in (0, 2b]$, and determine a function $u_j'(x, t)$ as the solution of the Cauchy problem

$$\left(\frac{\partial}{\partial t} + (-1)^b \Delta^b\right)^{r_j} u_j'(x, t) = 0,$$
$$\left.\frac{\partial^{i-1} u_j'}{\partial t^{i-1}}\right|_{t=0} = \tilde{\varphi}_j^{(i)}(x), \quad i = 1, \ldots, r_j, \tag{12}$$

where $\tilde{\varphi}_j^{(i)}(x)$ is an extension of $\varphi_j^{(i)}$ to the whole space R^n, effected so that

$$|\tilde{\varphi}_j^{(i)}(x)|_{R^n}^{(2br_j-k-2b(i-1))} \leqslant c_1 |\varphi_j^{(i)}|^{(2br_j-k-2b(i-1))}.$$

We show that

$$|u_j'(x, t)|_{2br_j-k, R^n}^{(2br_j+\mu)} \leqslant c_2 \sum_{i=1}^{r_j} |\tilde{\varphi}_j^{(i)}|_{R^n}^{(2br_j-k-2b(i-1))}. \tag{13}$$

As was shown in [1], the function u_j' is found by the formula

$$u_j'(x, t) = \sum_{i=1}^{r_j} M_i^{(j)}\left(\frac{\partial}{\partial x}, \frac{\partial}{\partial t}\right) \int_{R^n} \Gamma^{(j)}(x-y, t) \tilde{\varphi}_j^{(i)}(y)\, dy = \sum_{i=1}^{r_j} u_{ij}(x, t),$$

where $M_i^{(j)}$ is an operator of order $2b(r_j - i)$ and $\Gamma^{(j)}$ is a fundamental solution of the equation $(\partial/\partial t + (-1)^b\Delta^b)^{r_j} u = 0$ for which

$$|D_t^k D_x^\alpha \Gamma^{(j)}(x, t)| \leqslant c_3 t^{-(n+|\alpha|)/2b-k+r_j-1} \exp\left(-c_4 \frac{|x|^{2b/(2b-1)}}{t^{1/(2b-1)}}\right).$$

We estimate the derivatives $D_t^l D_x^\beta u_{ij}$. If $l < i - 1$, from the formula

$$D_t^l D_x^\beta u_{ij} = D_t^l M_i^{(j)}\left(\frac{\partial}{\partial x}, \frac{\partial}{\partial t}\right) \int_{R^n} D_x^{\beta-\gamma} \Gamma^{(j)}(x-y, t) D^\gamma \tilde{\varphi}_j^{(i)}(y)\, dy,$$

where $|\gamma| < \min(|\beta|, [2b(r_j - i + 1) - k])$, and

$$D_t^l D_x^\beta u_{ij} = D_t^l M_i^{(j)} \int_{R^n} D_x^{\beta-\gamma} \Gamma^{(j)}(x-y, t) [D^\gamma \tilde{\varphi}_j^{(i)}(y) - D^\gamma \tilde{\varphi}_j^{(i)}(x)]\, dy,$$

where $|\gamma| = [2b(r_j - i + 1) - k] < |\beta|$, follow the inequalities

$$|D_t^l D_x^\beta u_{ij}(x, t)| \leqslant c_5 t^{-l+i-1-(|\beta|-|\gamma|)/2b} |D^\gamma \tilde{\varphi}_j^{(i)}|_{R^n}, \tag{14}$$

$$|D_t^l D_x^\beta u_{ij}(x, t)| \leqslant c_6 t^{-l-|\beta|/2b+r_j-k/2b} [\tilde{\varphi}_j^{(i)}]_{R^n}^{(2b(r_j-i+1)-k)}. \tag{15}$$

In the case $l \geqslant i$, the equation $(\partial/\partial t + (-1)^b\Delta^b)^{r_j}\Gamma^{(j)} = 0$ allows one to express $D_t^l D_x^\beta u_{ij}$ in the form of a linear combination of functions

$$D_x^{\alpha+\beta} Q_\alpha\left(\frac{\partial}{\partial x}, \frac{\partial}{\partial t}\right) \int_{R^n} \Gamma^{(j)}(x-y, t) \tilde{\varphi}_j^{(i)}(y)\, dy$$
$$= D_x^{\alpha+\beta-\gamma} Q_\alpha \int_{R^n} \Gamma^{(j)}(x-y, t) D^\gamma \varphi_j^{(i)}(y)\, dy, \tag{16}$$

where Q_α is an operator of order $r_j - 1$,

$$|\alpha| = 2b(l - i + 1),$$

and

$$|\gamma| \leqslant \min(|\beta| + 2b[l - i + 1), \ [2b(r_j - i + 1) - k]).$$

Estimating the right side of (16), we again arrive at (14) and (15), which are thereby proved under the conditions

$$|\gamma| \leqslant \min\{|\beta| + \max(0, 2b(l - i + 1)), \ [2b(r_j - i + 1) - k]\}$$

and

$$|\beta| + \max(0, 2b(l - i + 1)) > [2b(r_j - i + 1) - k]$$

respectively. It is easy to verify, with the help of (14) and (15), that for $2bl + |\beta| > 2br_j - k$ we have

$$|D_t^l D_x^\beta u_j(x, t)| \leqslant c_7 t^{-l - |\beta|/2b + r_j - k/2b} \sum_{j=1}^{r_j} [\bar{\varphi}_j^{(i)}]_{R^n}^{(2b(r_j - i + 1) - k)}$$

and, moreover,

$$\sum_{\substack{2bl + |\beta| \leqslant \\ \leqslant 2br_j - k}} |D_t^l D_x^\beta u_j|_{Q_T} + [u_j]_{Q_T}^{(2br_j - k)} \leqslant c_8(T) \sum_{i=1}^{r_j} [\bar{\varphi}_j^{(i)}]_{R^n}^{(2b(r_j - i + 1) - k)},$$

which implies (13).

For the new unknown vector $v = u - u'$, one obtains the problem

$$\mathcal{L}v = f - \mathcal{L}u' = g, \quad \mathcal{B}v = \Phi - \mathcal{B}u'|_{x \in S} = \psi, \quad \mathcal{C}\left(\frac{\partial}{\partial t}\right)v = 0, \tag{17}$$

which for $k \in (0, 2b]$ is analogous to (6). Here

$$g_j \in \overset{\circ}{C}{}_{-k}^{\mu, \mu/2b}(Q_T), \quad \psi_q \in \overset{\circ}{C}{}_{-k-\sigma_q}^{\mu - \sigma_q, (\mu - \sigma_q)/2b}(S_T)$$

(by virtue of (3)) and

$$\sum_{j=1}^{m} |g_j|_{-k, Q_T}^{(\mu)} + \sum_{q=1}^{R} |\psi_q|_{-k-\sigma_q, S_T}^{(\mu - \sigma_q)} \leqslant c_9 \left\{ \sum_{j=1}^{m} |f_j|_{-k, Q_T}^{(\mu)} + \sum_{q=1}^{R} |\Phi_q|_{-k-\sigma_q, S_T}^{(\mu - \sigma_q)} + \sum_{j=1}^{m} \sum_{i=1}^{r_j} |\varphi_j^{(i)}|_{\Omega}^{(2b(r_j - i + 1) - k)} \right\}. \tag{18}$$

From this and Theorem 3, applied to (17), we obtain the conclusion of Theorem 1.

In the case $k \in (-\mu, 0]$, for u_j' one should take the solution of the problem

$$\left(\frac{\partial}{\partial t} + (-1)^b \Delta^b\right)^{r_j + 1} u_j' = 0,$$

$$\left.\frac{\partial^{i-1} u_j'}{\partial t^{i-1}}\right|_{t=0} = \bar{\varphi}_j^{(i)}(x), \quad i = 1, \ldots, r_j,$$

$$\left.\frac{\partial^{r_j} u_j'}{\partial t^{r_j}}\right|_{t=0} = \bar{f}_j(x) - \sum_{k=1}^{m} \bar{A}_{jk}\left(x, \frac{\partial}{\partial x}, \frac{\partial}{\partial t}\right) u_k'|_{t=0},$$

where $\tilde{f}_j(x)$ is the extension of the function $f_j(x, 0)$ onto the entire space, with class retained, i.e.

$$|\tilde{f}_j|_{R^n}^{(-k)} \leqslant c_{10} |f_j(x, 0)|_{\Omega}^{(-k)},$$

and $\tilde{A}_{jk}$ is an operator whose coefficients are obtained by extension within the class of the corresponding coefficients of the operators $A_{jk}(x, 0, \partial/\partial x, \partial/\partial t)$. With such a definition of the vector u', we shall again have

$$g_j \in \overset{\circ}{C}{}_{-k}^{\mu, \mu/2b}(Q_T), \quad \psi_q \in \overset{\circ}{C}{}_{-k-\sigma_q}^{\mu-\sigma_q, (\mu-\sigma_q)/2b}(S_T)$$

in problem (17), and (18) will be preserved.

Theorem 2 is proved by the same argument.

Bibliography

1. V. A. Solonnikov, *On boundary value problems for linear parabolic systems of differential equations of general form*, Trudy Mat. Inst. Steklov. 83 (1965); English transl., Proc. Steklov Math. Inst. 83 (1965).

2. ______, *On an estimate for the maxima of the absolute values of the derivatives of the solution of a homogeneous parabolic initial-boundary value problem*, Preprint P-2-77, Leningrad. Otdel. Mat. Inst. Steklov., Leningrad, 1977. (Russian)

3. V. S. Belonosov and T. I. Zelenjak, *Nonlocal problems in the theory of quasilinear parabolic equations*, Novosibirsk. Gos. Univ., Novosibirsk, 1975. (Russian)

4. S. D. Ivasišen, *A study of the Green matrix of a general inhomogeneous parabolic boundary-value problem*, Methods of Functional Anal. in Problems of Math. Phys., Inst. Mat. Akad. Nauk Ukrain. SSR, Kiev, 1971, pp. 5–74. (Russian)

Translated by P. C. FIFE

Труды Матем. Инст. Стеклов.
147(1980)

Proceedings of the Steklov
Institute of Mathematics
1981, Issue 2

A NONSTATIONARY PROBLEM OF MAGNETOHYDRODYNAMICS FOR THE CASE OF TWO SPACE VARIABLES

UDC 517.946

L. I. STUPJALIS

ABSTRACT. This paper studies a two-dimensional nonlinear nonstationary problem of magnetohydrodynamics in which displacement currents cannot be neglected. The unique solvability of an initial-boundary value problem for this system of equations is proved.

Bibliography: 15 titles.

Introduction

The problem to be studied in this paper is a continuation of the problem presented in [1]. Here we consider two-dimensional nonlinear nonstationary problems of magnetohydrodynamics in which displacement currents cannot be neglected in comparison with conduction currents. By two-dimensional we mean problems in which the vectors $\mathbf{v}$ and $\mathbf{H}$ have x_3 components equal to zero and in which all the data of the problem and the unknown quantities are independent of x_3. In this case the motion of a conducting viscous incompressible fluid, placed in an electromagnetic field, is described by the following set of equations:*

$$\frac{\partial \mathbf{v}}{\partial t} - \nu\Delta\mathbf{v} + \sum_{k=1}^{2} v_k \frac{\partial \mathbf{v}}{\partial x_k} - \frac{\mu}{\rho}\sum_{k=1}^{2} H_k \frac{\partial \mathbf{H}}{\partial x_k} + \frac{1}{\rho}\operatorname{grad}\left(p + \frac{\mu H^2}{2}\right) = \mathbf{f}, \tag{1}$$

$$\operatorname{div}\mathbf{v} = 0, \tag{2}$$

$$\varepsilon\frac{\partial \mathbf{E}}{\partial t} - \operatorname{rot}\mathbf{H} + \sigma[\mathbf{E} + \mu(\mathbf{v}\times\mathbf{H})] = \mathbf{j}, \tag{3}$$

$$\mu\frac{\partial \mathbf{H}}{\partial t} + \operatorname{rot}\mathbf{E} = 0, \tag{4}$$

$$\operatorname{div}\mathbf{H} = 0, \tag{5}$$

where

$$\mathbf{E} = E_3\mathbf{e}_3,\quad \operatorname{rot}\mathbf{E} = \frac{\partial E_3}{\partial x_2}\mathbf{e}_1 - \frac{\partial E_3}{\partial x_1}\mathbf{e}_2,\quad \operatorname{rot}\mathbf{H} = \left(\frac{\partial H_2}{\partial x_1} - \frac{\partial H_1}{\partial x_2}\right)\mathbf{e}_3,\quad \mathbf{j} = j\mathbf{e}_3.$$

Here $\mathbf{v}(x, t)$ is the fluid velocity; $\mathbf{H}(x, t)$ and $\mathbf{E}(x, t)$ are the magnetic and electric field intensities; $p(x, t)$ is the pressure; $\mathbf{f}(x, t)$ and $\mathbf{j}(x, t)$ are specified external hydrodynamic forces and currents; ε is the dielectric constant; μ is the magnetic permeability; σ is the conductivity of the medium; ν is the kinematic viscosity of the fluid; and, finally, ρ is the fluid density.

1980 *Mathematics Subject Classification*. Primary 35A05, 35Q99, 35D05, 76W05.

* *Editor's note*. In the Russian literature, rot is used in place of curl.

We assume that the fluid is enclosed in a bounded container Ω, the surface S of which is an ideal conductor subjected to the action of known hydrodynamic and electromagnetic forces **f** and **j**. The problem consists in determining **v**, p, **E**, and **H** in the domain $Q_T = \Omega \times [0, T]$ ($x \in \Omega$, $t \in [0, T]$) from the system of equations (1)–(5), the boundary conditions

$$\mathbf{v}|_S = 0, \quad H_n|_S = 0, \quad E_3|_S = 0, \quad t \in [0, T], \tag{6}$$

where $H_n = \mathbf{H} \cdot \mathbf{n} = H_1 n_1 + H_2 n_2 + H_3 n_3$ are the normal components of the vector **H** on S, and **n** is the normal to S, exterior to Ω; and the initial conditions

$$\mathbf{v}|_{t=0} = \mathbf{v}_0(x), \quad \mathbf{E}|_{t=0} = \mathbf{E}_0(x), \quad \mathbf{H}|_{t=0} = \mathbf{H}_0(x). \tag{7}$$

In this paper we prove the solvability of the initial-boundary value problem (1)–(7).

§1. Definitions and auxiliary propositions

We now introduce the function spaces we need to study the initial-boundary value problem (1)–(7).

$L_p(\Omega)$, $p > 1$, is the Banach space consisting of all functions (or vector-valued functions), defined and Lebesgue measurable on the bounded domain Ω of euclidean space R_2, and having the finite norm

$$\|u\|_{p,\Omega} = \left(\int_\Omega |u(x)|^p\,dx\right)^{1/p} \quad \left(\|\mathbf{u}\|_{p,\Omega} = \left(\int_\Omega |\mathbf{u}(x)|^p\,dx\right)^{1/p}\right).$$

$W_p^{(l)}(\Omega)$ ($p > 1$, l integral) is the space of functions (or vector-valued functions) belonging to $L_p(\Omega)$ and having generalized derivatives belonging to $L_p(\Omega)$ to order l inclusive; the norm in $W_p^l(\Omega)$ is given by

$$\|u\|_{p,\Omega}^{(l)} = \sum_{|\alpha| \leq l} \|D^\alpha u\|_{p,\Omega}.$$

$W_p^{2,1}(Q_T)$ ($p > 1$, $Q_T = \Omega \times [0, T]$) is the space of functions (or vector-valued functions) defined in Q_T which, together with their generalized derivatives of the first order with respect to t and their generalized derivatives of the first and second orders with respect to x_k, belong to $L_p(Q_T)$; the norm in $W_p^{2,1}(Q_T)$ is defined as follows:

$$\|u\|_{p,Q_T}^{(2,1)} = \|u_t\|_{p,Q_T} + \|u_{xx}\|_{p,Q_T} + \|u_x\|_{p,Q_T} + \|u\|_{p,Q_T}.$$

$W_{p,0}^{2,1}(Q_T)$ is the subspace of $W_p^{2,1}(Q_T)$ consisting of all vector-valued functions which vanish on S_T, where $S_T = \{(x, t)\colon x \in S, t \in [0, T]\}$ is the lateral surface of Q_T.

$W_p^{2-2/p}(\Omega)$ is the space of L. N. Slobodeckiĭ, V. A. Solonnikov, and O. V. Besov, and consists of all vector-valued functions, measurable on Ω, with the finite norm

$$\|\mathbf{u}\|_{p,\Omega}^{(2-2/p)} = \|\mathbf{u}\|_{p,\Omega} + \left(\int_\Omega\int_\Omega \frac{|\mathbf{u}(x) - \mathbf{u}(y)|^p}{|x - y|^{2p}}\,dx\,dy\right)^{1/p},$$

where $p \in (1, 2)$. It is known (see [2], Theorem 2, and [3], Theorem 2.1) that the traces of vector-valued functions $\mathbf{u}(x, t)$ from $W_p^{2,1}(Q_T)$ are defined on each section Ω_{t_1} of the cylinder Q_{T_2} by the plane $t = t_1 \in [0, T]$ as vector-valued functions from $W_p^{2-2/p}(\Omega_{t_1})$, and that they vary continuously with respect to t in the $W_p^{2-2/p}(\Omega)$ norm as t varies in $[0, T]$.

We also define the norm

$$|\mathbf{u}|_{p,\Omega}^{(2-2/p)} = \begin{cases} \|\mathbf{u}\|_{p,\Omega}^{(2-2/p)} & \text{for } p \neq \frac{3}{2}, \\ \|\mathbf{u}\|_{3/2,\Omega}^{(2/3)} + \left(\int_\Omega |\mathbf{u}(x)|^{3/2} \rho^{-1}(x)\, dx\right)^{2/3} & \text{for } p = \frac{3}{2}, \end{cases}$$

where $\rho(x)$ is the distance from x to $\partial\Omega$, and we denote by $W_{p,0}^{2-2/p}(\Omega)$ the closure of the set of vector-valued functions from the class $C^2(\bar{\Omega})$ which vanish on S in this norm. In the limiting case $p = 3/2$, in the definition of the norm $|\mathbf{u}|_{p,\Omega}^{(2-2/p)}$, the term $(\int_\Omega |\mathbf{u}(x)|^{3/2}\rho^{-1}(x)\, dx)^{2/3}$ expresses the conditions of compatibility (for $p < 3/2$ there are no compatibility conditions). This is consistent with the fact that for an arbitrary vector-valued function $\mathbf{u}(x, t) \in W_{3/2,0}^{2,1}(Q_T)$ we have

$$\left(\int_\Omega |\mathbf{u}(x, 0)|^{3/2} \rho^{-1}(x)\, dx\right)^{2/3} \leqslant C_1 \|\mathbf{u}\|_{3/2, Q_T}^{(2,1)}$$

(see [4], the proof of Theorem 1). Conversely, an arbitrary vector-valued function $\mathbf{v}(x) \in W_{3/2,0}^{(2/3)}(\Omega)$ can be extended onto the cylinder Q_T, so that $\mathbf{u}(x, t) \in W_{3/2,0}^{2,1}(Q_T)$, $\mathbf{u}|_{t=0} = \mathbf{v}(x)$, and $\|\mathbf{u}\|_{3/2,Q_T}^{(2,1)} < C_2 |\mathbf{v}|_{3/2,\Omega}^{(2/3)}$, with the constant C_2 being independent of $\mathbf{v}$ (see [2], §5).

It is known (see, for example, [5], Chapter I, §2.2) that $L_2(\Omega) = G(\Omega) \oplus \mathring{J}(\Omega) = \mathring{G}(\Omega) \oplus J(\Omega)$, where $G(\Omega)$ and $\mathring{G}(\Omega)$ are spaces of vector-valued functions representable in the form $\mathbf{u} = \operatorname{grad} p$, $p(x) \in W_2^1(\Omega)$, and $\mathbf{v} = \operatorname{grad} q$, $q(x) \in \mathring{W}_2^1(\Omega)$, respectively; $J(\Omega)$ is the closure in $L_2(\Omega)$ of the set of smooth solenoidal vector-valued functions, while $\mathring{J}(\Omega)$ is the closure of the set of smooth solenoidal vector-valued functions satisfying the condition

$$u_n|_S = 0. \tag{8}$$

$J^l(\Omega)$ is the subspace of $W_2^l(\Omega)$ consisting of all solenoidal vector-valued functions.

$\mathring{J}_n^l(\Omega)$ is the subspace of vector-valued functions from $J^l(\Omega)$ which satisfy the condition (8) on S.

The following results are known (see [6], p. 138).

THEOREM 1. *If $S \in C^2$, then the operator* rot *establishes a one-to-one correspondence between the spaces $\mathring{W}_2^1(\Omega)$ and $\mathring{J}(\Omega)$, under which for all $V \in \mathring{W}_2^1(\Omega)$*

$$\|\operatorname{rot} V\|_{2,\Omega} \leqslant \|V\|_{2,\Omega}^{(1)} \leqslant \left(\sqrt{2} + \frac{1}{\sqrt{\mu_1}}\right) \|\operatorname{rot} V\|_{2,\Omega}, \tag{9}$$

where the number μ_1 is the smallest eigenvalue of the operator $-\Delta$ in the domain Ω subject to the zero boundary condition.

THEOREM 2. *If $S \in C^2$, then the operator* rot *establishes a one-to-one correspondence between the spaces $\mathring{J}_n^1(\Omega)$ and $L_2(\Omega)$, under which for all $\psi = (\psi_1, \psi_2) \in \mathring{J}_n^1(\Omega)$*

$$\|\operatorname{rot} \psi\|_{2,\Omega} \leqslant \|\psi\|_{2,\Omega}^{(1)} \leqslant C_3 \|\operatorname{rot} \psi\|_{2,\Omega}, \tag{10}$$

where the constant C_3 does not depend on ψ.

We consider the following auxiliary boundary value problem:

$$\operatorname{rot}\operatorname{rot} \mathbf{h} = \mathbf{g}, \quad \operatorname{div} \mathbf{h} = 0, \tag{11}$$

$$h_n|_S = 0, \quad \operatorname{rot} \mathbf{h}|_S = 0. \tag{12}$$

The problem (11), (12) reduces to the two problems

$$\operatorname{rot} V = \mathbf{g}, \quad V|_S = 0,$$

$$\operatorname{rot} \mathbf{h} = V, \quad \operatorname{div} \mathbf{h} = 0, \quad h_n|_S = 0.$$

If $S \in C^2$ and $\mathbf{g} \in \mathring{J}(\Omega)$, then, by virtue of Theorems 1 and 2, both problems are uniquely solvable. Further, following the reasoning from [6] (see the proof of Theorem 11, p.138) and relying on Theorem 1 of [7] (p. 68; see also [8], pp. 23–29), it is not difficult to show that if $S \in C^3$, then the operator rot establishes a one-to-one correspondence between the spaces $\mathring{J}_n^2(\Omega)$ and $W_2^1(\Omega)$, and we then have, for $\psi \in \mathring{J}_n^2(\Omega)$, the two-sided estimate

$$\frac{1}{2}\|\operatorname{rot} \psi\|_{2,\Omega}^{(1)} \leqslant \|\psi\|_{2,\Omega}^{(2)} \leqslant C_4 \|\operatorname{rot} \psi\|_{2,\Omega}^{(1)},$$

where the positive constant C_4 does not depend on ψ. Consequently

$$\begin{aligned} \|\mathbf{g}\|_{2,\Omega} = \|\operatorname{rot} V\|_{2,\Omega} = \|\operatorname{rot}\operatorname{rot} \mathbf{h}\|_{2,\Omega} \leqslant \|\mathbf{h}\|_{2,\Omega}^{(2)} \\ \leqslant C_4 \|V\|_{2,\Omega}^{(1)} \leqslant C_4\left(\sqrt{2} + \frac{1}{\sqrt{\mu_1}}\right)\|\mathbf{g}\|_{2,\Omega}. \end{aligned} \tag{13}$$

Thus, the operator $\mathfrak{L} = \operatorname{rot}\operatorname{rot}$ establishes a one-to-one correspondence between $\hat{J}(\Omega)$ and $\mathring{J}(\Omega)$, where $\hat{J}(\Omega)$ is the subspace of vector-valued functions from $J^2(\Omega)$ which satisfy (12) on S. From the equations

$$(\operatorname{rot}\operatorname{rot} \mathbf{h}, \mathbf{h}')_{2,\Omega} = (\operatorname{rot} \mathbf{h}, \operatorname{rot} \mathbf{h}')_{2,\Omega} = (\mathbf{h}, \operatorname{rot}\operatorname{rot} \mathbf{h}')_{2,\Omega}, \tag{14}$$

valid for all $\mathbf{h}, \mathbf{h}' \in \hat{J}(\Omega)$, and (13) and (10), we see that the operator is selfadjoint and positive definite.

In addition, it is evident from (14) and Theorem 5 from [9] (§205) that the domain of definition $D(\mathfrak{L}^{\frac{1}{2}})$ of the operator $\mathfrak{L}^{\frac{1}{2}}$ is the closure of $\hat{J}(\Omega)$ in the norm $\|\operatorname{rot} \mathbf{h}\|_{2,\Omega}$, which, by virtue of Theorem 2, is equivalent to the norm of $W_2^1(\Omega)$. It is obvious that $D(\mathfrak{L}^{\frac{1}{2}}) \subset \mathring{J}_n^1(\Omega)$. Reasoning exactly as was done in [10] (see Russian p. 87, English p. 418), we conclude that $D(\mathfrak{L}^{\frac{1}{2}}) = \mathring{J}_n^1(\Omega)$. Indeed, let all the $\mathbf{h}$ belong to $\mathring{J}_n^1(\Omega)$. Then $\operatorname{rot} \mathbf{h} = V \in L_2(\Omega)$. An arbitrary function $V \in L_2(\Omega)$ can be approximated by functions $V^{(k)} \in \mathring{W}_2^1(\Omega)$ in the $L_2(\Omega)$ norm. Let $\mathbf{h}^{(k)}$ be a solution of the problem

$$\operatorname{rot} \mathbf{h}^{(k)} = V^{(k)}, \quad \operatorname{div} \mathbf{h}^{(k)} = 0, \quad h_n^{(k)}|_S = 0.$$

It is clear that $\mathbf{h}^{(k)} \in \hat{J}(\Omega)$. Since $V^{(k)} \Rightarrow V$ (in the $L_2(\Omega)$ norm), by virtue of Theorem 2 it follows that $\mathbf{h}^{(k)} \Rightarrow \mathbf{h}$ in the $W_2^1(\Omega)$ norm, which is what we wished to prove. Since $W_2^l(\Omega)$, $l \geqslant 1$, is compactly imbedded in $L_2(\Omega)$, the operators $\mathfrak{L}^{-1}$ and $\mathfrak{L}^{-\frac{1}{2}}$ are completely continuous. In view of the above, and also of the coercive estimates (13), the spectra of the operators $\mathfrak{L}$ and $\mathfrak{L}^{\frac{1}{2}}$ are discrete and positive, and the system $\{\varphi_k(x)\}$, $k = 1, 2, \ldots,$ of eigenelements forms a basis in $\mathring{J}(\Omega)$, $\mathring{J}_n^1(\Omega)$, and $\hat{J}(\Omega)$.

The orthogonal decompositions of $L_2(\Omega)$, which were introduced above, induce decompositions of $L_2(Q_T)$:

$$L_2(Q_T) = G(Q_T) \oplus \mathring{J}(Q_T) = \mathring{G}(Q_T) \oplus J(Q_T),$$

where $G(Q_T)$, $\mathring{J}(Q_T)$, $\mathring{G}(Q_T)$ and $J(Q_T)$ are subspaces of $L_2(Q_T)$ consisting of vector-valued functions belonging to $G(\Omega)$, $\mathring{J}(\Omega)$, $\mathring{G}(\Omega)$ and $J(\Omega)$ for almost all $t \in [0, T]$, respectively.

$B(Q_T)$ is the Banach space obtained by the closure of the set of all solenoidal vector-valued functions of the class $C_2(\overline{Q}_T)$ which satisfy (8) for all $t \in [0, T]$ in the norm

$$\|\mathbf{u}\|_{B(Q_T)} = \max_{0 \leqslant t \leqslant T} \left(\|\mathbf{u}_t(x, t)\|_{2,\Omega}^2 + \|\mathbf{u}(x, t)\|_{2,\Omega}^{(1)2} \right)^{1/2}.$$

The problem (1)–(7) can be reformulated, eliminating the vector-valued function $\mathbf{E}(x, t)$. Doing this, we obtain the following initial-boundary value problem for the vector-valued functions $\mathbf{v}$ and $\mathbf{H}$ and the function p:

$$\frac{\partial \mathbf{v}}{\partial t} - \nu\Delta\mathbf{v} + \sum_{k=1}^{2} v_k \frac{\partial \mathbf{v}}{\partial x_k} - \frac{\mu}{\rho} \sum_{k=1}^{2} H_k \frac{\partial \mathbf{H}}{\partial x_k} + \frac{1}{\rho} \operatorname{grad}\left(p + \frac{\mu \mathbf{H}^2}{2}\right) = \mathbf{f}, \tag{15}$$

$$\varepsilon\mu \frac{\partial^2 \mathbf{H}}{\partial t^2} + \operatorname{rot}\operatorname{rot}\mathbf{H} + \sigma\mu \frac{\partial \mathbf{H}}{\partial t} - \sigma\mu \operatorname{rot}(\mathbf{v} \times \mathbf{H}) = -\operatorname{rot}\mathbf{j}, \tag{16}$$

$$\operatorname{div}\mathbf{v} = 0, \quad \operatorname{div}\mathbf{H} = 0, \tag{17}$$

$$\mathbf{v}|_{t=0} = \mathbf{v}_0(x), \quad \mathbf{H}|_{t=0} = \mathbf{H}_0(x), \quad \left.\frac{\partial \mathbf{H}}{\partial t}\right|_{t=0} = -\frac{1}{\mu} \operatorname{rot}\mathbf{E}_0, \tag{18}$$

$$\mathbf{v}|_S = 0, \quad H_n|_S = 0, \quad \operatorname{rot}\mathbf{H}|_S = -\mathbf{j}|_S, \quad t \in [0, T]. \tag{19}$$

After this, we can determine $\mathbf{E}(x, t)$ as the solution of the problem

$$\operatorname{rot}\mathbf{E} = -\mu \frac{\partial \mathbf{H}}{\partial t},$$

$$\mathbf{E}|_S = 0, \quad t \in [0, T].$$

§2. Linear problem

If in (15) and (16) we discard the nonlinear terms, the problem (15)–(19) breaks up into two initial-boundary value problems:

$$\frac{\partial \mathbf{v}}{\partial t} - \nu\Delta\mathbf{v} + \frac{1}{\rho} \operatorname{grad} p = \mathbf{f}, \quad \operatorname{div}\mathbf{v} = 0, \tag{20}$$

$$\mathbf{v}|_S = 0, \quad \mathbf{v}|_{t=0} = \mathbf{v}_0(x) \tag{21}$$

and

$$\varepsilon\mu\frac{\partial^2 \mathbf{H}}{\partial t^2}+\operatorname{rot}\operatorname{rot}\mathbf{H}+\sigma\mu\frac{\partial \mathbf{H}}{\partial t}=-\operatorname{rot}\mathbf{j},\ \operatorname{div}\mathbf{H}=0, \tag{22}$$

$$H_n|_S=0,\ \operatorname{rot}\mathbf{H}|_S=-\mathbf{j}|_S,\ \mathbf{H}|_{t=0}=\mathbf{H}_0(x),\ \left.\frac{\partial \mathbf{H}}{\partial t}\right|_{t=0}=-\frac{1}{\mu}\operatorname{rot}\mathbf{E}_0. \tag{23}$$

We first consider the problem (22), (23). Let $\mathring{J}_n^1(Q_T)$ denote the subspace of $W_2^1(Q_T)$ in which the smooth solenoidal vector-valued functions satisfying the condition $u_n|_{S_T}=0$ are dense. Following O. A. Ladyženskaja, by a generalized solution of the problem (22), (23) we shall mean a vector-valued function $\mathbf{H}(x,t)$ which belongs to $\mathring{J}_n^1(Q_T)$, takes on the initial value $\mathbf{H}_0(x)$, and satisfies the integral identity

$$\int_{Q_T}\left(-\varepsilon\mu\frac{\partial \mathbf{H}}{\partial t}\cdot\frac{\partial \psi}{\partial t}+\operatorname{rot}\mathbf{H}\cdot\operatorname{rot}\psi+\sigma\mu\frac{\partial \mathbf{H}}{\partial t}\cdot\psi+\mathbf{j}\cdot\operatorname{rot}\psi\right)dx\,dt$$
$$+\varepsilon\int_\Omega \operatorname{rot}\mathbf{E}_0\cdot\psi\,dx\,|_{t=0}=0 \tag{24}$$

for an arbitrary vector-valued function $\psi(x,t)$ from $\mathring{J}_n^1(Q_T)$, equal to zero for $t=T$.

LEMMA 1. *The problem* (22), (23) *cannot have two distinct generalized solutions from* $\mathring{J}_n^1(Q_T)$.

This lemma may be proved in precisely the same way as was Theorem 1 in [11].

We call $\varphi_k(x)$ a generalized vector-valued eigenfunction of the problem

$$\operatorname{rot}\operatorname{rot}\varphi_k=\lambda_k\varphi_k,\ \operatorname{div}\varphi_k=0, \tag{25}$$

$$\varphi_{kn}|_S=0,\quad \operatorname{rot}\varphi_k|_S=0, \tag{26}$$

corresponding to the eigenvalue λ_k, if it is not identically zero, belongs to $\mathring{J}_n^1(\Omega)$, and for all vector-valued functions $\psi(x)$ from $\mathring{J}_n^1(\Omega)$ satisfies the integral identity

$$\int_\Omega \operatorname{rot}\varphi_k\cdot\operatorname{rot}\psi\,dx=\lambda_k\int_\Omega\varphi_k\cdot\psi\,dx. \tag{27}$$

We show now that if the generalized vector-valued eigenfunction $\varphi_k(x)\in J^2(\Omega)$, then (25) is satisfied almost everywhere in Ω and the conditions (26) are satisfied in the sense of imbedding theorems. Actually, if in (27) we substitute smooth solenoidal functions ψ having compact support in Ω, and then integrate by parts, we obtain

$$\int_\Omega(\operatorname{rot}\operatorname{rot}\varphi_k-\lambda_k\varphi_k)\cdot\psi\,dx=0;$$

consequently $\operatorname{rot}\operatorname{rot}\varphi_k-\lambda_k\varphi_k\in G(\Omega)$. Now let ψ be an arbitrary vector-valued function from $\mathring{J}_n^1(\Omega)$. Taking note of the fact that $\operatorname{rot}\operatorname{rot}\varphi_k-\lambda_k\varphi_k\in G(\Omega)$, and again integrating by parts in (27), we obtain

$$\int_S(\operatorname{rot}\varphi_k\times\mathbf{n})\cdot\psi\,dS=0,\quad \psi\in\mathring{J}_n^1(\Omega). \tag{28}$$

It follows from this that $\operatorname{rot} \varphi_k|_S = 0$. Actually, as shown in [5] (see Chapter I, §2.1), we can construct a vector-valued function $\psi(x) \in \mathring{J}_n^1(\Omega)$, which takes on the value $\operatorname{rot} \varphi_k \times \mathbf{n}$ on the boundary S. For this vector-valued function the relation (28) becomes

$$\int_S |(\operatorname{rot} \varphi_k \times \mathbf{n})|^2 \, dS = 0.$$

Further, by virtue of Theorems 1 and 2, we have $\operatorname{rot}\operatorname{rot} \varphi_k - \lambda_k \varphi_k \in \mathring{J}(\Omega)$. Consequently

$$\operatorname{rot}\operatorname{rot} \varphi_k - \lambda_k \varphi_k = 0.$$

The smoothness of the generalized vector-valued eigenfunctions is determined by the smoothness of the surface S.

The problem of finding these eigenfunctions is equivalent to the variational problem for the functional

$$f(\varphi) = \frac{\| \operatorname{rot} \varphi \|_{2,\Omega}^2}{\| \varphi \|_{2,\Omega}^2}.$$

Just as was done in [11] (see pp. 44–46; see also [12], pp. 195–198), we can establish the validity of the following propositions: If $S \in C^2$, then the variational problem has the infinite set of eigenvalues[1]

$$\frac{1}{C_3^2} \leqslant \lambda_1 \leqslant \lambda_2 \leqslant \ldots \leqslant \lambda_n \ldots; \quad \lim_{n \to \infty} \lambda_n = +\infty,$$

where C_3 is the constant from (10). The corresponding generalized vector-valued eigenfunctions are orthogonal in $L_2(\Omega)$ and in the new inner product $[\varphi, \psi] = (\operatorname{rot} \varphi, \operatorname{rot} \psi)_{2,\Omega}$ of the space $\mathring{J}_n^1(\Omega)$. The system $\{\varphi_k\}$ of eigenfunctions is complete in $\mathring{J}_n^1(\Omega)$ and in $\mathring{J}(\Omega)$. In view of all we have said, the solution of the nonstationary problem (22), (23) can be found with the aid of the Fourier method.

LEMMA 2. *Let $S \in C^2$, $\mathbf{j}(x, t) \in W_{2,0}^{1,0}(Q_T)$,[2] $\mathbf{H}_0(x) \in \mathring{J}_n^1(\Omega)$ and $\mathbf{E}_0(x) \in \mathring{W}_2^1(\Omega)$. Then the problem* (22), (23) *has a generalized solution which belongs to the space $B(Q_T)$ and satisfies the inequality*

$$\| \mathbf{H} \|_{B(Q_T)} \leqslant C_5 \left(\| \mathbf{j} \|_{2, Q_T}^{(1,0)} + \| \mathbf{H}_0 \|_{2,\Omega}^{(1)} + \| \mathbf{E}_0 \|_{2,\Omega}^{(1)} \right) \tag{29}$$

with the constant C_5 depending only on ε, μ, σ, and the domain Ω.

This lemma is proved in exactly the same way as Theorem 2 of [11] (see pp. 75–83).

We shall assume that $\mathbf{j}|_S = 0$.

---

(1) If a spectral value has multiplicity $m < \infty$, i.e. to it there correspond m linearly independent solutions, it is repeated m times in the sequence $\{\lambda_k\}$.

(2) $W_{2,0}^{1,0}(Q_T)$ is the Hilbert space consisting of the elements $u(x, t)$ of $L_2(Q_T)$ having generalized derivatives $\partial u / \partial x_i$, $i = 1, 2$, square-summable with respect to Q_T that satisfy the condition $u|_{S_T} = 0$. The inner product and the norm in this space are given by

$$(u, v)_{2,Q_T}^{(1,0)} = \int_{Q_T} (uv + u_x v_x) \, dx dt, \quad \| u \|_{2,Q_T}^{(1,0)} = \sqrt{(u, u)_{2,Q_T}^{(1,0)}}.$$

A generalized solution from $B(Q_T)$ (energy class) of the problem (22), (23) can be defined somewhat differently. Namely, $\mathbf{H}(x, t)$ is a generalized solution from $B(Q_T)$ of the problem (22), (23) if $\mathbf{H}(x, t)$ belongs to $B(Q_T)$, assumes the initial conditions by continuity in $\mathring{J}_n^1(\Omega)$ and $\mathring{J}(\Omega)$, respectively, and satisfies the identity

$$\varepsilon\mu \int_\Omega \frac{\partial \mathbf{H}}{\partial t} \cdot \psi dx \Big|_{t=t_1} + \int_{Q_{t_1}} \Big(-\varepsilon\mu \frac{\partial \mathbf{H}}{\partial t} \cdot \frac{\partial \psi}{\partial t} + \operatorname{rot} \mathbf{H} \cdot \operatorname{rot} \psi + \sigma\mu \frac{\partial \mathbf{H}}{\partial t} \cdot \psi$$
$$+ \operatorname{rot} \mathbf{j} \cdot \psi \Big) dx dt + \varepsilon \int_\Omega \operatorname{rot} \mathbf{E}_0 \cdot \psi dx \Big|_{t=0} = 0, \quad \forall t_1 \in [0, T],$$

where ψ is an arbitrary element of $\mathring{J}_n^1(Q_T)$ and $Q_{t_1} = \Omega \times [0, t_1]$, and the energy relation

$$\frac{1}{2} \int_\Omega \Big[\varepsilon\mu \Big(\frac{\partial \mathbf{H}}{\partial t} \Big)^2 + (\operatorname{rot} \mathbf{H})^2 \Big] dx \Big|_{t=0}^{t=t_1} + \int_{Q_{t_1}} \Big[\sigma\mu \Big(\frac{\partial \mathbf{H}}{\partial t} \Big)^2 + \operatorname{rot} \mathbf{j} \cdot \frac{\partial \mathbf{H}}{\partial t} \Big] dx dt = 0, \forall t_1 \in [0, T].$$

From the last equation we get the "energy estimate"

$$\| \mathbf{H} \|_{B(Q_T)} \leqslant \delta \Big\{ \int_\Omega \Big[\frac{(\operatorname{rot} \mathbf{E}_0)^2}{\mu^2} + (\operatorname{rot} \mathbf{H}_0)^2 \Big] dx \Big\}^{1/2} + \frac{2}{\gamma} \int_0^T \| \operatorname{rot} \mathbf{j} \|_{2,\Omega} dt, \tag{30}$$

where $\delta = 1/\varepsilon\mu$ for $\varepsilon\mu \leqslant 1$, and $\delta = \varepsilon\mu$ for $\varepsilon\mu > 1$; $\gamma = \min(\varepsilon\mu, 1)$. We denote by $W_{2,1,0}^{1,0}(Q_T)$ the Banach space obtained from the closure of the set of all smooth functions satisfying the condition $u|_{S_T} = 0$ in the norm

$$\| u \|_{2,1,Q_T}^{(1,0)} = \int_0^T \| u \|_{2,\Omega}^{(1)} dt.$$

LEMMA 2′. *Let* $S \in C^2$, $\mathbf{j}(x, t) \in W_{2,1,0}^{1,0}(Q_T)$, $\mathbf{H}_0 \in \mathring{J}_n^1(\Omega)$ *and* $\mathbf{E}_0 \in \mathring{W}_2^1(\Omega)$. *Then the problem* (22), (23) *has a unique generalized solution from* $B(Q_T)$ *such that*

$$\| \mathbf{H} \|_{B(Q_T)} \leqslant C_6 \big(\| \mathbf{j} \|_{2,1,Q_T}^{(1,0)} + \| \mathbf{H}_0 \|_{2,\Omega}^{(1)} + \| \mathbf{E}_0 \|_{2,\Omega}^{(1)} \big) \tag{31}$$

with the constant C_6 *depending only on* ε *and* μ.

This lemma is an immediate consequence of Lemmas 1 and 2 and the inequality (30).

We denote by $G_p(\Omega)$ the set of all vector-valued functions grad $\boldsymbol{\varphi} \in L_p(\Omega)$, and by $\mathring{J}_p(\Omega)$ we denote the set of all vector-valued functions $\mathbf{u}(x)$ satisfying the identity

$$\int_\Omega \mathbf{u} \cdot \operatorname{grad} \Phi dx = 0$$

for all grad $\Phi \in L_{p'}(\Omega)$. Let $G_p(Q_T)$ and $\mathring{J}_p(Q_T)$ be subspaces of $L_p(Q_T)$ consisting of vector-valued functions which belong for almost all t to $G_p(\Omega)$ and $\mathring{J}_p(\Omega)$, and let P_G and P_J be projections onto G_p and $\mathring{J}_p$. It follows from Theorem 2.2 of [13] that

$$L_p(\Omega) = G_p(\Omega) \dotplus \mathring{J}_p(\Omega), \quad L_p(Q_T) = G_p(Q_T) \dotplus \mathring{J}_p(Q_T),$$

where

$$\| P_J \mathbf{f} \|_{p,\Omega} \leqslant C_7 \| \mathbf{f} \|_{p,\Omega}. \tag{32}$$

On the set of solenoidal vector-valued functions $\mathbf{v}(x)$ equal to zero on S and belonging to $W_p^2(\Omega)$, we define an operator A, putting

$$A\mathbf{v} = \nu P_J \Delta \mathbf{v}.$$

It is known (see [5], Chapter III, §5 and Chapter II, §4, [14], §4, and [15], §2) that the operator A is closed, that its spectrum consists of eigenvalues which are all negative and can be enumerated in nondecreasing order of their moduli: $|\lambda_1| \leqslant |\lambda_2| \leqslant \cdots$.

We denote by $\mathring{J}_p^{2-2/p}(\Omega)$ the subspace of $W_{p,0}^{2-2/p}(\Omega)$ consisting of all solenoidal vector-valued functions.

The following theorem was proved in [13] (see §4) for the problem (20), (21).

THEOREM 3. *Let $S \in C^2$, $\mathbf{f}(x, t) \in L_p(Q_T)$ and $\mathbf{v}_0(x) \in \mathring{J}_p^{2-2/p}(\Omega)$. Then the problem* (20), (21) *has a unique solution $\mathbf{v}(x, t) \in W_p^{2,1}(Q_T)$, $\operatorname{grad} p \in L_p(Q_T)$, and*

$$\|\mathbf{v}\|_{p,Q_T}^{(2,1)} + \|\operatorname{grad} p\|_{p,Q_T} \leqslant C_8\left(\|\mathbf{f}\|_{p,Q_T} + |\mathbf{v}_0|_{p,\Omega}^{(2-2/p)}\right) \tag{33}$$

with the constant C_8 being independent of T.(3)

§3. Nonlinear problem

In this section we consider the problem (15)–(19), which we can also write in the form(4)

$$\frac{d\mathbf{v}}{dt} - A\mathbf{v} + K\mathbf{v} - Q\mathbf{H} = \mathbf{f}, \tag{34}$$

$$\mathbf{v}|_{t=0} = \mathbf{v}_0, \tag{35}$$

$$\varepsilon\mu \frac{\partial^2 \mathbf{H}}{\partial t^2} + \operatorname{rot}\operatorname{rot}\mathbf{H} + \sigma\mu \frac{\partial \mathbf{H}}{\partial t} - \sigma\mu \operatorname{rot}(\mathbf{v} \times \mathbf{H}) = -\operatorname{rot}\mathbf{j}, \quad \operatorname{div}\mathbf{H} = 0, \tag{36}$$

$$\mathbf{H}|_{t=0} = \mathbf{H}_0(x), \quad \left.\frac{\partial \mathbf{H}}{\partial t}\right|_{t=0} = -\frac{1}{\mu}\operatorname{rot}\mathbf{E}_0, \tag{37}$$

$$H_n|_S = 0, \quad \operatorname{rot}\mathbf{H}|_S = 0, \quad t \in [0, T], \tag{38}$$

where

$$K\mathbf{v} = P_J \sum_{k=1}^{2} v_k \frac{\partial \mathbf{v}}{\partial x_k}, \quad Q\mathbf{H} = \frac{\mu}{\rho} P_J \sum_{k=1}^{2} H_k \frac{\partial \mathbf{H}}{\partial x_k}.$$

We introduce the following notation:

$L_{q,r}(Q_T)$, for $1 \leqslant q, r \leqslant \infty$, is the space of vector-valued functions with the norm

$$\|\mathbf{u}\|_{q,r,Q_T} = \left(\int_0^T \|\mathbf{u}(x, t)\|_{q,\Omega}^r \, dt\right)^{1/r}.$$

$\mathring{J}_q^{2,1}(Q_T)$ is the subspace of $W_q^{2,1}(Q_T)$ consisting of all solenoidal vector-valued functions $\mathbf{v}(x,t)$ which are equal to zero on S_T.

LEMMA 3. *Let $u(x, t) \in W_q^{2,1}(Q_T)$. If $p, r \geqslant q > 1$ are such that*

$$\rho = 1 - \frac{2}{q} + \frac{1}{p} + \frac{1}{r} \geqslant 0$$

(3)Moreover, we have learned that the spectrum of the operator A is negative (see [13], §5, Remark 2).

(4)With no loss of generality we assume that in (15) $\mathbf{f}$ belongs to $\mathring{J}_p(Q_T)$.

or

$$\rho_1 = \frac{1}{2} - \frac{2}{q} + \frac{1}{p} + \frac{1}{r} \geqslant 0,$$

then, correspondingly,

$$\|\mathbf{u}\|_{p,r,Q_T} \leqslant C_9 \|\mathbf{u}\|^{(2,1)}_{q,Q_T} \tag{39}$$

or

$$\sum_{i=1}^{2} \|\mathbf{u}_{x_i}\|_{p,r,Q_T} \leqslant C_{10} \|\mathbf{u}\|^{(2,1)}_{q,Q_T} \tag{40}$$

with the constants C_9 and C_{10} being independent of T and **u**.

LEMMA 4. *For any $q \in [4/3, 2)$ and any* $\mathbf{v}, \mathbf{w} \in \mathring{J}^{2,1}_q(Q_T)$(5)

$$\|K(\mathbf{v}, \mathbf{w})\|_{q,Q_T} \leqslant C_{11} \min(1, T^{1/q-1/\sigma}) \big(\|\mathbf{v}\|^{(2,1)}_{q,Q_T} + |\mathbf{v}(x, 0)|^{(2-2/q)}_{q,\Omega}\big)\big(\|\mathbf{w}\|^{(2,1)}_{q,Q_T} + |\mathbf{w}(x, 0)|^{(2-2/q)}_{q,\Omega}\big), \tag{41}$$

where $0 < 1/\sigma < 1/q$, C_{11} is a constant independent of T, **v** *and* **w**, *and* $K(\mathbf{v}, \mathbf{w}) = P_J \Sigma^2_{k=1} v_k \partial \mathbf{w}/\partial x_k$ *is a bilinear form in* **v** *and* **w**.

Lemmas 3 and 4 are proved in the very same manner as in [13] (see the proof of Theorem 2.1 and the proof of Lemma 10.1).

It follows from imbedding theorems and from (32) that for any $q \in (1, 2)$ and **H**, $\mathbf{H}' \in B(Q_T)$

$$\|Q(\mathbf{H}', \mathbf{H})\|_{q,Q_T} \leqslant C_{12} \left\{ \int_0^T \|\mathbf{H}'\|^q_{2q/(2-q),\Omega} (\|\mathbf{H}\|^{(1)}_{2,\Omega})^q \, dt \right\}^{1/q} \leqslant C_{13} T^{1/q} \|\mathbf{H}'\|_{B(Q_T)} \|\mathbf{H}\|_{B(Q_T)}. \tag{42}$$

Here

$$Q(\mathbf{H}', \mathbf{H}) = \frac{\mu}{\rho} P_J \sum_{k=1}^{2} H'_k \frac{\partial \mathbf{H}}{\partial x_k}$$

is a bilinear form in **H**′ and **H**.

Let $\mathbf{v} \in \mathring{J}^{2,1}_q(Q_T)$ and $\mathbf{H} \in B(Q_T)$. Then $\mathbf{v} \times \mathbf{H}$ belongs to $W^{1,0}_{2,1,0}(Q_T)$, whence we have

$$\begin{aligned}
\|(\mathbf{v} \times \mathbf{H})\|^{(1,0)}_{2,1,Q_T} &\leqslant \left(\sqrt{2} + \frac{1}{\sqrt{\mu_1}}\right) \int_0^T \|\operatorname{rot}(\mathbf{v} \times \mathbf{H})\|_{2,\Omega} \, dt \\
&\leqslant 2\left(\sqrt{2} + \frac{1}{\sqrt{\mu_1}}\right) \left(\sum_{k=1}^{2} \int_0^T \left\| \frac{\partial \mathbf{v}}{\partial x_k} \right\|_{2q/(2-q),\Omega} \|\mathbf{H}\|_{2q/2q-2,\Omega} \, dt \right. \\
&\quad \left. + \sum_{k=1}^{2} \int_0^T \operatorname*{vrai\,max}_{x \in \Omega} |\mathbf{v}| \left\| \frac{\partial \mathbf{H}}{\partial x_k} \right\|_{2,\Omega} dt \right) \leqslant C_{14} \int_0^T \|\mathbf{v}\|^2_{q,\Omega} \|\mathbf{H}\|^1_{2,\Omega} \, dt \\
&\leqslant C_{14} T^{1/q'} \|\mathbf{v}\|^{(2,1)}_{q,Q_T} \|\mathbf{H}\|_{B(Q_T)}.
\end{aligned} \tag{43}$$

(5)For $T > 1$ the terms $|\mathbf{v}(x, 0)|^{(2-2/q)}_{q,\Omega}$ and $|\mathbf{w}(x, 0)|^{(2-2/q)}_{q,\Omega}$ are redundant in (41).

Here we have used Theorem 1, Hölder's inequality, and imbedding theorems.

The following lemma is known (see [13], Lemma 10.2).

LEMMA 5. *Let the numbers* $\xi_n \geqslant 0$, $n = 0, 1, \ldots$, *satisfy the recursion inequality* $\xi_{n+1} \leqslant a\xi_n^2 + b\xi_n + c$. *If* $a, c > 0$, $0 < b < 1$, $(1-b)^2 > 4ac$ *and*

$$\xi_0 \leqslant 2c\left(1-b+\sqrt{(1-b)^2-4ac}\right)^{-1},$$

then

$$\xi_n \leqslant 2c\left(1-b+\sqrt{(1-b)^2-4ac}\right)^{-1}$$

for all $n > 0$.

By a generalized solution of the problem (34)–(38) we mean a pair of vector-valued functions $\mathbf{v}$ and $\mathbf{H}$ having the following properties: $\mathbf{v}$ belongs to $\mathring{J}_q^{2,1}(Q_T)$ and satisfies the initial condition (35); $\mathbf{H}$ belongs to $B(Q_T)$ and assumes the initial value $\mathbf{H}_0(x)$; finally, $\mathbf{v}$ and $\mathbf{H}$ satisfy (34) and the identity

$$\int\limits_{Q_T}\Big[-\varepsilon\mu\frac{\partial \mathbf{H}}{\partial t}\cdot\frac{\partial \psi}{\partial t} + \operatorname{rot}\mathbf{H}\cdot\operatorname{rot}\psi + \sigma\mu\frac{\partial \mathbf{H}}{\partial t}\cdot\psi - \sigma\mu(\mathbf{v}\times\mathbf{H})\cdot\operatorname{rot}\psi + \operatorname{rot}\mathbf{j}\cdot\psi\Big]dx\,dt + \varepsilon\int\limits_{\Omega}\operatorname{rot}\mathbf{E}_0\cdot\psi\,dx\,\Big|_{t=0} = 0 \qquad (44)$$

for arbitrary $\psi(x, t)$ from $B(Q_T)$, equal to zero for $t = T$.

To prove a theorem for the existence of generalized solutions of the problem (34)–(38), we use the method of successive approximations in the following form:

$$\begin{gathered}
\frac{d\mathbf{v}^{(m)}}{dt} - A\mathbf{v}^{(m)} + K\mathbf{v}^{(m-1)} - Q\mathbf{H}^{(m-1)} = \mathbf{f}, \quad \mathbf{v}^{(m)}\big|_{t=0} = \mathbf{v}_0, \\
\varepsilon\mu\frac{\partial^2\mathbf{H}^{(m)}}{\partial t^2} + \operatorname{rot}\operatorname{rot}\mathbf{H}^{(m)} + \sigma\mu\frac{\partial\mathbf{H}^{(m)}}{\partial t} - \sigma\mu\operatorname{rot}(\mathbf{v}^{(m-1)}\times\mathbf{H}^{(m-1)}) = -\operatorname{rot}\mathbf{j}, \quad \operatorname{div}\mathbf{H}^{(m)} = 0, \\
\mathbf{H}^{(m)}\big|_{t=0} = \mathbf{H}_0(x), \quad \frac{\partial\mathbf{H}^{(m)}}{\partial t}\Big|_{t=0} = -\frac{1}{\mu}\operatorname{rot}\mathbf{E}_0, \\
H_n^{(m)}\big|_S = 0, \quad \operatorname{rot}\mathbf{H}^{(m)}\big|_S = 0, \quad t\in[0, T], \\
m = 1, 2, \ldots; \quad \mathbf{v}^{(0)} = \mathbf{H}^{(0)} = 0.
\end{gathered} \qquad (45)$$

Let

$$\begin{gathered}
S\in C^2, \quad \mathbf{j}(x, t)\in \mathring{J}_q(Q_T), \quad \mathbf{v}_0(x)\in \mathring{J}_q^{2-2/q}(\Omega), \quad 4/3\leqslant q<2, \\
\mathbf{j}(x, t)\in W_{2,1,0}^{1,0}(Q_T), \quad \mathbf{H}_0(x)\in \mathring{J}_n^1(\Omega), \quad \mathbf{E}_0(x)\in \mathring{W}_2^1(\Omega).
\end{gathered}$$

Then the problems (45) are uniquely solvable in $\mathring{J}_q^{2,1}(Q_T)\times B(Q_T)$. This follows from (41)–(43), Lemma 2′, and Theorem 3.

The inequalities (41)–(43), (33), and (31) guarantee the following estimates:

$$\|\mathbf{v}^{(m)}\|_{q,Q_T}^{(2,1)} \leqslant C_8\Big[\|\mathbf{f}\|_{q,Q_T} + |\mathbf{v}_0|_{q,\Omega}^{(2-2/q)} + C_{11}\min(1, T^{1/q-1/2})\big(\|\mathbf{v}^{(m-1)}\|_{q,Q_T}^{(2,1)} + |\mathbf{v}_0|_{q,\Omega}^{(2-2/q)}\big)^2 + C_{13}T^{1/q}\|\mathbf{H}^{(m-1)}\|_{B(Q_T)}^2\Big], \qquad (46)$$

$$\|\mathbf{H}^{(m)}\|_{B(Q_T)} \leqslant C_6\Big[\|\mathbf{j}\|_{2,1,Q_T}^{(1,0)} + \|\mathbf{H}_0\|_{2,\Omega}^{(1)} + \|\mathbf{E}_0\|_{2,\Omega}^{(1)} + C_{14}\sigma\mu T^{1/q'}\|\mathbf{v}^{(m-1)}\|_{q,Q_T}^{(2,1)}\|\mathbf{H}^{(m-1)}\|_{B(Q_T)}\Big]. \qquad (47)$$

Let $V_{m,T}$ denote the norm. It follows from (46) and (47) that

$$V_{m,T} \leqslant F + \{C_8[C_{11}\min(1,\ T^{1/q-1/\sigma}) + C_{13}T^{1/q}] + C_6C_{14}\sigma\mu T^{1/q'}\}\, V^2_{m-1,T},$$

where

$$F = C_8\|\mathbf{f}\|_{q,Q_T} + (C_8+1)\,|\mathbf{v}_0|^{(2-2/q)}_{q,\Omega} + C_6\left(\|\mathbf{j}\|^{(1,0)}_{2,1,Q_T} + \|\mathbf{H}_0\|^1_{2,\Omega} + \|\mathbf{E}_0\|^1_{2,\Omega}\right).$$

If

$$F\{C_8[C_{11}\min(1,\ T^{1/q-1/\sigma}) + C_{13}T^{1/q}] + C_6C_{14}\sigma\mu T^{1/q'}\} < \frac{1}{4},$$

then, on the basis of Lemma 5,

$$V_{m,T} \leqslant \frac{2F}{1+\sqrt{1-4F\{C_8[C_{11}\min(1,\ T^{1/q-1/\sigma}) + C_{13}T^{1/q}] + C_6C_{14}\sigma\mu T^{1/q'}\}}}. \tag{48}$$

We now prove that $\mathbf{v}^{(m)}$ and $\mathbf{H}^{(m)}$ converge in the corresponding norms. Thanks to (45), the differences $\mathbf{w}^{(m)} = \mathbf{v}^{(m)} - \mathbf{v}^{(m-1)}$ and $\mathbf{h}^{(m)} = \mathbf{H}^{(m)} - \mathbf{H}^{(m-1)}$ are a solution of the linear problem

$$\frac{d\mathbf{w}^{(m)}}{dt} - A\mathbf{w}^{(m)} + K(\mathbf{w}^{(m-1)},\ \mathbf{v}^{(m-1)}) + K(\mathbf{v}^{(m-2)},\ \mathbf{w}^{(m-1)})$$
$$- Q(\mathbf{h}^{(m-1)},\ \mathbf{H}^{(m-1)}) - Q(\mathbf{H}^{(m-2)},\ \mathbf{h}^{(m-1)}) = 0,$$
$$\mathbf{w}^{(m)}|_{t=0} = 0,$$

$$\varepsilon\mu\frac{\partial^2\mathbf{h}^{(m)}}{\partial t^2} + \operatorname{rot}\operatorname{rot}\mathbf{h}^{(m)} + \sigma\mu\frac{\partial\mathbf{h}^{(m)}}{\partial t} - \sigma\mu\operatorname{rot}(\mathbf{w}^{(m-1)}\times\mathbf{H}^{(m-1)})$$
$$- \sigma\mu\operatorname{rot}(\mathbf{v}^{(m-2)}\times\mathbf{h}^{(m-1)}) = 0,$$

$$\operatorname{div}\mathbf{h}^{(m)} = 0,$$

$$\mathbf{h}^{(m)}|_{t=0} = 0,\quad \left.\frac{\partial\mathbf{h}^{(m)}}{\partial t}\right|_{t=0} = 0,\quad \mathbf{h}^{(m)}_n|_{S_T} = 0,\quad \operatorname{rot}\mathbf{h}^{(m)}|_{S_T} = 0. \tag{49}$$

We denote the norm $\|\mathbf{w}^{(m)}\|^{(2,1)}_{q,Q_\tau} + \|\mathbf{h}^{(m)}\|_{B(Q_\tau)}$ by $W_{m,\tau}$, assuming that $\tau < T$. By virtue of (41)–(43), (33), and (31), we have

$$W_{m,\tau} \leqslant \{C_8[C_{11}\min(1,\ \tau^{1/q-1/\sigma})(\|\mathbf{v}^{(m-1)}\|^{(2,1)}_{q,Q_\tau} + \|\mathbf{v}^{(m-2)}\|^{(2,1)}_{q,Q_\tau} + 2|\mathbf{v}_0|^{(2-2/q)}_{q,\Omega})$$
$$+ C_{13}\tau^{1/q}(\|\mathbf{H}^{(m-1)}\|_{B(Q_\tau)} + \|\mathbf{H}^{(m-2)}\|_{B(Q_\tau)})] + C_6C_{14}\sigma\mu\tau^{1/q'}(\|\mathbf{H}^{(m-1)}\|_{B(Q_\tau)}$$
$$+ \|\mathbf{v}^{(m-2)}\|^{(2,1)}_{q,Q_\tau})\}\, W_{m-1,\tau}.$$

Since the uniform boundedness of the norms $V_{m,T}$ has already been proved, the convergence of $\Sigma_{m=1}^{\infty} W_{m,\tau}$ follows from this estimate, i.e. the sequences $\{\mathbf{v}^{(m)}, \mathbf{H}^{(m)}\}$ in the norm of $\mathring{J}^{2,1}_q(Q_\tau) \times B(Q_\tau)$ with some $\tau \leqslant T$ depending on the magnitude of the right side of (48). In this way the convergence of $\{\mathbf{v}^{(m)}, \mathbf{H}^{(m)}\}$ in $\mathring{J}^{2,1}_q(Q_{2\tau}) \times B(Q_{2\tau})$ is proved, etc. In a similar way we may also establish the uniqueness of the solution in $\mathring{J}^{2,1}_q(Q_T) \times B(Q_T)$.

We observe that the just proven convergence of $\{\mathbf{v}^{(m)}, \mathbf{H}^{(m)}\}$ in the norm of $\mathring{J}^{2,1}_q(Q_T) \times B(Q_T)$ makes it possible to prove that the desired generalized solution of our problem is the limit of $\{\mathbf{v}^{(m)}, \mathbf{H}^{(m)}\}$. We have thus proved the following theorem.

THEOREM 4. *Let*

$$S \in C^2,\ \mathbf{f}(x,\ t) \in \mathring{J}_q(Q_T),\ \mathbf{v}_0(x) \in \mathring{J}_q^{2-2/q}(\Omega),\ 4/3 \leqslant q < 2,$$
$$\mathbf{j} \in W_{2,1,0}^{1,0}(Q_T),\ \mathbf{H}_0(x) \in \mathring{J}_n^1(\Omega),\ \mathbf{E}_0(x) \in \mathring{W}_2^1(\Omega).$$

If

$$F\omega(T) < \frac{1}{4},$$

then the problem (34)–(38) *is uniquely solvable for* $0 \leqslant t \leqslant T$ *in* $\mathring{J}_q^{2,1}(Q_T) \times B(Q_T)$, *and for this solution*

$$\| \mathbf{v} \|_{q, Q_T}^{(2,1)} + \| \mathbf{H} \|_{B(Q_T)} \leqslant \frac{2F}{1 + \sqrt{1 - 4F\omega(T)}},$$

where

$$\omega(T) = C_8 [C_{11} \min(1,\ T^{1/q - 1/\sigma}) + C_{13} T^{1/q}] + C_6 C_{14} \sigma\mu T^{1/q'}.$$

The author thanks V. A. Solonnikov for a discussion of the results.

BIBLIOGRAPHY

1. L. Stupjalis, *Nonstationary problems in magnetohydrodynamics*, Zap. Naučn. Sem. Leningrad. Otdel. Mat. Inst. Steklov. (LOMI) **52** (1975), 175–217; English transl. in J. Soviet Math. **9** (1978), no. 6.

2. V. A. Solonnikov, *A priori estimates for second-order parabolic equations*, Trudy Mat. Inst. Steklov. **70** (1964), 133–212; English transl. in Amer. Math. Soc. Transl. (2) **65** (1967).

3. O. V. Besov, *Investigation of a family of function spaces in connection with theorems of imbedding and extension*, Trudy Mat. Inst. Steklov. **60** (1961), 42–81; English transl. in Amer. Math. Soc. Transl. (2) **40** (1964).

4. Š. Sahaev and V. A. Solonnikov, *Estimates for solutions of a boundary-value problem in magnetohydrodynamics*, Trudy Mat. Inst. Steklov. **127** (1975), 76–92; English transl. in Proc. Steklov Inst. Math. **127** (1975).

5. O. A. Ladyženskaja, *Mathematical problems in the dynamics of a viscous incompressible fluid*, 2nd rev. aug. ed., "Nauka", Moscow, 1970; English transl. of 1st ed., *The mathematical theory of viscous incompressible flow*, Gordon and Breach, New York, 1963; rev. 1969.

6. O. A. Ladyženskaja and V. A. Solonnikov, *Solution of some nonstationary problems of magnetohydrodynamics for a viscous incompressible fluid*, Trudy Mat. Inst. Steklov. **59** (1960), 115–173. (Russian)

7. O. A. Ladyženskaja, *On integral estimates, convergence, approximate methods, and solutions in functionals for linear elliptic operators*, Vestnik Leningrad. Univ. **1958**, no. 7 (Ser. Mat. Meh. Astr. vyp. 2), 60–69. (Russian)

8. ______, *A simple proof of the solvability of the fundamental boundary-value problems and of a problem on eigenvalues for linear elliptic equations*, Vestnik Leningrad. Univ. **1955**, no. 11, 23–29. (Russian)

9. V. I. Smirnov, *Course in higher mathematics*, Vol. 5, rev. ed., "Nauka", Moscow, 1959; English transl., Pergamon Press, Oxford, and Addison-Wesley, Reading, Mass., 1964.

10. O. A. Ladyženskaja and V. A. Solonnikov, *On the linearization principle and invariant manifolds for problems in magnetohydrodynamics*, Zap. Naučn. Sem. Leningrad. Otdel. Mat. Inst. Steklov. (LOMI) **38** (1973), 46–93; English transl. in J. Soviet Math. **8** (1977), no. 4.

11. O. A. Ladyženskaja, *The mixed problem for a hyperbolic equation*, GITTL, Moscow, 1953. (Russian)

12. S. G. Mihlin, *Variational methods in mathematical physics*, GITTL, Moscow, 1957; English transl., Macmillan, 1964.

13. V. A. Solonnikov, *Estimates for solutions of nonstationary Navier-Stokes equations*, Zap. Naučn. Sem. Leningrad. Otdel. Mat. Inst. Steklov. (LOMI) **38** (1973), 153–231; English transl. in J. Soviet Math. **8** (1977), no. 4.

14. S. G. Kreĭn, *On functional properties of operators of vector analysis and hydrodynamics*, Dokl. Akad. Nauk SSSR **93** (1953), 969–972. (Russian)

15. V. I. Judovič, *On the stability of stationary flows of a viscous incompressible fluid*, Dokl. Akad. Nauk SSSR **161** (1965), 1037–1040; English transl. in Soviet Phys. Dokl. **10** (1965/66).

Translated by J. F. HEYDA

Труды Матем. Инст. Стеклов.
147(1980)

Proceedings of the Steklov
Institute of Mathematics
1981, Issue 2

ON A BOUNDARY VALUE PROBLEM FOR THE STATIONARY SYSTEM OF EQUATIONS OF MAGNETOHYDRODYNAMICS

UDC 517.946

L. I. STUPJALIS

ABSTRACT. In this paper a boundary value problem is considered for a nonlinear stationary system of equations of magnetohydrodynamics. The solvability of this problem is studied, and the dependence of the smoothness of the solutions on the smoothness of the data of the problem is examined. A model problem concerning the motion of a metal in the presence of a free surface is also examined.
Bibliography: 18 titles.

§1. Introduction

In this paper we consider the following boundary value problem for a stationary system of equations of magnetohydrodynamics in a bounded domain $\Omega \subset R^3$ with the boundary $S = S_1 \cup S_2$:*

$$-\nu\Delta \mathbf{v} + v_k \frac{\partial \mathbf{v}}{\partial x_k} - \frac{\mu}{\rho} H_k \frac{\partial \mathbf{H}}{\partial x_k} = -\frac{1}{\rho}\operatorname{grad}\left(p + \frac{\mu \mathbf{H}^2}{2}\right) + \mathbf{f}, \tag{1}$$

$$\operatorname{div} \mathbf{v} = 0, \quad \operatorname{div} \mathbf{H} = 0, \quad \operatorname{rot} \mathbf{E} = 0, \tag{2}$$

$$\operatorname{rot} \mathbf{H} = \sigma(\mathbf{E} + \mu[\mathbf{v} \times \mathbf{H}]) + \mathbf{j}_0, \tag{3}$$

$$\mathbf{v}\cdot\mathbf{n}|_{S_1} = 0, \quad \mathbf{t} - \mathbf{n}(\mathbf{t}\cdot\mathbf{n})|_{S_1} = 0, \quad \mathbf{v}|_{S_2} = 0, \tag{4}$$

$$\mathbf{H}\cdot\mathbf{n}|_S = \beta(x), \quad \mathbf{E}_\tau|_S = 0. \tag{5}$$

Here ν is the kinematic viscosity, ρ is the density, μ is the magnetic permeability of the fluid, σ is the conductivity, p is the pressure, $\mathbf{v}$ is the fluid velocity, $\mathbf{H}$ is the magnetic field intensity, $\mathbf{E}$ is the electric field intensity, $\mathbf{f}$ denotes the external forces acting on the fluid, $\mathbf{j}_0$ represents specified currents, $\mathbf{n}$ is the unit exterior normal vector to the surface S, and $\mathbf{t}$ is the vector with the components

$$t_i = \sum_{k=1}^{3} t_{ik}(x) n_k(x), \quad \mathbf{t}\cdot\mathbf{n} = \sum_{i=1}^{3} t_i(x) n_i(x), \quad t_{ik} = -\delta_i^k \frac{p}{\rho} + \nu\left(\frac{\partial v_i}{\partial x_k} + \frac{\partial v_k}{\partial x_i}\right);$$

$$\mathbf{E}_\tau = \mathbf{E} - \mathbf{n}(\mathbf{E}\cdot\mathbf{n}), \quad \mathbf{H}^2 = \mathbf{H}\cdot\mathbf{H}.$$

We shall assume that the following condition is satisfied:

$$\int_S \beta(x)\, dS = 0.$$

1980 *Mathematics Subject Classification*. Primary 35A05, 35J60, 35Q10; Secondary 76W05.

* *Editor's note*. In the Russian literature, rot is used instead of curl.

In [1] we studied a nonlinear stationary system of equations of magnetohydrodynamics subject to the first boundary condition on the velocity **v** and the conditions (5) on **H** and **E**. In the present paper we examine the same system of equations subject to the conditions (4). The investigation of the problem (1)–(5) requires a number of auxiliary propositions, the proof of which constitutes the main obstacle. We present these propositions in §2. Solvability of the problem (1)–(5) is studied in §3. Next, we show in §4 that smoothness properties of the solutions improve with improvement of the smoothness of the data of the problem, and that this improvement of the smoothness properties is of a local nature. Finally, in §5 we study a model problem concerning the motion of a molten metal under the action of a moving magnetic field in the presence of a free surface. The two-dimensional stationary problem with a free boundary for the Navier-Stokes equations was considered in [2].

§2. Auxiliary propositions

Let $\Omega \subset R^3$ be a bounded domain with boundary S. We recall that the norms in the function spaces $L_p(\Omega)$ and $W_p^l(\Omega)$ with integral $l > 0$ and $p > 1$ are defined by

$$\|u\|_{p,\Omega}=\left(\int_\Omega |u(x)|^p\,dx\right)^{1/p}, \quad \|u\|_{p,\Omega}^{(l)}=\sum_{j=0}^{l}\sum_{(j)}\|D^j u\|_{p,\Omega}.$$

In the case of nonintegral l the norm in $W_p^l(\Omega)$ is defined by

$$\|u\|_{p,\Omega}^{(l)}=\langle\!\langle u\rangle\!\rangle_{p,\Omega}^{(l)}+\|u\|_{p,\Omega}^{([l])},$$

where

$$\langle\!\langle u\rangle\!\rangle_{p,\Omega}^{(l)}=\sum_{j=0}^{[l]}\sum_{(j)}\left(\int_\Omega dx\int_\Omega \frac{|D^j u(x)-D^j u(y)|^p}{|x-y|^{n+p(l-[l])}}\,dy\right)^{1/p}.$$

We now define the spaces $W_p^l(S)$ with nonintegral l on the surface S. Let $S \in C^l$. It is well known (see [3], §4) that S can be covered by a finite number of sets $S^{(k)} \subset S$ $(k = 1, \dots, N)$, possessing the following properties:

1) For an arbitrary point $x \in S$ we can find an $S^{(k)}$ such that $x \in S^{(k)}$ and the distance $\rho(x, S \setminus S^{(k)}) > \delta$, where δ is a positive number independent of x.

2) Each $S^{(k)}$ is mapped onto the unit square σ by an invertible transformation of class C^l such that the ratio of the distance between two arbitrary points of S in the old and the new coordinates lies between two positive numbers.

Under the mapping of $S^{(k)}$ onto σ let the function $u(x)$ $(x \in S^{(k)})$ go over into $u^{(k)}(z)$ $(z \in \sigma)$. We put

$$\|u\|_{p,S}^{(l)}=\left\{\sum_{k=1}^{N}\left(\|u^{(k)}(z)\|_{p,\sigma}^{(l)}\right)^p\right\}^{1/p}.$$

It is known (see [3], §4) that the norms corresponding to various decompositions of S into the sets $S^{(k)}$, where these sets satisfy the properties indicated above, are mutually equivalent.

We shall employ the same symbols even in the case when $\mathbf{u}(x)$ is a vector-valued function with components $u_i(x)$, $i = 1, 2, 3$.

We consider the following auxiliary problem:

$$\operatorname{div} \mathbf{v} = \psi(x), \tag{6}$$

$$\mathbf{v} \cdot \mathbf{n}|_S = \alpha(x), \quad \mathbf{t} - \mathbf{n}(\mathbf{t} \cdot \mathbf{n})|_S = \mathbf{b}(x). \tag{7}$$

We assume the following conditions to be satisfied:

$$\int_\Omega \psi(x)\,dx = \int_S \alpha(x)\,dS, \quad \mathbf{b} \cdot \mathbf{n}|_S = 0. \tag{8}$$

This problem has infinitely many solutions; we show that there exists at least one solution of the problem (6), (7).

LEMMA 1. *Assume that the conditions* (8) *are satisfied; assume also that* $\alpha(x) \in W_2^{3/2}(S)$, $\mathbf{b}(x) \in W_2^{1/2}(S)$, $\psi(x) \in W_2^1(\Omega)$ *and* $S \in C^3$. *Then the problem* (6), (7) *has at least one solution from* $W_2^2(\Omega)$. *For this solution the inequality*

$$\|\mathbf{v}\|_{2,\Omega}^{(2)} \leqslant C_1(\|\psi\|_{2,\Omega}^{(1)} + \|\alpha\|_{2,S}^{(3/2)} + \|\mathbf{b}\|_{2,S}^{(1/2)}) \tag{9}$$

holds, where the constant C_1 *does not depend on* ψ, α *or* $\mathbf{b}$.

PROOF. We seek a solution of the problem (6), (7) in the form $\mathbf{v} = \operatorname{grad} \varphi + \operatorname{rot} \mathbf{d}$, where φ is a solution of the problem

$$\Delta\varphi = \psi(x), \quad \left.\frac{\partial\psi}{\partial n}\right|_S = \alpha(x). \tag{10}$$

By virtue of the first of the conditions (8) the problem (10) is solvable to within a constant term, which we fix by requiring that $\int_\Omega \varphi(x)\,dx = 0$. Further, under the conditions of the lemma the solution of this problem belongs to $W_2^3(\Omega)$ (see [4], and also [3], §2, Theorem 2 and §4) and satisfies the inequality

$$\|\varphi\|_{2,\Omega}^{(3)} \leqslant C_2(\|\psi\|_{2,\Omega}^{(1)} + \|\alpha\|_{2,S}^{(3/2)}) \tag{11}$$

where the constant C_2 does not depend on ψ or α.

Let $\omega^{(1)}$ and $\omega^{(2)}$ be unit vectors directed along the y_1 and y_2 axes of the local $\{y\}$ coordinate system with origin at the point $\xi \in S$. Then the conditions (7) can be rewritten in the form

$$\mathbf{v} \cdot \mathbf{n}|_{\xi \in S} = \alpha(\xi), \quad \mathbf{T} \cdot \boldsymbol{\omega}^{(l)}|_{\xi \in S} = \mathbf{b} \cdot \boldsymbol{\omega}^{(l)}|_{\xi \in S}, \quad l = 1, 2, \tag{12}$$

where $\mathbf{T}$ is a vector with the components

$$T_i = \nu \sum_{j=1}^{3} \left(\frac{\partial v_i}{\partial x_j} + \frac{\partial v_j}{\partial x_i}\right) n_j$$

along the coordinate axes x_i, $i = 1, 2, 3$.

We construct a vector-valued function $\mathbf{d}$ satisfying the following conditions:

$$\operatorname{rot} \mathbf{d} \cdot \mathbf{n}|_{\xi \in S} = 0, \quad \tilde{\mathbf{T}} \cdot \boldsymbol{\omega}^{(l)}|_{\xi \in S} = \tilde{\mathbf{b}} \cdot \boldsymbol{\omega}^{(l)}|_{\xi \in S}, \quad l = 1, 2, \tag{13}$$

where $\tilde{\mathbf{b}} = \mathbf{b} - \mathbf{h}$ and $\tilde{\mathbf{T}}$ and $\mathbf{h}$ are vectors with respective components

$$\tilde{T}_i = \nu \sum_{j=1}^{3} \left(\frac{\partial c_i}{\partial x_j} + \frac{\partial c_j}{\partial x_i} \right) n_j \text{ and } h_i = \nu \sum_{j=1}^{3} \left(\frac{\partial g_i}{\partial x_j} + \frac{\partial g_j}{\partial x_i} \right) n_j; \ \mathbf{g} = \operatorname{grad} \varphi, \ \mathbf{c} = \operatorname{rot} \mathbf{d}.$$

In S we construct a sufficiently fine partition of unity,

$$1 = \sum_{k=1}^{N} \psi_k(x) \tag{14}$$

(see, for example, [5], formula (1.8)). Let $\mathbf{b}^{(k)} = \psi_k \tilde{\mathbf{b}}$ ($k = 1, \ldots, N$). Next, we construct in Ω a vector $\mathbf{d}^{(k)}$ such that $\operatorname{rot} \mathbf{d}^{(k)} \equiv \mathbf{c}^{(k)}(x)$ will satisfy the conditions

$$\operatorname{rot} \mathbf{d}^{(k)} \cdot \mathbf{n} |_{\xi \in S} = 0, \quad \mathbf{T}^{(k)} \cdot \boldsymbol{\omega}^{(l)} |_{\xi \in S} = \mathbf{b}^{(k)} \cdot \boldsymbol{\omega}^{(l)} |_{\xi \in S}, \quad l = 1, 2, \tag{15}$$

where $\mathbf{T}^{(k)}$ is a vector with the components

$$T_i^{(k)} = \nu \sum_{j=1}^{3} \left(\frac{\partial c_i^{(k)}}{\partial x_j} + \frac{\partial c_j^{(k)}}{\partial x_i} \right) n_j.$$

Then $\sum_{k=1}^{N} \mathbf{c}^{(k)}(x)$ gives us the desired vector $\mathbf{c}(x)$. We now consider how it is possible to select $\mathbf{d}^{(k)}(x)$ on S in order to satisfy the conditions (15). If $\xi \in S \setminus S^{(k)}$, then in it $\mathbf{b}^{(k)} = 0$ and we can take $\mathbf{d}^{(k)}$, $\partial \mathbf{d}^{(k)}/\partial x_l$ and $\partial^2 \mathbf{d}^{(k)}/\partial x_l \partial x_m$ to be equal to zero. (Here $\{S^{(k)}\}$ ($k = 1, \ldots, N$) is a covering of S corresponding to the decomposition of unity (14).) Let $\xi \in S^{(k)}$. The conditions (15) in the local $\{y\}$ coordinate system with origin at ξ have the same form:

$$\frac{\partial \hat{d}_2^{(k)}}{\partial y_1} - \frac{\partial \hat{d}_1^{(k)}}{\partial y_2} = 0, \quad \frac{\partial^2 \hat{d}_3^{(k)}}{\partial y_2 \partial y_3} - \frac{\partial^2 \hat{d}_2^{(k)}}{\partial y_3^2} + \frac{\partial^2 \hat{d}_2^{(k)}}{\partial y_1^2} - \frac{\partial^2 \hat{d}_1^{(k)}}{\partial y_1 \partial y_2} = \hat{b}_1^{(k)},$$

$$\frac{\partial^2 \hat{d}_1^{(k)}}{\partial y_3^2} - \frac{\partial^2 \hat{d}_3^{(k)}}{\partial y_3 \partial y_1} + \frac{\partial^2 \hat{d}_2^{(k)}}{\partial y_1 \partial y_2} - \frac{\partial^2 \hat{d}_1^{(k)}}{\partial y_2^2} = \hat{b}_2^{(k)},$$

where $\hat{d}_m^{(k)}$ ($\hat{b}_m^{(k)}$) are the components of $\mathbf{d}^{(k)}$ ($\mathbf{b}^{(k)}$) along the coordinate axes y_m ($m = 1, 2, 3$). We satisfy these equations at the point ξ, putting

$$\hat{d}_m^{(k)} = 0, \quad \frac{\partial \hat{d}_m^{(k)}}{\partial y_p} = 0, \quad \frac{\partial^2 \hat{d}_m^{(k)}}{\partial y_p \partial y_q} = 0 \ (q < 3 \text{ for } m = 1, 2),$$

$$\frac{\partial^2 \hat{d}_1^{(k)}}{\partial y_3^2} = \hat{b}_2^{(k)}, \quad \frac{\partial^2 \hat{d}_2^{(k)}}{\partial y_3^2} = -\hat{b}_1^{(k)}.$$

It is easy to see that the values of $\mathbf{d}^{(k)}$, $\partial \mathbf{d}^{(k)}/\partial y_p$ and $\partial^2 \mathbf{d}^{(k)}/\partial y_p \partial y_q$ ($p, q = 1, 2, 3$), calculated at each point $\xi \in S$, do not contradict one another.

Once more we now go to the variables $\{x\}$. This leads us to equations at the point $\xi \in S$:

$$d_m^{(k)} = 0, \quad \frac{\partial d_m^{(k)}}{\partial n} = 0,$$

$$\frac{\partial^2 d_m^{(k)}}{\partial n^2} = \sum_{j=1}^{3} [\cos(y_1, x_m) \cos(y_2, x_j) - \cos(y_2, x_m) \cos(y_1, x_j)] b_j^{(k)} \tag{16}$$

$$(m = 1, 2, 3).$$

Here $d_m^{(k)}$ ($b_m^{(k)}$) are the components of $\mathbf{d}^{(k)}$ ($\mathbf{b}^{(k)}$) along the coordinate axes x_m ($m = 1, 2, 3$). By virtue of Theorem 2 proved in [3] (see §2 and also §4) we have $b_j^{(k)} \in W_2^{1/2}(S)$. Since the surface $S \in C^3$, we also have

$$\sum_{j=1}^{3} [\cos(y_1, x_m)\cos(y_2, x_j) - \cos(y_2, x_m)\cos(y_1, x_j)]\, b_j^{(k)} \in W_2^{1/2}(S).$$

Therefore there exist functions $d_m^{(k)}(x) \in W_2^3(\Omega)$ (see [3], loc. cit.) satisfying the boundary conditions (16) and the inequality

$$\sum_{m=1}^{3} \| d_m^{(k)} \|_{2,\Omega}^{(3)} \leqslant C_3 \sum_{j=1}^{3} \| b_j^{(k)} \|_{2,S}^{(1/2)}, \tag{17}$$

where C_3 depends only on Ω. The sum $\Sigma_{k=1}^{N} \mathbf{d}^{(k)}(x)$ gives the desired vector-valued function $\mathbf{d}(x)$.

Further, we have

$$\| \mathbf{d}(x) \|_{2,\Omega}^{(3)} \leqslant \sum_{k=1}^{N} \| \mathbf{d}^{(k)}(x) \|_{2,\Omega}^{(3)} \leqslant C_4 \sum_{j=1}^{3} (\| b_j \|_{2,S}^{(1/2)} + \| h_j \|_{2,S}^{(1/2)})$$
$$\leqslant C_5 (\| \psi \|_{2,\Omega}^{(1)} + \| \alpha \|_{2,S}^{(3/2)} + \| \mathbf{b} \|_{2,S}^{(1/2)}). \tag{18}$$

The estimate (9) follows from (11) and (18). This completes the proof of Lemma 1.

We consider the following boundary value problem for the linearized Navier-Stokes system of equations in the bounded domain $\Omega \subset R^3$ with the boundary S:

$$-\nu\Delta \mathbf{v} + \frac{1}{\rho} \operatorname{grad} p = \mathbf{f}, \quad \operatorname{div} \mathbf{v} = \psi(x), \tag{19}$$

$$\mathbf{v} \cdot \mathbf{n} |_S = \alpha(x), \quad \mathbf{t} - \mathbf{n}(\mathbf{t} \cdot \mathbf{n}) |_S = \mathbf{b}(x). \tag{20}$$

It is well known (see [6], §1) that the linearized stationary Navier-Stokes system of equations is elliptic in the sense of Douglis-Nirenberg. If we reason as in [6] (Russian pp. 672–673, English pp. 200–201) it is easily seen that the matrix of boundary conditions (20) satisfies the complementarity condition.

With no loss of generality we assume that Ω is not a domain of revolution about any constant vector $\mathbf{B}$.

We recall now the definition of the Hölder space $C^l(\Omega)$. Here l is always a nonintegral positive number. $C^l(\Omega)$ is a Banach space whose elements are functions $u(x)$ which in $\bar{\Omega}$ are continuous and have continuous derivatives to the order $[l]$ inclusive; in addition, the quantity

$$\| u \|_{\Omega}^{(l)} = \langle u \rangle_{\Omega}^{(l)} + \sum_{j=0}^{[l]} \langle u \rangle_{\Omega}^{(j)}, \tag{21}$$

where

$$\langle u \rangle_{\Omega}^{(j)} = \sum_{(j)} \max_{\Omega} | D^j u |, \quad \langle u \rangle_{\Omega}^{(l)} = \sum_{([l])} \sup_{x, x' \in \Omega} \frac{| D_x^{[l]} u(x) - D_{x'}^{[l]} u(x') |}{| x - x' |^{l-[l]}},$$

has a finite value. Equation (21) defines the norm $\| u \|_{\Omega}^{(l)}$ in $C^l(\Omega)$.

THEOREM 1. *Let $l \geqslant 0$ be an integer, let $p > 1$, and let*

$$S \in C^{l+3},\quad \mathbf{f}(x) \in W_p^l(\Omega),\quad \psi(x) \in W_p^{l+1}(\Omega),$$
$$\alpha(x) \in W_p^{l+2-1/p}(S),\quad \mathbf{b}(x) \in W_p^{l+1-1/p}(S).$$

If the conditions (8) *are satisfied, the problem* (19), (20) *has a unique solution such that*

$$\|\mathbf{v}\|_{p,\Omega}^{(l+2)} + \|p\|_{p,\Omega}^{(l+1)} \leqslant C_6(\|\mathbf{f}\|_{p,\Omega}^{(l)} + \|\psi\|_{p,\Omega}^{(l+1)} + \|\alpha\|_{p,S}^{(l+2-1/p)} + \|\mathbf{b}\|_{p,S}^{(l+1-1/p)}). \quad (22)$$

THEOREM 2. *Let $l > 0$ be nonintegral, and let $S \in C^{l+3}$. If*

$$\mathbf{f}(x) \in C^l(\Omega),\quad \psi(x) \in C^{l+1}(\Omega),\quad \alpha(x) \in C^{l+2}(S),\quad \mathbf{b}(x) \in C^{l+1}(S)$$

and if the conditions (8) *are satisfied, the problem* (19), (20) *has a unique solution such that $\mathbf{v}(x) \in C^{l+2}(\Omega)$, $p(x) \in C^{l+1}(\Omega)$ and*

$$\|\mathbf{v}\|_\Omega^{(l+2)} + \|p\|_\Omega^{(l+1)} \leqslant C_7(\|\mathbf{f}\|_\Omega^{(l)} + \|\psi\|_\Omega^{(l+1)} + \|\alpha\|_S^{(l+2)} + \|\mathbf{b}\|_S^{(l+1)}). \quad (23)$$

We first prove Theorem 1 for $p = 2$ and $l = 0$. In view of Lemma 1 there exists, under the conditions of Theorem 1, a vector-valued function $\mathbf{u}(x) \in W_2^2(\Omega)$ satisfying (6), (7), and the inequality

$$\|\mathbf{u}(x)\|_{2,\Omega}^{(2)} \leqslant C_8(\|\psi\|_{2,\Omega}^{(1)} + \|\alpha\|_{2,S}^{(3/2)} + \|\mathbf{b}\|_{2,S}^{(1/2)}). \quad (24)$$

We introduce the new unknown vector-valued function $\mathbf{V}(x) = \mathbf{v}(x) - \mathbf{u}(x)$. It must be a solution of the problem

$$\begin{gathered} -\nu\Delta\mathbf{V} + \frac{1}{\rho}\operatorname{grad} p = \mathbf{f} + \nu\Delta\mathbf{u},\quad \operatorname{div}\mathbf{V} = 0, \\ \mathbf{V}\cdot\mathbf{n}|_S = 0,\quad \mathbf{t}' - \mathbf{n}(\mathbf{t}'\cdot\mathbf{n})|_S = 0, \end{gathered} \quad (25)$$

where $\mathbf{t}'$ is the vector with components

$$t_i' = \sum_{k=1}^3 \left[-\delta_i^k \frac{p}{\rho} + \nu\left(\frac{\partial V_i}{\partial x_k} + \frac{\partial V_k}{\partial x_i}\right)\right] n_k.$$

On the basis of Theorems 1 and 2 in [7], for the case $\psi(x) = 0$, $\alpha(x) = 0$, $\mathbf{b}(x) = 0$, $p = 2$, $l = 0$, the solution of problem (25) exists and

$$\|\mathbf{V}\|_{2,\Omega}^{(2)} + \|p\|_{2,\Omega}^{(1)} \leqslant C_9\|\mathbf{f} + \nu\Delta\mathbf{u}\|_{2,\Omega}, \quad (26)$$

whence, by virtue of (24), Theorem 1 follows for these same p and l and for the derivatives of ψ, α, and $\mathbf{b}$.

We introduce the spaces

$$H_{p,l} = W_p^l(\Omega) \times W_p^{l+1}(\Omega) \times W_p^{l+2-1/p}(S) \times W_p^{l+1-1/p}(S),$$
$$V_{p,l} = W_p^{l+2}(\Omega) \times W_p^{l+1}(\Omega)$$

and we write the problem (19), (20) in the form

$$Aw = h, \quad (27)$$

where $w = [\mathbf{v}, p] \in V_{p,l}$, $h = [\mathbf{f}, \psi, \alpha, \mathbf{b}] \in H_{p,l}$, and the operator A makes correspond to each element $w \in V_{p,l}$ the element

$$[-\nu\Delta\mathbf{v} + \rho^{-1}\operatorname{grad} p, \operatorname{div}\mathbf{v}, \mathbf{v}\cdot\mathbf{n}|_S, \mathbf{t} - \mathbf{n}(\mathbf{t}\cdot\mathbf{n})|_S] \in H_{p,l}.$$

Under the conditions of Theorem 1, A is a bounded operator acting from $V_{p,l}$ into $H_{p,l}$; its domain is the whole space $V_{p,l}$. In view of Theorems 2.2 and 2.4 in [5], for general boundary value problems for systems elliptic in the Douglis-Nirenberg sense, A is a normally solvable operator with a finite defect number and with a finite dimensionality of the space of zeros. The same also is valid for Hölder spaces (see [5], Theorems 2.1 and 2.3).

It follows from the above that the dimensionality of the space of zeros of the operator A in the space $V_{2,0}$ is equal to zero, and the defect number of this operator in $H_{2,0}$ is equal to one since the functions ψ and α must be subject to the single condition

$$\int_\Omega \psi(x)\,dx = \int_S \alpha(x)\,dS, \tag{28}$$

defining a subspace in $H_{2,0}$. By virtue of Theorems 2.12, 2.13, and 2.14 given in [5], the dimensionality of the space of zeros of A and its defect number do not depend on what space $V_{p,l}$ or Hölder space is considered. Therefore, subject to the conditions of Theorem 1 or Theorem 2, the problem (19), (20) will have a solution in the corresponding Banach spaces if the condition (28) is satisfied. The estimates (22) and (23) are consequences of Theorems 2.1, 2.2, and 2.8 in [5]. This completes the proofs of Theorems 1 and 2.

In the domain Ω we consider the boundary value problems:

$$\begin{gathered} \operatorname{rot}\mathbf{v} = \boldsymbol{\psi}(x), \quad \operatorname{div}\mathbf{v} = f(x), \\ \mathbf{v}_\tau|_S = \mathbf{a}(x), \end{gathered} \tag{29}$$

$$\begin{gathered} \operatorname{rot}\mathbf{u} = \mathbf{j}(x), \quad \operatorname{div}\mathbf{u} = \varphi(x), \\ u_n|_S = \alpha(x). \end{gathered} \tag{30}$$

We assume that

$$\int_\Omega \boldsymbol{\psi}\cdot\operatorname{grad} p\,dx = \int_S [\mathbf{n}\times\mathbf{a}]\cdot\operatorname{grad} p\,dS \tag{31}$$

for an arbitrary function $p \in W_2^2(\Omega)$, and also that

$$\int_\Omega \varphi(x)\,dx = \int_S \alpha(x)\,dS. \tag{32}$$

The vector $\mathbf{a}$, given on the surface S, must satisfy the compatibility condition $\mathbf{a}\cdot\mathbf{n} = 0$.

It is well known (see [8], Chapter I, §2.2) that

$$L_2(\Omega) = \mathring{J}(\Omega) \oplus G(\Omega),$$

where $G(\Omega)$ is the subspace of vectors of the form $\mathbf{u} = \operatorname{grad} p$, $p \in W_2^1(\Omega)$, and $\mathring{J}(\Omega)$ is the closure of the set $J(\Omega)$ of compactly supported solenoidal vectors infinitely differentiable in Ω. The space $L_2(\Omega)$ can also be represented in the form

$$L_2(\Omega) = J(\Omega) \oplus \mathring{G}(\Omega),$$

where $\mathring{G}(\Omega)$ is the subspace of vectors of the form $\mathbf{u} = \operatorname{grad} q$, $q \in \mathring{W}_2^1(\Omega)$, and $J(\Omega)$ is the space in which the set $\tilde{J}(\Omega)$ of all continuously differentiable solenoidal vectors is dense.

We introduce the following subspaces of the space $W_2^1(\Omega)$: $J^1(\Omega) = W_2^1(\Omega) \cap J(\Omega)$ is the set of all solenoidal vectors from $W_2^1(\Omega)$, and $\mathring{J}_\tau^1(\Omega)$ ($\mathring{J}_n^1(\Omega)$) are subspaces of vectors from $J^1(\Omega)$ which satisfy on S the condition

$$\mathbf{u}_\tau|_S = 0 \ (u_n|_S = 0).$$

THEOREM 3. *Let*

$$S \in C^2, \ \psi(x) \in L_2(\Omega), \ f(x) \in L_2(\Omega), \ \mathbf{a}(x) \in W_2^{1/2}(S).$$

If condition (31) *is satisfied, the problem* (29) *has a unique solution such that*

$$\|\mathbf{v}\|_{2,\Omega}^{(1)} \leqslant C_{10}(\|\psi\|_{2,\Omega} + \|f\|_{2,\Omega} + \|\mathbf{a}\|_{2,S}^{(1/2)}). \tag{33}$$

PROOF. We first consider the following auxiliary problem:

$$\operatorname{div} \mathbf{w} = f(x), \quad \mathbf{w}_\tau|_S = \mathbf{a}(x). \tag{34}$$

This problem has an infinite number of solutions. We consider one of them. We seek $\mathbf{w}$ in the form $\mathbf{w} = \operatorname{grad} p + \mathbf{h}$, where

$$\Delta p = f(x), \quad p|_S = 0, \tag{35}$$

$$\operatorname{div} \mathbf{h} = 0, \quad \mathbf{h}_\tau|_S = \mathbf{a}(x). \tag{36}$$

It is well known (see, for example, [9], Chapter III, Theorem 9.1) that the problem (35) has a unique solution in $W_2^2(\Omega)$, and that

$$\|p\|_{2,\Omega}^{(2)} \leqslant C_{11}\|f\|_{2,\Omega}. \tag{37}$$

Problem (36) was considered in [8] (see Chapter I, §2.1), where a solution was shown to exist satisfying the inequality

$$\|\mathbf{h}\|_{2,\Omega}^{(1)} \leqslant C_{12}\|\mathbf{a}\|_{2,S}^{(1/2)}. \tag{38}$$

In place of $\mathbf{v}$ we introduce the vector-valued function $\mathbf{v}'(x)$:

$$\mathbf{v}'(x) = \mathbf{v}(x) - \mathbf{w}(x).$$

Then for it we obtain from (29) the following:

$$\operatorname{rot} \mathbf{v}' = \psi - \operatorname{rot} \mathbf{w}, \quad \operatorname{div} \mathbf{v}' = 0, \qquad \mathbf{v}'_\tau|_S = 0. \tag{39}$$

By virtue of the condition (31), $\psi - \operatorname{rot} \mathbf{w} \in \mathring{J}(\Omega)$. Therefore, on the basis of Theorem 3.2 of [10], the problem (39) has a unique solution from $\mathring{J}_\tau^1(\Omega)$ for which

$$\| \mathbf{v}' \|_{2,\Omega}^{(1)} \leqslant C_{13} \| \psi - \operatorname{rot} \mathbf{w} \|_{2,\Omega}. \tag{40}$$

Validity of (33) follows from (37), (38), and (40). This completes the proof of the theorem.

THEOREM 4. *Let*

$$S \in C^2, \quad \mathbf{j}(x) \in J(\Omega), \quad \varphi(x) \in L_2(\Omega), \quad \alpha(x) \in W_2^{1/2}(S).$$

If condition (32) *is satisfied, the problem* (30) *has a unique solution such that* $\mathbf{u}(x) \in W_2^1(\Omega)$ *and*

$$\| \mathbf{u} \|_{2,\Omega}^{(1)} \leqslant C_{14} (\| \mathbf{j} \|_{2,\Omega} + \| \varphi \|_{2,\Omega} + \| \alpha \|_{2,S}^{(1/2)}). \tag{41}$$

The proof of this theorem is close to the proof of Theorem 3. In proving it, it is necessary to make use of Theorem 4.2 of [10] concerning the single-valued representation of an arbitrary vector-valued function $\mathbf{j}(x) \in J(\Omega)$ in the form $\mathbf{j} = \operatorname{rot} \mathbf{d}$, where $\mathbf{d} \in \mathring{J}_n^1(\Omega)$.

LEMMA 2. *Let* $S \in C^2$ *and* $p > 1$, *and let* $\mathbf{u}(x)$ *be a smooth vector-valued solenoidal function satisfying the condition* $\mathbf{u}_\tau|_S = 0$ *or* $u_n|_S = 0$. *Then*

$$\| \mathbf{u} \|_{p,\Omega}^{(1)} \leqslant C_{15} \| \operatorname{rot} \mathbf{u} \|_{p,\Omega}. \tag{42}$$

To prove this lemma we need to use Lemmas 1.2 and 2.2 of [10] concerning the single-valued representation of an arbitrary smooth solenoidal vector-valued function satisfying the condition $\mathbf{u}_\tau|_S = 0$ or $u_n|_S = 0$ across electrostatic potentials and the solutions of the Dirichlet and Neumann problems for the Laplace operator, and to establish estimates for these solutions in $L_p(\Omega)$ norms in a known way (see, for example, [8], Chapter III, §5).

THEOREM 5. *Let* $p > 1$, *let* l *be a nonnegative integer, and let* Ω *be a bounded domain with boundary* $S \in C^{l+2}$. *If*

$$\psi(x) \in W_p^l(\Omega), \quad f(x) \in W_p^l(\Omega), \quad \mathbf{a}(x) \in W_p^{l+1-1/p}(S)$$

and if condition (31) *is satisfied, then the problem* (29) *has a unique solution such that* $\mathbf{v}(x) \in W_p^{l+1}(\Omega)$ *and*

$$\| \mathbf{v} \|_{p,\Omega}^{(l+1)} \leqslant C_{16} (\| \psi \|_{p,\Omega}^{(l)} + \| f \|_{p,\Omega}^{(l)} + \| \mathbf{a} \|_{p,S}^{(l+1-1/p)}). \tag{43}$$

PROOF. It is convenient for the study of problem (29) to write it in the form

$$A\mathbf{v} = h, \tag{44}$$

where

$$A\mathbf{v} = [\operatorname{rot} \mathbf{v}, \ \operatorname{div} \mathbf{v}, \ \mathbf{v}_\tau|_S], \ h = [\psi, \ f, \ \mathbf{a}].$$

We define the operator A on $W_p^{l+1}(\Omega)$, where $l \geqslant 0$ is an integer and $p > 1$. Under the conditions of Theorem 5, A is a bounded operator from $W_p^{l+1}(\Omega)$ into

$$B_p^l = W_p^l(\Omega) \times W_p^l(\Omega) \times W_p^{l+1-1/p}(S).$$

It is well known (see [11], Russian p. 119, English p. 482) that the system of equations

$$\text{rot}\, \mathbf{v} = \boldsymbol{\psi}(x), \quad \text{div}\, \mathbf{v} = f(x)$$

is an overdetermined elliptic system and that the operator $\mathbf{v}_\tau|_S$ covers this system of equations. Therefore, on the basis of Theorem 1.1 of [11] the following assertions hold (these assertions are equivalent):

1) The operator A has a left regularizer, i.e. there exists a bounded operator R: $B_p^l \to W_p^{l+1}(\Omega)$ such that $RA = I - T$ and T is completely continuous in $W_p^{l+1}(\Omega)$.

2) The following a priori estimate is satisfied:

$$\|\mathbf{v}\|_{p,\Omega}^{(l+1)} \leqslant C_{17}(\|\boldsymbol{\psi}\|_{p,\Omega}^{(l)} + \|f\|_{p,\Omega}^{(l)} + \|\mathbf{a}\|_{p,S}^{(l+1-1/p)} + \|\mathbf{v}\|_{p,\Omega}). \tag{45}$$

In addition, we shall use the following assertions proved in [12] (see Theorems 1 and 2).

Let A be a bounded linear operator acting from the Banach space B_1 into the Banach space B_2 and having a left regularizer R: $B_2 \to B_1$. Then the following statements are true:

1) We can construct a bounded operator R_0 such that $R_0 A = I - P$, where P is a projection operator ($P^2 = P$) and $R(P) = N(I - P) = N(A)$.

2) For the solvability of the equation

$$Au = h \tag{46}$$

it is necessary and sufficient that

$$Qh \equiv (I - AR_0)h = 0. \tag{47}$$

Subject to this condition, the element $u = R_0 h$ is the unique solution of (46) for which $Pu = 0$.

The condition (31) defines a subspace in B_p^l, which we denote by R_p^l. On the basis of Theorem 3 for the case $p = 2$ and $l = 0$ the problem (29) has a unique solution in $W_2^1(\Omega)$ if (31) is satisfied. Therefore h satisfies (47) for these same p and l. It is obvious that h will satisfy the same condition even for $p \geqslant 2$ and $l = 0$. In the case $p \in (1, 2)$ and $l = 0$ satisfaction of (47) follows from the compactness of R_2^0 in R_p^0 and the boundedness of the operator Q. Uniqueness of the solution of (44) in $W_p^1(\Omega)$ is a consequence of (42). In view of all this, h will be subject to the condition (47) even when $l > 0$ and for arbitrary $p > 1$. Consequently, the problem (29) will have a unique solution in $W_p^{l+1}(\Omega)$ for arbitrary $l \geqslant 0, p > 1, \boldsymbol{\psi}, f$, and $\mathbf{a}$ if condition (31) is satisfied. The absence on the right of (43) of a term of the form $C\|\mathbf{v}\|_{p,\Omega}$ explains the uniqueness of a solution of (29) in $W_p^{l+1}(\Omega)$. This completes the proof of the theorem.

We denote by $J_p^l(\Omega)$ the subspace of the Banach space $W_p^l(\Omega)$ consisting of all solenoidal vector-valued functions.

THEOREM 6. *Let $l \geqslant 0$ be an integer, let $p > 1$, and let Ω be a bounded domain with the boundary $S \in C^{l+2}$. If*

$$\mathbf{j}(x) \in J_p^l(\Omega), \quad \varphi(x) \in W_p^l(\Omega), \quad \alpha(x) \in W_p^{l+1-1/p}(S)$$

and if condition (32) *is satisfied, then the problem* (30) *has a unique solution such that*

$$\|\mathbf{u}\|_{p,\Omega}^{(l+1)} \leqslant C_{18}(\|\mathbf{j}\|_{p,\Omega}^{(l)} + \|\varphi\|_{p,\Omega}^{(l)} + \|\alpha\|_{p,S}^{(l+1-1/p)}). \tag{48}$$

The general scheme of the proof is the same as that for Theorem 5 since the operator $u_n|_S$ also covers the system of equations

$$\operatorname{rot}\mathbf{u} = \mathbf{j}(x), \quad \operatorname{div}\mathbf{u} = \varphi(x),$$

and, therefore, Theorem 1.1 of [11], which was used in the proof of Theorem 5, also applies to the problem (30).

Results analogous to those in Theorems 5 and 6 are obtained also when the problems (29) and (30) are investigated in the Hölder spaces $C^l(\Omega)$ with nonintegral l.

We consider the following auxiliary problem:

$$\begin{gathered}\operatorname{rot}\operatorname{rot}\mathbf{H} = \mathbf{F}(x), \quad \operatorname{div}\mathbf{H} = \rho(x), \\ H_n|_S = \alpha(x), \quad \operatorname{rot}_\tau \mathbf{H}|_S = \mathbf{b}(x).\end{gathered} \tag{49}$$

We assume that the conditions

$$\int_\Omega \mathbf{F}\cdot\operatorname{grad} p\,dx = \int_S [\mathbf{n}\times\mathbf{b}]\cdot\operatorname{grad} p\,dS \tag{50}$$

are satisfied for an arbitrary function $p(x) \in W_2^2(\Omega)$, and also that

$$\int_\Omega \rho(x)\,dx = \int_S \alpha(x)\,dS. \tag{51}$$

THEOREM 7. *Let $p > 1$, let l be a nonnegative integer, and let Ω be a bounded domain with the boundary $S \in C^{l+3}$. If*

$$\mathbf{F}(x) \in W_p^l(\Omega), \quad \rho(x) \in W_p^{l+1}(\Omega), \quad \alpha(x) \in W_p^{l+2-1/p}(S), \quad \mathbf{b}(x) \in W_p^{l+1-1/p}(S)$$

and if conditions (50) *and* (51) *are satisfied, then the problem* (49) *has a unique solution such that* $\mathbf{H}(x) \in W_p^{l+2}(\Omega)$ *and*

$$\|\mathbf{H}\|_{p,\Omega}^{(l+2)} \leqslant C_{19}(\|\mathbf{F}\|_{p,\Omega}^{(l)} + \|\rho\|_{p,\Omega}^{(l+1)} + \|\alpha\|_{p,S}^{(l+2-1/p)} + \|\mathbf{b}\|_{p,S}^{(l+1-1/p)}). \tag{52}$$

PROOF. The problem (49) reduces to the two problems

$$\begin{gathered}\operatorname{rot}\mathbf{u} = \mathbf{F}(x), \quad \operatorname{div}\mathbf{u} = 0, \\ \mathbf{u}_\tau|_S = \mathbf{b}(x)\end{gathered} \tag{53}$$

$$\begin{gathered}\operatorname{rot}\mathbf{H} = \mathbf{u}, \quad \operatorname{div}\mathbf{H} = \rho(x), \\ H_n|_S = \alpha(x).\end{gathered} \tag{54}$$

By virtue of Theorem 5, under the conditions of Theorem 7, the problem (53) has a unique solution from $J_p^{l+1}(\Omega)$ and

$$\|\mathbf{u}\|_{p,\Omega}^{(l+1)}\leqslant C_{20}(\|\mathbf{F}\|_{p,\Omega}^{(l)}+\|\mathbf{b}\|_{p,S}^{(l+1-1/p)}). \tag{55}$$

Then, according to Theorem 6, problem (54) has a unique solution in $W_p^{l+2}(\Omega)$ such that

$$\|\mathbf{H}\|_{p,\Omega}^{(l+2)}\leqslant C_{21}(\|\mathbf{u}\|_{p,\Omega}^{(l+1)}+\|\rho\|_{p,\Omega}^{(l+1)}+\|\alpha\|_{p,S}^{(l+2-1/p)}). \tag{56}$$

Inequality (52) then follows from (55) and (56). This completes the proof of the theorem.

An analogous result holds also for the spaces $C^l(\Omega)$.

§3. Solvability of the problem (1)–(5)

As in [1] and [13], we replace the classical statement of problem (1)–(5) by a generalized formulation in which the whole problem is reduced to finding vector-valued functions $\mathbf{v}$ and $\mathbf{H}$ satisfying certain integral identities and belonging to well-defined Hilbert function spaces.

We now give the definitions of these spaces which we shall need in what follows.

$H(\Omega)$ is the subspace of the Hilbert space $W_2^1(\Omega)$ consisting of all the vector-valued functions from $W_2^1(\Omega)$ satisfying the boundary conditions

$$u_n|_{S_1}=0,\quad \mathbf{u}|_{S_2}=0.$$

Here $u_n|_{S_1}=(\mathbf{u},\mathbf{n})|_{S_1}$ is the normal component of the vector-valued function $\mathbf{u}$ on S_1. We introduce in $H(\Omega)$ the new inner product

$$(\mathbf{u},\ \mathbf{v})_H=E(\mathbf{u},\ \mathbf{v}),$$

where

$$E(\mathbf{u},\ \mathbf{v})=\int_\Omega\sum_{i,k=1}^{3}\left(\frac{\partial u_i}{\partial x_k}+\frac{\partial u_k}{\partial x_i}\right)\left(\frac{\partial v_i}{\partial x_k}+\frac{\partial v_k}{\partial x_i}\right)dx.$$

Without loss of generality we assume that for $S_1=S$ the domain Ω is not a domain of revolution about any constant vector $\mathbf{B}$. The following lemma furnishes the basis for introducing such an inner product (see [7], §2).

LEMMA 3. *For any vector-valued function* $\mathbf{u}(x)\in H(\Omega)$

$$\|\mathbf{u}\|_{2,\Omega}^{(1)}\leqslant C_{22}\|\mathbf{u}\|_H, \tag{57}$$

where $\|\mathbf{u}\|_H=\sqrt{(\mathbf{u},\mathbf{u})_H}$.

Lemma 3 holds provided that the surface $S\in C^2$. Throughout this section we assume that this condition is satisfied.

It is obvious that $\|\cdot\|_H$ is equivalent to the norm of $W_2^1(\Omega)$. $\hat{J}^1(\Omega)$ is the subspace of $H(\Omega)$ consisting of all solenoidal vector-valued functions. $\hat{J}^{-1}$ is the space of linear functionals over $\hat{J}^1(\Omega)$.

On the basis of Theorem 4.2 of [10] the operator rot establishes a one-to-one correspondence between the spaces $\mathring{J}_N^1(\Omega)$ and $J(\Omega)$, whereby for all $\mathbf{u}(x)\in\mathring{J}_n^1(\Omega)$ we have the inequalities

$$C_{23}\|\operatorname{rot}\mathbf{u}\|_{2,\Omega}\leqslant\|\mathbf{u}\|_{2,\Omega}^{(1)}\leqslant C_{24}\|\operatorname{rot}\mathbf{u}\|_{2,\Omega} \tag{58}$$

where the constants C_{23} and C_{24} do not depend on $\mathbf{u}(x)$. This allows us to introduce in $\mathring{J}_n^1(\Omega)$ the new inner product

$$[\mathbf{u}, \mathbf{v}] = \int_\Omega \operatorname{rot} \mathbf{u} \cdot \operatorname{rot} \mathbf{v} dx,$$

which generates in $\mathring{J}_n^1(\Omega)$ a norm $|\cdot|$ equivalent to the old norm.

We come now to the study of problem (1)–(5). First of all, we reduce the nonhomogeneous condition

$$H_n|_S = \beta(x)$$

to a homogeneous one. To do this, we construct the vector-valued function $\mathbf{h}_0(x)$ from $W_2^1(\Omega)$ to have the following properties (see §2, Theorem 4):

$$\begin{gathered} \operatorname{rot} \mathbf{h}_0 = 0, \quad \operatorname{div} \mathbf{h}_0 = 0, \quad h_{0n}|_S = \beta(x), \\ \|\mathbf{h}_0\|_{2,\Omega}^{(1)} \leqslant C_{25} \|\beta\|_{2,S}^{(1/2)}. \end{gathered} \tag{59}$$

We introduce the new unknown vector $\mathbf{h} = \mathbf{H} - \mathbf{h}_0$. Then the system (1)–(3) may be written in the form

$$\begin{gathered} -\nu\Delta\mathbf{v} + v_k \frac{\partial \mathbf{v}}{\partial x_k} - \frac{\mu}{\rho}\left(h_k \frac{\partial \mathbf{h}}{\partial x_k} + h_{0k}\frac{\partial \mathbf{h}}{\partial x_k} + h_k \frac{\partial \mathbf{h}_0}{\partial x_k}\right) \\ = -\frac{1}{\rho}\operatorname{grad}\left[p + \frac{\mu}{2}(h^2 + 2(\mathbf{h}\cdot\mathbf{h}_0))\right] + \mathbf{f}, \end{gathered} \tag{60}$$

$$\operatorname{div} \mathbf{v} = 0, \quad \operatorname{div} \mathbf{h} = 0, \tag{61}$$

$$\operatorname{rot} \mathbf{h} = \sigma(\mathbf{E} + \mu[\mathbf{v} \times (\mathbf{h} + \mathbf{h}_0)]) + \mathbf{j}_0, \quad \operatorname{rot} \mathbf{E} = 0. \tag{62}$$

The problem here is to determine in the bounded domain Ω a solution of the system (60)–(62) satisfying the conditions

$$v_n|_{S_1} = 0, \quad \mathbf{t} - \mathbf{n}(\mathbf{t}\cdot\mathbf{n})|_{S_1} = 0, \quad \mathbf{v}|_{S_2} = 0, \tag{63}$$

$$h_n|_S = 0, \quad \mathbf{E}_\tau|_S = 0. \tag{64}$$

By a generalized solution of the problem (60)–(64) we mean a pair of vector-valued functions $\mathbf{v}$ and $\mathbf{h}$, $\mathbf{v} \in \hat{J}^1(\Omega)$, $\mathbf{h} \in \mathring{J}_n^1(\Omega)$, satisfying the identities

$$\frac{\nu}{2} E(\mathbf{v}, \boldsymbol{\varphi}) - \int_\Omega \left\{\left[v_k \mathbf{v} - \frac{\mu}{\rho} h_k(\mathbf{h} + \mathbf{h}_0)\right]\cdot\frac{\partial \boldsymbol{\varphi}}{\partial x_k} + \frac{\mu}{\rho} h_{0k}\frac{\partial \mathbf{h}}{\partial x_k}\cdot\boldsymbol{\varphi}\right\} dx = \int_\Omega \mathbf{f}\cdot\boldsymbol{\varphi} dx, \tag{65}$$

$$\int_\Omega (\operatorname{rot}\mathbf{h} - \sigma\mu[\mathbf{v}\times(\mathbf{h}+\mathbf{h}_0)])\cdot\operatorname{rot}\boldsymbol{\psi} dx = \int_\Omega \mathbf{j}_0\cdot\operatorname{rot}\boldsymbol{\psi} dx \tag{66}$$

for arbitrary vector-valued functions $\boldsymbol{\varphi} \in \hat{J}^1(\Omega)$ and $\boldsymbol{\psi} \in \mathring{J}_n^1(\Omega)$.

In order to verify that every classical solution of (60)–(64) is a generalized solution, it is necessary to use the following Green's formula, which is valid for smooth solenoidal vector-valued functions $\mathbf{v}$ and $\boldsymbol{\varphi}$, and the function $p(x) \in C^1(\Omega)$:

$$\int_\Omega \left(\nu\Delta\mathbf{v} - \frac{1}{\rho}\operatorname{grad} p\right)\cdot\boldsymbol{\varphi} dx = -\frac{\nu}{2}E(\mathbf{v}, \boldsymbol{\varphi}) + \int_S \mathbf{t}\cdot\boldsymbol{\varphi} dS, \tag{67}$$

and to take into account the fact that

$$\int_{\Omega} \mathbf{E} \cdot \operatorname{rot} \psi dx = 0$$

by virtue of the equation $\operatorname{rot} \mathbf{E} = 0$ and the second of the conditions (64), for an arbitrary vector-valued function $\psi \in \mathring{J}_n^1(\Omega)$.

We assume now, relative to the generalized solution $\mathbf{v}$, $\mathbf{h}$, that it is known that $\mathbf{v}(x)$ and $\mathbf{h}(x)$ belong to $W_2^2(\Omega')$, where Ω' is an arbitrary subdomain of Ω, and $\mathbf{f}(x) \in L_2(\Omega')$. Then (65) can be transformed to the form

$$\int_{\Omega'} \left[-\nu\Delta\mathbf{v} + v_k \frac{\partial \mathbf{v}}{\partial x_k} - \frac{\mu}{\rho}\left(h_k \frac{\partial \mathbf{h}}{\partial x_k} + h_{0k} \frac{\partial \mathbf{h}}{\partial x_k} + h_k \frac{\partial \mathbf{h}_0}{\partial x_k}\right) - \mathbf{f}\right] \cdot \varphi dx = 0$$

for $\varphi(x) \in \dot{J}(\Omega')$, where $\dot{J}(\Omega')$ is the set of vector-valued functions that are infinitely differentiable, compactly supported and solenoidal in Ω'. Since $\dot{J}(\Omega')$ is dense in $\mathring{J}(\Omega')$ (see, for example, [8], Chapter I, §2.2), $\varphi(x)$ is an arbitrary element from $\dot{J}(\Omega')$, and

$$-\nu\Delta\mathbf{v} + v_k \frac{\partial \mathbf{v}}{\partial x_k} - \frac{\mu}{\rho}\left(h_k \frac{\partial \mathbf{h}}{\partial x_k} + h_{0k} \frac{\partial \mathbf{h}}{\partial x_k} + h_k \frac{\partial \mathbf{h}_0}{\partial x_k}\right) - \mathbf{f} \in L_2(\Omega')$$

by virtue of the differential properties of a generalized solution $\mathbf{v}$ and $\mathbf{h}$, it then follows from the last identity that

$$-\nu\Delta\mathbf{v} + v_k \frac{\partial \mathbf{v}}{\partial x_k} - \frac{\mu}{\rho}\left(h_k \frac{\partial \mathbf{h}}{\partial x_k} + h_{0k} \frac{\partial \mathbf{h}}{\partial x_k} + h_k \frac{\partial \mathbf{h}_0}{\partial x_k}\right) - \mathbf{f}$$

is the gradient of a function $q(x)$. Putting $p = -\rho q - (\mu/2)(h^2 + 2\mathbf{h} \cdot \mathbf{h}_0)$, we confirm that (60) is satisfied.

Further, if we follow the reasoning employed in [7] (see §3), we arrive at the result that $\mathbf{t}_\rho|_{S_1'} \equiv \mathbf{t} - \mathbf{n}(\mathbf{t} \cdot \mathbf{n})|_{S_1'} = 0$, where $S_1' = S_1 \cap \partial\Omega'$.

Let $\mathbf{j}_0(x) \in W_2^1(\Omega')$. We define the vector-valued function $\mathbf{E}(x)$. The identity (66) can be written in the form

$$\int_{\Omega'} (\operatorname{rot} \mathbf{h} - \sigma\mu [\mathbf{v} \times (\mathbf{h} + \mathbf{h}_0)] - \mathbf{j}_0) \cdot \operatorname{rot} \psi dx = 0$$

for an arbitrary $\psi \in \dot{J}(\Omega')$. Since

$$\operatorname{rot} \mathbf{h} - \sigma\mu [\mathbf{v} \times (\mathbf{h} + \mathbf{h}_0)] - \mathbf{j}_0 \in L_2(\Omega')$$

by virtue of the differential properties of a generalized solution $\mathbf{v}$ and $\mathbf{h}$, it then follows in accordance with Theorem 1 in Chapter I, §2.2 of [8] that

$$\operatorname{rot} \mathbf{h} - \sigma\mu [\mathbf{v} \times (\mathbf{h} + \mathbf{h}_0)] - \mathbf{j}_0 = \operatorname{grad} q.$$

We put

$$\mathbf{E} = \frac{1}{\sigma} \operatorname{grad} q.$$

The vector-valued function $\mathbf{E}(x)$ satisfies (62). We show that $\mathbf{E}(x)$ satisfies the boundary condition

$$\mathbf{E}_\tau|_{S'} = 0,$$

where $S' = S \cap \partial\Omega'$. Let $\mathring{J}^1_{n,\sigma}(\Omega')$ be a lineal of vector-valued functions $\psi(x) \in \mathring{J}^1_n(\Omega)$, which are equal to zero outside of Ω' and take on zero values on S outside of σ, where σ is an arbitrary smooth submanifold of S'. Integrating by parts in (66), we obtain

$$\int_\sigma [\mathbf{n} \times \mathbf{E}] \cdot \psi dS = 0$$

for arbitrary $\psi(x) \in \mathring{J}^1_{n,\sigma}(\Omega')$. Further, reasoning precisely as in [7] (see §3), we find that $\mathbf{E}_\tau|_\sigma = 0$.

THEOREM 8. *Let*

$$\mathbf{j}_0(x) \in L_2(\Omega), \quad \mathbf{f}(x) \in \hat{J}^{-1}(\Omega), \quad \beta(x) \in W_2^{1/2}(S).$$

If the condition

$$\int_\Omega \beta(x)\, dS = 0$$

is satisfied, it follows that the problem (60)–(64) *has at least one generalized solution.*

THEOREM 9. *Let ν, μ, ρ, σ, f, j_0, and Ω satisfy the condition*

$$\frac{C_{26}(2\mu+\rho)}{\beta^2}\sqrt{\rho^2\|\mathbf{f}\|^2_{\hat{J}^{-1}(\Omega)} + \frac{1}{\sigma^2}\|\mathbf{j}_0\|^2_{2,\Omega}} < 1, \tag{68}$$

where the constant C_{26} depends on Ω, and

$$\beta = \min(\nu\rho/2,\ 1/\sigma), \quad a\|\mathbf{f}\|_{\hat{J}^{-1}(\Omega)} = \sup_{\mathbf{v}\in \hat{J}^1(\Omega)} \frac{\left|\int_\Omega \mathbf{f}\mathbf{v}\,dx\right|}{\|\mathbf{v}\|_H}.$$

Then the problem (60)–(64) *has at most one generalized solution.*

Theorems 8 and 9 are proved in precisely the same way as in [8] (see Chapter V, §2) and in [1] (see §2).

REMARK 1. The problem (60)–(64) is handled in a similar way even in the case of the nonhomogeneous boundary condition

$$\mathbf{E}_\tau|_S = \mathbf{a}.$$

Here it is necessary to assume that $\mathbf{a}(x) \in W_2^{-1/2}(S)$, where $W_2^{-1/2}(S)$ is the space of linear functionals over $W_2^{1/2}(S)$; also, in (68) it is necessary to replace $\|\mathbf{j}_0\|^2_{2,\Omega}/\sigma^2$ by $(C_{27}\|\mathbf{a}\|^{(-1/2)}_{2,S} + \|\mathbf{j}_0\|_{2,\Omega}/\sigma)^2$. Here

$$\|\mathbf{a}\|^{(-1/2)}_{2,S} = \sup_{\varphi\in W_2^{1/2}(S)} \frac{\left|\int_S \mathbf{a}\varphi\, dS\right|}{\|\varphi\|^{(1/2)}_{2,S}},$$

and C_{27} is a constant that depends only on Ω.

REMARK 2. Following the same reasoning as in [7] (see §3) and relying on Lemma 3 of [7], §2, it is easy to show that (65) is equivalent to

$$\frac{\nu}{2}E(\mathbf{v}, \varphi) - \int_\Omega \left\{\left[v_k\mathbf{v} - \frac{\mu}{\rho}h_k(\mathbf{h}+\mathbf{h}_0)\right]\cdot\frac{\partial\varphi}{\partial x_k} + \frac{\mu}{\rho}\mathbf{h}_{0k}\frac{\partial\mathbf{h}}{\partial x_k}\cdot\varphi - \mathbf{f}\cdot\varphi\right\}dx$$
$$= \frac{1}{\rho}\int_\Omega \left[p + \frac{\mu}{2}(h^2 + 2\mathbf{h}\cdot\mathbf{h}_0)\right]\operatorname{div}\varphi\, dx$$

for an arbitrary vector-valued function $\varphi \in H(\Omega)$, where $p(x) \in L_2(\Omega) \ominus \{1\}$ and

$$\|p\|_{2,\Omega} \leqslant C_{28}(\|\mathbf{v}\|_H + \|\mathbf{v}\|_H^2 + |\mathbf{h}|^2 + \|\mathbf{h}_0\|_{2,\Omega}^{(1)2} + \|\mathbf{f}\|_{\mathring{J}^{-1}(\Omega)})$$
$$\leqslant C_{29}\left(\sqrt{\|\mathbf{f}\|^2_{\mathring{J}^{-1}(\Omega)} + \|\mathbf{j}_0\|^2_{2,\Omega}} + \|\mathbf{f}\|^2_{\mathring{J}^{-1}(\Omega)} + \|\mathbf{j}_0\|^2_{2,\Omega} + (\|\beta\|^{(1/2)}_{2,S})^2\right). \quad (69)$$

§4. Study of the differential properties of generalized solutions

We devote this section to local estimates of the solution of the problem (60)–(64) in the W_p^l norms and also in the Hölder norms.

We introduce the following notation. Let $S' = S \cap \bar{\Omega}'$, where $\Omega' \subseteq \Omega$, and let $\Omega \supseteq \Omega'' \supseteq \Omega'$, where the distance from Ω' to $\Omega \setminus \Omega''$ is positive. Let $S'' = S \cap \bar{\Omega}''$. We assume that the surface S'' belongs in its entirety either to S_1 or to S_2. If S' is empty, we select Ω'' so that S'' will also be empty.

THEOREM 10. *Let* $\mathbf{v}$ *and* $\mathbf{h}$ *be a generalized solution of the problem* (60)–(64). *If*

$$S'' \in C^{l+3},\ \mathbf{f}(x) \in W_p^l(\Omega''),\ \mathbf{j}_0(x) \in W_p^{l+1}(\Omega''),\ \beta(x) \in W_p^{l+2-1/p}(S'') \cap W_2^{1/2}(S''),$$

where $l \geqslant 0$ *is an integer and* $p > 1$, *then*

$$\mathbf{v}(x) \in W_p^{l+2}(\Omega'),\ p(x) \in W_p^{l-1}(\Omega'),\ \mathbf{h}(x) \in W_p^{l+2}(\Omega'),\ \mathrm{E}(x) \in W_p^{l+1}(\Omega').$$

PROOF. We note that it is sufficient to prove the theorem for domains Ω'' and Ω' of canonical form (see, for example, [5], §1).

Let the domain Ω'', adjacent to the boundary S_1, be specified at the point $\xi^{(k)} \in S_1$ in the local system of coordinates by the inequalities

$$|y_\alpha| < \lambda\ (\alpha = 1, 2),\quad -2\lambda < y_3 - F(y_1, y_2) < 0.$$

Along with Ω'' we consider the domains Ω_m, $m = 0, 1, \ldots, N$, defined by

$$|y_\alpha| < \left(1 - \frac{m}{2N}\right)\lambda\ (\alpha = 1, 2),\quad -2\left(1 - \frac{m}{2N}\right)\lambda < y_3 - F(y_1, y_2) < 0,$$

where $N \geqslant 0$ is an integer determined by p and l; $\Omega_0 = \Omega''$.

Let $\zeta_m(x)$ be smooth cutoff functions possessing the properties

$$\zeta_m(x) = \begin{cases} 1 & \text{for } x \in \Omega_m, \\ 0 & \text{for } x \in \Omega \setminus \Omega_{m-1}, \end{cases}$$
$$|D_x^s \zeta_m| \leqslant \frac{C_s}{\lambda^s},$$

where $m = 1, \ldots, N$. We denote by S_m the intersection $S \cap \bar{\Omega}_m$ $(m = 0, \ldots, N)$. The functions $\mathbf{v}_m(x) = \zeta_m \mathbf{v}(x)$ and $p_m(x) = \zeta_m p(x)$ are defined in Ω_{m-1} (they are equal to zero outside of Ω_{m-1}) and are a solution of the following boundary value problem:

$$-\nu \Delta \mathbf{v}_m + \frac{1}{\rho} \operatorname{grad} p_m = \boldsymbol{\xi}_m,\quad \operatorname{div} \mathbf{v}_m = \rho_m, \quad (70)$$

$$v_{mn}|_{S_m} = 0,\quad \mathbf{t}^{(m)} - \mathbf{n}(\mathbf{t}^{(m)}, \mathbf{n})|_{S_m} = \mathbf{b}_m(x). \quad (71)$$

Here

$$\xi_m = \left[\mathbf{f} - \frac{\mu}{2\rho}\operatorname{grad}(h^2 + 2\mathbf{h}\cdot\mathbf{h}_0) + \frac{\mu}{\rho}\left(h_k\frac{\partial\mathbf{h}}{\partial x_k} + h_{0k}\frac{\partial\mathbf{h}}{\partial x_k} + h_k\frac{\partial\mathbf{h}_0}{\partial x_k}\right) - v_k\frac{\partial\mathbf{v}}{\partial x_k}\right]$$
$$\times\zeta_m - \nu\Delta\zeta_m\mathbf{v} - 2\nu\frac{\partial\zeta_m}{\partial x_k}\frac{\partial\mathbf{v}}{\partial x_k} + \frac{P}{\rho}\operatorname{grad}\zeta_m,\quad \rho_m = \mathbf{v}\cdot\operatorname{grad}\zeta_m,$$

$\mathbf{t}^{(m)}$ is a vector with the components

$$t_i^{(m)} = \sum_{k=1}^{3}\left[-\delta_i^k\frac{p_m}{\rho} + \nu\left(\frac{\partial v_{mi}}{\partial x_k} + \frac{\partial v_{mk}}{\partial x_i}\right)\right]n_k$$

and, finally, $\mathbf{b}_m(x) = \boldsymbol{\tau}^{(m)} - \mathbf{n}(\boldsymbol{\tau}^{(m)}\cdot\mathbf{n})|_{S_m}$, where $\boldsymbol{\tau}^{(m)}$ is a vector with the components

$$\tau_i^{(m)} = \nu\sum_{k=1}^{3}\left(\frac{\partial\zeta_m}{\partial x_k}v_i + \frac{\partial\zeta_m}{\partial x_i}v_k\right)n_k.$$

The vector-valued function $\mathbf{h}_m(x) = \zeta_m\mathbf{h}(x)$ is defined in Ω_{m-1} and can be considered as the solution of the boundary value problem

$$\operatorname{rot}\operatorname{rot}\mathbf{h}_m = \boldsymbol{\eta}_m,\quad \operatorname{div}\mathbf{h}_m = \delta_m, \tag{72}$$
$$h_{mn}|_{S_m} = 0,\quad \operatorname{rot}_\tau\mathbf{h}_m|_{S_m} = \boldsymbol{\omega}_m, \tag{73}$$

where

$$\boldsymbol{\eta}_m = (\sigma\mu\operatorname{rot}[\mathbf{v}\times(\mathbf{h}+\mathbf{h}_0)] + \operatorname{rot}\mathbf{j}_0)\zeta_m + \operatorname{rot}[\operatorname{grad}\zeta_m\times\mathbf{h}] + [\operatorname{grad}\zeta_m\times\operatorname{rot}\mathbf{h}],$$

$$\delta_m = \mathbf{h}\cdot\operatorname{grad}\zeta_m,\quad \boldsymbol{\omega}_m = [\operatorname{grad}\zeta_m\times\mathbf{h}]_\tau|_{S_m} + \sigma\mu[\mathbf{v}\times\beta\mathbf{n}]_\tau\zeta_m|_{S_m} + \mathbf{j}_{0\tau}\zeta_m|_{S_m}.$$

We first prove the theorem for $l = 0$. Since $\mathbf{v}$, $\mathbf{h}$, and $\mathbf{h}_0$ belong to $W_2^1(\Omega)$, it follows from a theorem of S. L. Sobolev concerning the boundedness of the operator of imbedding of the space $W_2^1(\Omega)$ in $L_q(\Omega)$ for $q < 6$ (see, for example, [14], Paragraph 117) that

$$\left[-\frac{\mu}{2\rho}\operatorname{grad}(h^2 + 2\mathbf{h}\cdot\mathbf{h}_0) + \frac{\mu}{\rho}\left(h_k\frac{\partial\mathbf{h}}{\partial x_k} + h_{0k}\frac{\partial\mathbf{h}}{\partial x_k} + h_k\frac{\partial\mathbf{h}_0}{\partial x_k}\right) - v_k\frac{\partial\mathbf{v}}{\partial x_k}\right]\zeta_1 \in L_p(\Omega'')$$

for $p < 3/2$. Then, taking note of (69), we find that $\xi_1(x) \in L_p(\Omega'')$ $(p < 3/2)$. It is clear that $\rho_1(x) \in W_p^1(\Omega'')$ and that

$$\int_{\Omega''}\rho_1(x)\,dx = 0.$$

Further, by virtue of the Theorem 1 of [15] we have $\mathbf{b}_1(x) \in W_p^{1-1/p}(S'')$, where $S'' = S_0$. Therefore, $\mathbf{v}$ will be an element of $W_p^2(\Omega_1)$ on the basis of Theorem 1, and $p(x)$ will be an element of $W_p^1(\Omega_1)$ $(p < 3/2)$.

In exactly the same way we confirm that $\boldsymbol{\eta}_1(x) \in L_p(\Omega'')$ $(p < 3/2)$, $\delta_1(x) \in W_p^1(\Omega'')$ and

$$\int_{\Omega''}\delta_1(x)\,dx = 0.$$

Further, since $[\mathbf{v}, \mathbf{h}_0]\zeta_1 \in W_p^1(\Omega'')$ $(p < 3/2)$, we then also have $\boldsymbol{\omega}_1(x) \in W_p^{1-1/p}(S'')$.

We verify that

$$\int_{\Omega''} \eta_1 \cdot \operatorname{grad} p\,dx = \int_{S''} [\mathbf{n} \times \boldsymbol{\omega}_1] \cdot \operatorname{grad} p\,dS$$

for an arbitrary function $p(x) \in W_2^2(\Omega'')$. Actually,

$$\int_{\Omega''} \{(\sigma_\mu \operatorname{rot} [\mathbf{v} \times (\mathbf{h} + \mathbf{h}_0)] + \operatorname{rot} \mathbf{j}_0)\zeta_1 + \operatorname{rot} [\operatorname{grad} \zeta_1 \times \mathbf{h}] + [\operatorname{grad} \zeta_1 \times \operatorname{rot} \mathbf{h}]\}$$

$$\times \operatorname{grad} p\,dx = \int_{\Omega''} \operatorname{rot} \operatorname{rot} \mathbf{h}_1 \cdot \operatorname{grad} p\,dx = -\int_{S''} [\operatorname{rot} \mathbf{h}_1 \times \mathbf{n}] \cdot \operatorname{grad} p\,dS$$

$$= \int_{S''} [\mathbf{n} \times \boldsymbol{\omega}_1] \operatorname{grad} p\,dS.$$

Consequently, the conditions of Theorem 7 are satisfied for $l = 0$ and $1 < p \leqslant 3/2$, and therefore $\mathbf{h}(x)$ belongs to $W_p^2(\Omega_1)$.

It follows from

$$\mathbf{E}(x) = \frac{1}{\sigma}(\operatorname{rot} \mathbf{h} - \sigma\mu [\mathbf{v} \times (\mathbf{h} + \mathbf{h}_0)] - \mathbf{j}_0) \tag{74}$$

that $\mathbf{E}(x)$ belongs to $W_p^1(\Omega_1)$. Thus, we have proved the theorem for $l = 0$ and $1 < p \leqslant 3/2$, since $\Omega_1 = \Omega'$ for $N = 1$.

Assume now that p is an arbitrary number from $(3/2, \infty)$ ($l = 0$). We take an arbitrary number p' from (r, s), where $r = 3/2$ for $p \in (3/2, 3)$ and $r = p/(1 + p/3)$ for $p \in [3, \infty)$, and $s = \min(3, p)$. The vector-valued functions $\mathbf{v}$ and $\mathbf{h}$, by virtue of what we have proved above, belong to $W_{3/2}^2(\Omega_1)$, and $p(x)$ belongs to $W_{3/2}^1(\Omega_1)$. However, according to the aforementioned imbedding theorem of Sobolev, the vector-valued functions $\mathbf{v}$ and $\mathbf{h}$ from $W_{3/2}^2(\Omega_1)$ are summable over Ω_1 to an arbitrary finite power, while their derivatives $\partial \mathbf{v}/\partial x_k$, $\partial \mathbf{h}/\partial x_k$ and $p(x)$ are summable to power 3. Consequently, $v_k \partial \mathbf{v}/\partial x_k$ and $h_k \partial \mathbf{h}/\partial x_k$ are summable over Ω_1 to the power p'.

We note, by virtue of Theorem 6, that under the conditions of the theorem being proved $\mathbf{h}_0$ belongs to $W_p^1(\Omega_1) \cap W_2^1(\Omega_1)$ (and even to $W_p^2(\Omega_1) \cap W_2^1(\Omega_1)$). We show that $h_{0k} \partial \mathbf{h}/\partial x_k$ and $h_k \partial \mathbf{h}_0/\partial x_k$ belong to $L_{p'}(\Omega_1)$. In fact,

$$\left\| h_{0k} \frac{\partial \mathbf{h}}{\partial x_k} \right\|_{p', \Omega_1} \leqslant \| \mathbf{h}_0 \|_{3p'/(3-p'), \Omega_1} \left\| \frac{\partial \mathbf{h}}{\partial x_k} \right\|_{3, \Omega_1} \leqslant C_{30} \| \mathbf{h}_0 \|^{(1)}_{p, \Omega_1} \| \mathbf{h} \|^{(2)}_{3/2, \Omega_1}$$

and

$$\left\| h_k \frac{\partial \mathbf{h}_0}{\partial x_k} \right\|_{p', \Omega_1} \leqslant \| \mathbf{h} \|_{p'p/(p-p'), \Omega_1} \left\| \frac{\partial \mathbf{h}_0}{\partial x_k} \right\|_{p, \Omega_1} \leqslant C_{31} \| \mathbf{h} \|^{(2)}_{3/2, \Omega_1} \| \mathbf{h}_0 \|^{(1)}_{p, \Omega_1}.$$

It follows from this that $\xi_2(x) \in L_{p'}(\Omega_1)$. It is obvious that $\rho_2(x) \in W_{p'}^1(\Omega_1)$ and $\mathbf{b}_2(x) \in W_{p'}^{1-1/p'}(S_1)$. Consequently, according to Theorem 1, $\mathbf{v}(x)$ belongs to $W_{p'}^2(\Omega_2)$ and $p(x)$ belongs to $W_{p'}^1(\Omega_2)$.

Precisely as was done above, we establish that $\boldsymbol{\eta}_2(x) \in L_{p'}(\Omega_1)$, $\delta_2(x) \in W_{p'}^1(\Omega_1) \omega_2(x) \in W_{p'}^{1-1/p'}(S_1)$, and, therefore, by virtue of Theorem 7, that $\mathbf{h}(x) \in W_{p'}^2(\Omega_2)$.

From what has been proved, and from Sobolev's imbedding theorems, it follows that $\mathbf{v}$ and $\mathbf{h}$ belong to $C(\Omega_2)$ and that $\partial \mathbf{v}/\partial x_k$, $\partial \mathbf{h}/\partial x_k$, and $p(x)$ belong to $L_q(\Omega_2)$ for $q \leqslant 3p'/(3 - p')$.

We show now that $\xi_3(x)$ belongs to $L_p(\Omega_2)$. Assume that $p \in (3/2, 3)$. Then

$$\left\| v_k \frac{\partial \mathbf{v}}{\partial x_k} \right\|_{p, \Omega_2} \leqslant \max_{x \in \Omega_2} |\mathbf{v}(x)| \left\| \frac{\partial \mathbf{v}}{\partial x_k} \right\|_{p, \Omega_2} \leqslant C_{32} (\|\mathbf{v}\|^{(2)}_{p', \Omega_2})^2.$$

In exactly the same way we prove that $h_k \partial \mathbf{h}/\partial x_k \in L_p(\Omega_2)$.

Further, we have

$$\left\| h_{0k} \frac{\partial \mathbf{h}}{\partial x_k} \right\|_{p, \Omega_2} \leqslant \|\mathbf{h}_0\|_{3p/(3-p), \Omega_2} \left\| \frac{\partial \mathbf{h}}{\partial x_k} \right\|_{3, \Omega_2} \leqslant C_{33} \|\mathbf{h}_0\|^{(1)}_{p, \Omega_2} \|\mathbf{h}\|^{(2)}_{p', \Omega_2},$$

$$\left\| h_k \frac{\partial \mathbf{h}_0}{\partial x_k} \right\|_{p, \Omega_2} \leqslant \max_{x \in \Omega_2} |\mathbf{h}(x)| \left\| \frac{\partial \mathbf{h}_0}{\partial x_k} \right\|_{p, \Omega_2} \leqslant C_{34} \|\mathbf{h}\|^{(2)}_{p', \Omega_2} \|\mathbf{h}_0\|^{(1)}_{p, \Omega_2}.$$

From this it follows that $\xi_3(x) \in L_p(\Omega_2)$ for $p \in (3/2, 3)$.

Assume now that $p \in [3, \infty)$. The terms $v_k \partial \mathbf{v}/\partial x_k$, $h_k \partial \mathbf{h}/\partial x_k$ are estimated in like manner; for example,

$$\left\| h_k \frac{\partial \mathbf{h}}{\partial x_k} \right\|_{p, \Omega_2} \leqslant \max_{x \in \Omega_2} |\mathbf{h}(x)| \left\| \frac{\partial \mathbf{h}}{\partial x_k} \right\|_{p, \Omega_2} \leqslant \max_{x \in \Omega_2} |\mathbf{h}(x)| \left\| \frac{\partial \mathbf{h}}{\partial x_k} \right\|_{3p'/(3-p'), \Omega_2} \leqslant C_{35} (\|\mathbf{h}\|^{(2)}_{p', \Omega_2})^2,$$

since $p < 3p'/(3 - p')$.

Further,

$$\left\| h_{0k} \frac{\partial \mathbf{h}}{\partial x_k} \right\|_{p, \Omega_2} \leqslant \|\mathbf{h}_0\|_{3p'p/(p'p-3(p-p')), \Omega_2} \left\| \frac{\partial \mathbf{h}}{\partial x_k} \right\|_{3p'/(3-p'), \Omega_2} \leqslant C_{36} \|\mathbf{h}_0\|^{(1)}_{p, \Omega_2} \|\mathbf{h}\|^{(2)}_{p', \Omega_2},$$

$$\left\| h_k \frac{\partial \mathbf{h}_0}{\partial x_k} \right\|_{p, \Omega_2} \leqslant \max_{x \in \Omega_2} |\mathbf{h}(x)| \left\| \frac{\partial \mathbf{h}_0}{\partial x_k} \right\|_{p, \Omega_2} \leqslant C_{37} \|\mathbf{h}\|^{(2)}_{p', \Omega_2} \|\mathbf{h}_0\|^{(1)}_{p, \Omega_2}.$$

Consequently, $\xi_3(x) \in L_p(\Omega_2)$. It is obvious that $\rho_3(x) \in W_p^1(\Omega_2)$ and $\mathbf{b}_3(x) \in W_p^{1-1/p}(S_2)$. It follows from this and from Theorem 1 that $\mathbf{v}(x) \in W_p^2(\Omega_3)$ and $p(x) \in W_p^1(\Omega_3)$. In a completely analogous way we may establish that $\mathbf{h}(x) \in W_p^2(\Omega_3)$. It follows from (74) that $\mathbf{E}(x)$ belongs to $W_p^1(\Omega_3)$.

Putting $N = 3$, we find that $\mathbf{v}$ and $\mathbf{h}$ belong to $W_p^2(\Omega')$ while $p(x)$ and $\mathbf{E}(x)$ belong to $W_p^1(\Omega')$. By the same token, the theorem is proved for $l = 0$ even for arbitrary $p > 1$. If $l > 0$, similar reasoning is repeated several times. If the boundary of Ω'' contains the portion S'' of the surface S_2, it is then necessary in proving the theorem to use Theorem 3.2 from [5]. This completes the proof of the theorem.

THEOREM 11. *Let* $\mathbf{v}$ *and* $\mathbf{h}$ *be a generalized solution of problem* (60)–(64). *If*

$$S'' \in C^{l+3}, \quad \mathbf{f}(x) \in C^l(\Omega''), \quad \mathbf{j}_0(x) \in C^{l+1}(\Omega''), \quad \beta(x) \in C^{l+2}(S''),$$

where $l > 0$ *is nonintegral, then*

$$\mathbf{v}(x) \in C^{l+2}(\Omega'), \quad p(x) \in C^{l+1}(\Omega'), \quad \mathbf{h}(x) \in C^{l+2}(\Omega'), \quad \mathbf{E} \in C^{l+1}(\Omega').$$

PROOF. We first prove the theorem assuming that $0 < l < 1$. According to a theorem of type 6 for Hölder spaces, $\mathbf{h}_0(x)$ belongs to $C^{l+2}(\Omega'')$. By virtue of Theorem 10, under the conditions of Theorem 11, $\mathbf{v}$ and $\mathbf{h}$ belong to $W_p^2(\Omega_3)$ and $p(x)$ belongs to $W_p^{-1}(\Omega_3)$, where $p > 1$ is an arbitrary number. Let $p = 3/(1 - l)$. Using a theorem concerning the imbedding of $W_p^2(\Omega_3)$ in $C^h(\Omega_3)$, $0 < h < 1$, with

$2p > 3$ (see, for example, [9], Chapter II, Theorem 2.1), we find that $\partial\mathbf{v}/\partial x_k$, $\partial\mathbf{h}/\partial x_k$ and $p(x)$ belong to $C^l(\Omega_3)$. Consequently, $\boldsymbol{\xi}_4(x) \in C^l(\Omega_3)$. It is clear that $\rho_4(x) \in C^{l+1}(\Omega_3)$ and $\mathbf{b}_4(x) \in C^{l+1}(S_3)$. Suppose that the boundary of Ω'' contains the portion S'' of S_1. Then, on the basis of Theorem 2, $\mathbf{v}(x)$ belongs to $C^{l+2}(\Omega_4)$ and $p(x)$ belongs to $C^{l+1}(\Omega_4)$. (In case Ω'' is adjacent to the boundary S_2, it is necessary to use Theorem 3.1 from [5].)

In a completely analogous way it may be established that $\mathbf{h}(x) \in C^{l+2}(\Omega_4)$ and that $\mathbf{E}(x) \in C^{l+1}(\Omega_4)$. By the same token, the theorem is proved for $0 < l < 1$, since $\Omega_4 = \Omega'$ for $N = 4$.

If $[l] > 0$, it is necessary to repeat similar reasoning several times. This completes the proof of the theorem.

§5. Some applications

We consider an electromagnetic trough–an apparatus designed for the transport of a molten metal (see [16], pp. 334–338). The metal moves under the action of a running periodic magnetic field generated by an inductor located below the trough, the trough being open from above.

In this section we restrict ourselves to considering plane-parallel steady flows in a curvilinear strip, the upper boundary of which is free while the lower boundary (the bottom) is a solid rectilinear wall. We assume that the magnetic field is located in the plane of the flow. Then the electric field intensity is constant throughout the flow region. With no loss of generality we assume that the mean depth of the fluid is equal to unity.

The equations (1)–(3) assume the following form:

$$-\nu\Delta\mathbf{v} + v_k\frac{\partial\mathbf{v}}{\partial x_k} - \frac{\mu}{\rho}H_k\frac{\partial\mathbf{H}}{\partial x_k} = -\frac{1}{\rho}\operatorname{grad}\left(p + \frac{\mu\mathbf{H}^2}{2}\right)\operatorname{div}\mathbf{v} = 0, \tag{75}$$

$$\frac{\partial H_2}{\partial x_1} - \frac{\partial H_1}{\partial x_2} = \sigma[E_0 + \mu(v_1H_2 - v_2H_1)], \quad \operatorname{div}\mathbf{H} = 0. \tag{76}$$

(All functions are specified in the strip $-\infty < x_1 < +\infty$, $-1 < x_2 < f(x_1)$; external forces and currents are absent.) Here $E_0 = \text{const}$ and $x_2 = f(x_1)$ is the equation of the free surface, which, we assume, projects single-valuedly onto the bottom $x_2 = -1$. We seek solutions $\mathbf{v}$, $\mathbf{H}$, p of the system (75), (76) and the form of the free boundary $f(x_1)$, periodic in x_1 with period l.

We introduce the following notation:

$$\Omega = \{x_1, x_2 \colon 0 < x_1 < l, \ -1 < x_2 < f(x_1)\},$$

$$\Sigma = \{x_1, x_2 \colon 0 < x_1 < l, \ x_2 = -1\}, \ \Gamma = \{x_1, x_2 \colon 0 < x_1 < l, \ x_2 = f(x_1)\}.$$

We require the following conditions to be satisfied: on the free surface,

$$v_n|_\Gamma = 0, \ t_\tau|_\Gamma = 0, \ H_n|_\Gamma = 0, \tag{77}$$

$$t_n|_\Gamma = \sigma_1\left(\frac{f'}{\sqrt{1+f'^2}}\right)', \tag{78}$$

on the bottom,

$$\mathbf{v}|_\Sigma = 0, \ H_2|_\Sigma = \beta_1(x_1) \tag{79}$$

and, finally,

$$\int_0^l f(x_1)\,dx_1 = 0. \tag{80}$$

Here the quantity $\sigma_1(f'/\sqrt{1+f'^2})'$, in accordance with Laplace's formula, is the capillary pressure, and the constant $\sigma_1 > 0$ is the coefficient of surface tension; $\beta(x_1)$ is a given l-periodic function. The condition (80) signifies that the mean depth of the fluid is equal to unity. This condition (or an analog of it) is necessary for the single-valued determination of the desired motion with a free boundary. In order to eliminate the possiblity of contact of the free boundary with the bottom, we require that $f(x_1) \geqslant -\delta > -1$ for all x_1 (δ = const > 0). Further, we assume that

$$\int_0^l \beta(x_1)\,dx_1 = 0. \tag{81}$$

Theorem 8 continues to hold for the boundary value problem (75)–(77), (79) (with $f(x_1)$ given and subject to periodicity conditions) if $\beta(x_1) \in W_2^{1/2}(0, l)$ and if the condition (81) is satisfied; while for the validity of Theorem 9 it is sufficient that $|E_0|$ be small. In the sequel we refer to this problem as the *auxiliary problem*.(1)

LEMMA 4. *If*

$$f(x_1) \in C^{3+\alpha}[0, l], \quad \beta(x_1) \in C^{2+\alpha}[0, l], \quad 0 < \alpha < 1,$$

then the following inequality holds for a generalized solution of the auxiliary problem:

$$\|\mathbf{v}\|_\Omega^{(2+\alpha)} + \|\mathbf{H}\|_\Omega^{(2+\alpha)} + \|\nabla p\|_\Omega^{(\alpha)} \leqslant C_{38}(|E_0| + \|\beta\|_{[0,l]}^{(2+\alpha)}), \tag{82}$$

where C_{38} is independent of f, β and E_0 if $\|f\|_{[0,l]}^{(3+\alpha)} \leqslant \delta$ and $|E_0| + \|\beta\|_{[0,1]}^{(2+\alpha)} \leqslant \varepsilon$.

For the proof of the lemma it is sufficient to use Schauder's method, which, applied to elliptic equations, is presented for Hölder norms in [17] (see §§6, 7) and in [9] (see Chapter III, §2), and also the reasoning employed in the proofs of Theorems 10 and 11.

We now investigate the continuous dependence of a solution of the auxiliary problem on the boundary Γ of Ω. We denote by $\hat{C}^{3+\alpha}$ the subspace of $C^{3+\alpha}(-\infty, \infty)$ consisting of all l-periodic functions satisfying (80), and by $S(\delta)$ the closed ball in $\hat{C}^{3+\alpha}$. We consider the family of domains $\{\Omega\}$ bounded from above by various curves $\Gamma: x_2 = f(x_1) \in S(\delta)$. The pressure $p(x)$ is fixed by the requirement $\int_\Omega p(x)\,dx = 0$, while the constant term to within which the pressure is

(1) In the proof of Theorem 8 we have used Theorem 4.2 from [10] concerning the single-valued representation of an arbitrary vector-valued function $\mathbf{w} \in J(\Omega)$ in the form $\mathbf{w} = \operatorname{rot} \mathbf{u}$, where $\mathbf{u} \in \mathring{J}_n^1(\Omega)$; it enables us to introduce a new inner product in the space $\mathring{J}_n^1(\Omega)$. Reasoning in the manner of Lemmas 2.2 and 5.2 in [10], we readily see that the theorem analogous to Theorem 4.2 also holds for vector-valued functions satisfying the condition of periodicity.

determined is chosen so that the mean value of the left side of (78) on $(0, l)$ is equal to zero. The need for such a choice of the constant term is a consequence of the periodicity of $f(x_1)$.

Let f and f_1 be two functions from the ball $S(\delta)$. To them correspond two auxiliary problems in the domains Ω and Ω_1, bounded from above by the curves Γ: $x_2 = f(x_1)$ and Γ_1: $x_2 = f_1(x_1)$. We rewrite the second auxiliary problem in the new notation, replacing x by y, and $\mathbf{v}$, $\mathbf{H}$, p, and $\mathbf{t}$ by $\mathbf{v}^{(1)}$, $\mathbf{H}^{(1)}$, $p^{(1)}$, and $\mathbf{t}^{(1)}$, respectively. We obtain

$$-\nu\Delta\mathbf{v}^{(1)} + v_k^{(1)}\frac{\partial\mathbf{v}^{(1)}}{\partial y_k} - \frac{\mu}{\rho}H_k^{(1)}\frac{\partial\mathbf{H}^{(1)}}{\partial y_k} = -\frac{1}{\rho}\operatorname{grad}\left(p^{(1)} + \frac{\mu\mathbf{H}^{(1)^2}}{2}\right),$$
$$\operatorname{div}\mathbf{v}^{(1)} = 0, \quad (83)$$

$$\frac{\partial H_2^{(1)}}{\partial y_1} - \frac{\partial H_1^{(1)}}{\partial y_2} = \sigma[E_0 + \mu(v_1^{(1)}H_2^{(1)} - v_2^{(1)}H_1^{(1)})], \operatorname{div}\mathbf{H}^{(1)} = 0, \quad (84)$$

$$v_n^{(1)}|_{\Gamma_1} = 0,\ t_\tau^{(1)}|_{\Gamma_1} = 0,\ H_n^{(1)}|_{\Gamma_1} = 0,\ \mathbf{v}^{(1)}|_\Sigma = 0,\ H_2^{(1)}|_\Sigma = \beta(x_1), \quad (85)$$

$$\mathbf{v}^{(1)}(y_1 + l,\ y_2) \equiv \mathbf{v}^{(1)}(y_1,\ y_2),\ \mathbf{H}^{(1)}(y_1 + l,\ y_2)$$
$$\equiv \mathbf{H}^{(1)}(y_1,\ y_2),\ p^{(1)}(y_1 + l,\ y_2) \equiv p^{(1)}(y_1,\ y_2), \quad (86)$$

$$\int_{\Omega_1} p^{(1)}(y)\,dy = 0. \quad (87)$$

In (83) and (84) we go over to the new variables x in accordance with the formulas

$$x_1 = y_1,\ x_2 = (1 + y_2)\frac{1 + f(y_1)}{1 + f_1(y_1)} - 1 \quad (88)$$

and to the new unknown functions $\mathbf{w}(x) = \mathbf{v}^{(1)}(y)$, $\mathbf{B}(x) = \mathbf{H}^{(1)}(y)$ and $q(x) = p^{(1)}(y)$. Moreover, Ω_1 goes over into Ω. From the transformed equations (83) and (84) we subtract the corresponding equations (75) and (76). This gives a linear system of equations in the functions $\mathbf{u} = \mathbf{w} - \mathbf{v}$, $\mathbf{h} = \mathbf{B} - \mathbf{H}$ and $r = q - p$ of the following form:

$$-\nu\Delta\mathbf{u} + w_k\frac{\partial\mathbf{u}}{\partial x_k} + u_k\frac{\partial\mathbf{v}}{\partial x_k} - \frac{\mu}{\rho}\left(h_k\frac{\partial\mathbf{B}}{\partial x_k} + H_k\frac{\partial\mathbf{h}}{\partial x_k}\right)$$
$$= -\frac{1}{\rho}\operatorname{grad}\left[r + \frac{\mu}{2}\mathbf{h}\cdot(\mathbf{B} + \mathbf{H})\right] + \zeta, \quad (89)$$

$$\operatorname{div}\mathbf{u} = \rho_1(x),\ \operatorname{div}\mathbf{h} = \vartheta(x), \quad (90)$$

$$\frac{\partial h_2}{\partial x_1} - \frac{\partial h_1}{\partial x_2} = \sigma\mu(u_1B_2 - u_2B_1 + v_1h_2 - v_2h_1) + \eta(x). \quad (91)$$

Here ζ is the vector with the components

$$\zeta_i = 2\nu\omega(1 + x_2)\frac{\partial^2 w_i}{\partial x_1\partial x_2} + \nu\left[\omega^2(1 + x_2)^2 - \frac{(2 + f + f_1)\lambda}{(1 + f_1)^2}\right]\frac{\partial^2 w_i}{\partial x_2^2}$$
$$+\left[\nu(\omega' + \omega^2)(1 + x_2) + \frac{\lambda w_2}{1 + f_1} - \omega(1 + x_2)w_1\right]\frac{\partial w_i}{\partial x_2} + \frac{\mu}{\rho}\left[\omega(1 + x_2)B_1\right.$$
$$\left.-\frac{\lambda B_2}{1 + f_1}\right]\frac{\partial B_i}{\partial x_2} - \frac{\varepsilon_i}{\rho}\frac{\partial}{\partial x_2}\left(q + \frac{\mu}{2}B^2\right),\ i = 1,\ 2,$$

where

$$\varepsilon_1 = \omega(1+x_2),\ \varepsilon_2 = -\frac{\lambda}{1+f_1},\quad \omega = \frac{\lambda f_1' - \lambda'(1+f_1)}{(1+f)(1+f_1)},\quad \lambda = f_1 - f;$$

$$\rho_1(x) = \frac{\lambda}{1+f_1}\frac{\partial w_2}{\partial x_2} - \omega(1+x_2)\frac{\partial w_1}{\partial x_2},\quad \eta(x) = -\frac{\lambda}{1+f_1}\frac{\partial B_1}{\partial x_2} - \omega(1+x_2)\frac{\partial B_2}{\partial x_2},$$

$$\vartheta(x) = \frac{a}{1+f_1}\frac{\partial B_2}{\partial x_2} - \omega(1+x_2)\frac{\partial B_1}{\partial x_2}.$$

Similarly, we obtain the boundary conditions

$$u_n|_\Gamma = \gamma_1(x),\ t_\tau(\mathbf{u})|_\Gamma = \chi(x),\ h_n|_\Gamma = \gamma_2(x),\ \mathbf{u}|_\Sigma = 0,\ h_2|_\Sigma = 0, \tag{92}$$

$$\mathbf{u}(x_1+l,\ x_2) \equiv \mathbf{u}(x_1,\ x_2),\ \mathbf{h}(x_1+l,\ x_2) \equiv \mathbf{h}(x_1,\ x_2),$$

$$r(x_1+l,\ x_2) \equiv r(x_1,\ x_2), \tag{93}$$

$$\int_\Omega r(x)\,dx = \varkappa, \tag{94}$$

where

$$\gamma_1(x) = \frac{\lambda' w_1}{\sqrt{1+f'^2}},\quad \gamma_2(x) = \frac{\lambda' B_1}{\sqrt{1+f'^2}},\quad \varkappa = -\int_\Omega \frac{\lambda q}{1+f}\,dx,$$

$$\chi(x) = \frac{\nu}{1+f'^2}\Big\{2\lambda'\frac{\partial w_1}{\partial x_1} + \Big[2f_1'\omega(1+x_2) + \frac{\lambda(1-f'^2)}{1+f_1} + \lambda'(f'+f_1')\frac{1+f}{1+f_1}\Big]$$
$$\times \frac{\partial w_1}{\partial x_2} + \lambda'(f'+f_1')\frac{\partial w_2}{\partial x_1} - \Big[\omega(1+x_2)(1-f_1'^2) + \frac{2\lambda' + 2(f\lambda' - \lambda f')}{1+f_1}\Big]\frac{\partial W_2}{\partial x_2}\Big\}.$$

First of all, we reduce the nonhomogeneous conditions to homogeneous ones. We do this by constructing functions $\mathbf{u}_0(x)$ and $r_0(x)$ possessing the following properties:

$$\nu\Delta\mathbf{u}_0 - \frac{1}{\rho}\operatorname{grad} r_0 = 0,\ \operatorname{div}\mathbf{u}_0 = \rho_1(x), \tag{95}$$

$$u_{0n}|_\Gamma = \gamma_1(x),\ t_\tau(\mathbf{u}_0)|_\Gamma = \chi(x),\ \mathbf{u}_0|_\Sigma = 0, \tag{96}$$

$$\mathbf{u}_0(x_1+l,\ x_2) \equiv \mathbf{u}_0(x_1,\ x_2),\ r_0(x_1+l,\ x_2) \equiv r_0(x_1,\ x_2), \tag{97}$$

$$\int_\Omega r_0(x)\,dx = \varkappa, \tag{98}$$

$$\|\mathbf{u}_0\|_\Omega^{(2+\alpha)} + \|r_0\|_\Omega^{(1+\alpha)} \leqslant C_{39}\|\mathbf{w}\|_\Omega^{(2+\alpha)}\|\lambda\|_{[0,\,l]}^{(3+\alpha)}. \tag{99}$$

In order that the problem (95)–(98) be solvable, it is necessary that

$$\int_\Omega \rho_1(x)\,dx = \int_\Gamma \gamma_1(x)\,d\Gamma. \tag{100}$$

We show that the condition (100) is satisfied. In fact, we transform the expressions $\rho_1(x)$ and $\gamma_1(x)$, making use of the equation $\operatorname{div}\mathbf{v}^{(1)} = 0$ and the condition $v_n^{(1)}|_{\Gamma_1} = 0$, which the vector-valued function $\mathbf{v}^{(1)}(y) = \mathbf{w}(x)$ satisfies. We obtain $\rho_1(x) = \operatorname{div}\mathbf{w}$ and $\gamma_1(x) = w_n|_\Gamma$, from whence we get (100).

We seek the function $r_0(x)$ in the form $r_0(x) = r_0'(x) + r_0''$, where $r_0'(x) \in L_2(\Omega) \ominus \{1\}$ and the constant r_0'' is equal to $\kappa/|\Omega|$.

Following reasoning analogous to that introduced in the proof of Theorem 2, we establish the unique solvability of the problem (95)–(97) in the space $C^{2+\alpha}(\Omega) \times C^{1+\alpha}(\Omega)$ subject to the condition

$$\int_\Omega r_0'(x)\,dx = 0,$$

where

$$\|\mathbf{u}_0\|_\Omega^{(2+\alpha)} + \|r_0'\|_\Omega^{(1+\alpha)} \leqslant C_{40}\left(\|\rho_1\|_\Omega^{(1+\alpha)} + \|\gamma_1\|_\Gamma^{(2+\alpha)} + \|\chi\|_\Gamma^{(1+\alpha)}\right);$$

from this, taking into account the form of the functions $\rho_1(x)$, $\gamma_1(x)$, $\chi(x)$, and κ, we obtain (99).

Next, we construct the vector-valued function $\mathbf{h}_0(x)$ possessing the following properties:

$$\frac{\partial h_{02}}{\partial x_1} - \frac{\partial h_{01}}{\partial x_2} = 0, \quad \operatorname{div} \mathbf{h}_0 = \vartheta(x), \tag{101}$$

$$h_{0n}|_\Gamma = \gamma_2(x), \quad h_{02}|_\Sigma = 0, \quad \mathbf{h}_0(x_1 + l, x_2) \equiv \mathbf{h}_0(x_1, x_2), \tag{102}$$

$$\|\mathbf{h}_0\|_\Omega^{(2+\alpha)} \leqslant C_{41}\|\mathbf{B}\|_\Omega^{(2+\alpha)} \|\lambda\|_{[0,l]}^{(3+\alpha)}. \tag{103}$$

The unique solvability of the problem (101), (102) and the validity of the estimate (103) can be established from considerations similar to those used above upon employing an analog of Theorem 6 for Hölder spaces.

We now introduce the new unknown functions

$$\mathbf{u}' = \mathbf{u} - \mathbf{u}_0, \quad r' = r - r_0, \quad \mathbf{h}' = \mathbf{h} - \mathbf{h}_0.$$

From (89)–(91) we then obtain

$$-\nu\Delta\mathbf{u}' + w_k \frac{\partial \mathbf{u}'}{\partial x_k} + u_k' \frac{\partial \mathbf{v}}{\partial x_k} - \frac{\mu}{\rho}\left(h_k' \frac{\partial \mathbf{B}}{\partial x_k} + H_k \frac{\partial \mathbf{h}'}{\partial x_k}\right) = -\frac{1}{\rho}\operatorname{grad}\left[r' + \frac{\mu}{2}\mathbf{h}' \cdot (\mathbf{B} + \mathbf{H})\right] + \zeta', \tag{104}$$

$$\operatorname{div} \mathbf{u}' = 0, \quad \operatorname{div} \mathbf{h}' = 0, \tag{105}$$

$$\frac{\partial h_2'}{\partial x_1} - \frac{\partial h_1'}{\partial x_2} = \sigma\mu\,(u_1' B_2 - u_2' B_1 + v_1 h_2' - v_2 h_1') + \eta'(x), \tag{106}$$

where

$$\zeta' = \zeta - w_k \frac{\partial \mathbf{u}_0}{\partial x_k} - u_{0k}\frac{\partial \mathbf{v}}{\partial x_k} + \frac{\mu}{\rho}\left(h_{0k}\frac{\partial \mathbf{B}}{\partial x_k} + H_k \frac{\partial \mathbf{h}_0}{\partial x_k}\right) - \frac{1}{\rho}\operatorname{grad}\left[\frac{\mu}{2}\mathbf{h}_0 \cdot (\mathbf{B} + \mathbf{H})\right],$$

$$\eta'(x) = \sigma\mu\,(u_{01}B_2 - u_{02}B_1 + v_1 h_{02} - v_2 h_{01}) + \eta(x),$$

and from (92)–(94) we obtain the conditions

$$u_n'|_\Gamma = 0, \quad t_\tau(\mathbf{u}')|_\Gamma = 0, \quad h_n'|_\Gamma = 0, \quad \mathbf{u}'|_\Sigma = 0, \quad h_2'|_\Sigma = 0, \tag{107}$$

$$\mathbf{u}'(x_1 + l, x_2) \equiv \mathbf{u}'(x_1, x_2), \quad \mathbf{h}'(x_1 + l, x_2) \equiv \mathbf{h}'(x_1, x_2), \quad r'(x_1 + l, x_2) \equiv r'(x_1, x_2), \tag{108}$$

$$\int_\Omega r'(x)\,dx = 0. \tag{109}$$

Let the conditions of Lemma 4 be satisfied. We note that if $f, f_1 \in S(\delta)$ $(\delta < 1)$, then the mapping (88) and the mapping inverse to it belong to $C^{3+\alpha}$. From this and from (82), valid for the solution $\mathbf{v}^{(1)}$, $p^{(1)}$, $\mathbf{H}^{(1)}$ of the problem (83)–(87), we obtain an estimate for $\mathbf{w}(x) = \mathbf{v}^{(1)}(y)$, $q(x) = p^{(1)}(y)$ and $\mathbf{B}(x) = \mathbf{H}^{(1)}(y)$:

$$\|\mathbf{w}\|_\Omega^{(2+\alpha)} + \|\mathbf{B}\|_\Omega^{(2+\alpha)} + \|\nabla q\|_\Omega^{(\alpha)} \leqslant C_{42}(|E_0| + \|\beta\|_{[0,l]}^{(2+\alpha)}). \tag{110}$$

In addition, the mapping (88) maps l-periodic functions of y_1 into l-periodic functions of x_1. This implies periodicity in x_1 of the functions $\zeta'(x)$ and $\eta'(x)$. Since $f, f_1 \in S(\delta)$, from (82), (99), (103), and (110) and the definition of these functions it follows that $\zeta'(x) \in C^\alpha(\Omega)$, $\eta'(x) \in C^{1+\alpha}(\Omega)$ and

$$\|\zeta'\|_\Omega^{(\alpha)} + \|\eta'\|_\Omega^{(1+\alpha)} \leqslant C_{43}(|E_0| + \|\beta\|_{[0,l]}^{(2+\alpha)})\|\lambda\|_{[0,l]}^{(3+\alpha)}. \tag{111}$$

Reasoning as we did in the proofs of Theorems 8, 10, and 11, we can easily show that the problem (104)–(109) has a solution $\mathbf{u}'$, r', $\mathbf{h}'$ from $C^{2+\alpha}(\Omega) \times C^{1+\alpha}(\Omega) \times C^{2+\alpha}(\Omega)$. It is important to note that for small ε the solution of this problem is unique in the class $C^{2+\alpha}(\Omega) \times C^{1+\alpha}(\Omega) \times C^{2k+\alpha}(\Omega)$ (see Theorem 9). Therefore, as a consequence of Theorem 2.1 from [5] and the analog of Theorem 1.1 from [11] for Hölder spaces, we have the estimate

$$\|\mathbf{u}'\|_\Omega^{(2+\alpha)} + \|\mathbf{h}'\|_\Omega^{(2+\alpha)} + \|\nabla r'\|_\Omega^{(\alpha)} \leqslant C_{44}(|E_0| + \|\beta\|_{[0,l]}^{(2+\alpha)})\|\lambda\|_{[0,l]}^{(3+\alpha)},$$

whence, taking (99), (103), and (110) into account, we obtain

$$\|\mathbf{u}\|_\Omega^{(2+\alpha)} + \|\mathbf{h}\|_\Omega^{(2+\alpha)} + \|\nabla r\|_\Omega^{(\alpha)} \leqslant C_{45}(|E_0| + \|\beta\|_{[0,l]}^{(2+\alpha)})\|\lambda\|_{[0,l]}^{(3+\alpha)}. \tag{112}$$

Thus, we have proved the following lemma.

LEMMA 5. *If $f, f_1 \in S(\delta)$ $(\delta < 1)$ and $|E_0| + \|\beta\|_{[0,l]}^{(2+\alpha)} < \varepsilon$, where $\varepsilon > 0$ is sufficiently small, then the estimate* (112) *is valid.*

We now study the operator

$$A(f) = -\frac{p}{\rho}\Big|_\Gamma + \nu\mathbf{T}(\mathbf{v})\cdot\mathbf{n}|_\Gamma - \frac{p_0}{\rho},$$

where $\mathbf{T}(\mathbf{v})$ is the vector with the components

$$T_i(\mathbf{v}) = \sum_{j=1}^{2}\left(\frac{\partial v_i}{\partial x_j} + \frac{\partial v_j}{\partial x_i}\right)n_j$$

along the coordinate axes x_i, $i = 1, 2$; $p(x) \in L_2(\Omega) \ominus \{1\}$ and

$$p_0 = \rho/l\int_0^l [-p/\rho|_\Gamma + \nu\mathbf{T}(\mathbf{v})\cdot\mathbf{n}|_\Gamma]\,dx_1.$$

We have the following lemma.

LEMMA 6. *Let the conditions of Lemma 5 be satisfied. Then $A(f) \in C^{1l+\alpha}[0, l]$ and*

$$\|A(f_1) - A(f)\|_{[0,l]}^{(1+\alpha)} \leqslant C_{46}(|E_0| + \|\beta\|_{[0,l]}^{(2+\alpha)})\|\lambda\|_{[0,l]}^{(3+\alpha)}. \tag{113}$$

PROOF. We have

$$A(f_1)-A(f)=-\frac{r}{\rho}\Big|_\Gamma+\nu\mathbf{T}(\mathbf{u})\mathbf{n}\big|_\Gamma+\nu R-\frac{1}{l}\int_0^l\Big[-\frac{r}{\rho}\Big|_\Gamma+\nu\mathbf{T}(\mathbf{u})\mathbf{n}\big|_\Gamma+\nu R\Big]dx_1,$$

where

$$R=\frac{2}{1+f_1'^2}\Big\{\frac{\lambda'(f'+f_1')f'^2}{1+f'^2}\frac{\partial w_1}{\partial x_1}-\frac{\lambda'(1-f'f_1')}{1+f'^2}\frac{\partial w_2}{\partial x_1}+\Big[\frac{\lambda f_1'}{1+f_1}-\frac{\lambda'(1-f'f_1')}{1+f'^2}$$
$$+\omega(1+f)\Big]\frac{\partial w_1}{\partial x_2}-\Big[\frac{\lambda}{1+f_1}+\frac{\lambda'(f'+f_1')}{1+f'^2}+f_1'\omega(1+f)\Big]\frac{\partial w_2}{\partial x_2}\Big\}.$$

The desired inequality (113) follows from this and from (110) and (112). This proves the lemma.

The relation (78) can now be written in the form

$$-f''+B(f)+\frac{1}{\sigma_1}A(f)=0, \tag{114}$$

where $B(f)=f''[1-(1+f'^2)^{3/2}]$. The equation (114) is equivalent to

$$f+L^{-1}B(f)+\frac{1}{\sigma_1}L^{-1}A(f)=0, \tag{115}$$

where L^{-1} is the operator inverse to $-d^2/dx_1^2$ and establishes a one-to-one correspondence between $\hat{C}^{3+\alpha}$ and $\hat{C}^{1+\alpha}$.

Applying Lemma 6, we find that there exist numbers $\delta_1\in(0,\delta]$ and $\varepsilon_1\in(0,\varepsilon]$ such that for $\varepsilon^*\in[0,\varepsilon_1]$ the operator $P(f)\equiv L^{-1}B(f)+(1/\sigma_1)L^{-1}A(f)$ maps the ball $S(\delta_1)$ of the space $\hat{C}^{3+\alpha}$ into itself, where

$$\|P(f_1)-P(f)\|_{[0,l]}^{(3+\alpha)}\leqslant C_{47}(\delta_1,\varepsilon_1)\|f_1-f\|_{[0,l]}^{(3+\alpha)},$$

where $C_{47}(\delta_1,\varepsilon_1)<1$ and $C_{47}(\delta_1,\varepsilon_1)\to 0$ as $\delta_1\to 0$ and $\varepsilon_1\to 0$ for arbitrary $f, f_1\in S(\delta_1)$. Consequently, $P(f)$ is a contraction operator. Therefore, on the basis of the Caccioppoli-Banach principle (see, for example, [18], Chapter XIV, §6.2) the equation (115) has a unique solution in the ball $S(\delta_1)$ of $\hat{C}^{3+\alpha}$.

Thus, we have the following theorem.

THEOREM 12. *Let the conditions of Lemma* 5 *be satisfied. Then there exists a number* $\varepsilon_1>0$ *such that for* $\varepsilon^*\in[0,\varepsilon_1]$ *the problem* (75)–(80) *has a unique solution* $\mathbf{v}, p, \mathbf{H}, f$ *in some neighborhood of the zero element of the space* $C^{2+\alpha}(\Omega)\times C^{1+\alpha}(\Omega)\times C^{2+\alpha}(\Omega)\times\hat{C}^{3+\alpha}$.

The author takes this occasion to express his sincere thanks to O. A. Ladyženskaja and V. A. Solonnikov for a number of valuable critical remarks.

BIBLIOGRAPHY

1. V. A. Solonnikov, *On some stationary boundary-value problems in magnetohydrodynamics*, Trudy Mat. Inst. Steklov. **59** (1960), 174–187. (Russian)

2. V. V. Puhnačev, *A plane steady-state free boundary problem for the Navier-Stokes equations*, Ž. Prikl. Meh. i Tehn. Fiz. **1972**, No. 3, 91–102; English transl. in J. Appl. Mech. Tech. Phys. **13** (1972).

3. V. A. Solonnikov, *A priori estimates for second-order parabolic equations*, Trudy Mat. Inst. Steklov. **70** (1964), 133–212; English transl. in Amer. Math. Soc. Transl. (2) **65** (1967).

4. O. A. Ladyženskaja, *On integral estimates, convergence, approximate methods and solutions in functionals for linear elliptic operators*, Vestnik Leningrad. Univ. **1958**, no. 7 (Ser. Mat. Meh. Astr. vyp. 2), 60–69. (Russian)

5. V. A. Solonnikov, *On general boundary value problems for systems which are elliptic in the sense of Douglis and Nirenberg*. II, Trudy Mat. Inst. Steklov **92** (1966), 233–297; English transl. in Proc. Steklov Inst. Math. **92** (1966).

6. V. A. Solonnikov, *On general boundary value problems for systems which are elliptic in the sense of Douglis and Nirenberg*. I, Izv. Akad. Nauk SSSR Ser. Mat. **28** (1964), 665–706; English transl. in Amer. Math. Soc. Transl. (2) **56** (1966).

7. V. A. Solonnikov and V. E. Ščadilov, *On a boundary problem for a stationary system of Navier-Stokes equations*, Trudy Mat. Inst. Steklov. **125** (1973), 196–210; English transl. in Proc. Steklov Inst. Math. **125** (1973).

8. O. A. Ladyženskaja, *Mathematical problems in the dynamics of a viscous imcompressible fluid*, 2nd rev. aug. ed., "Nauka", Moscow, 1970; English transl. of 1st ed., *The mathematical theory of viscous incompressible flow*, Gordon and Breach, New York, 1963; rev., 1969.

9. O. A. Ladyženskaja and N. N. Ural'ceva, *Linear and quasilinear elliptic equations*, 2nd ed., "Nauka", Moscow, 1974; English transl. of 1st ed., Academic Press, 1968.

10. E. B. Byhovskiĭ and N. V. Smirnov, *On orthogonal decomposition of the space of vector-valued functions square-summable on a given domain, and the operators of vector analysis*, Trudy Mat. Inst. Steklov. **59** (1960), 5–36. (Russian)

11. V. A. Solonnikov, *Overdetermined elliptic boundary-value problems*, Zap. Naučn. Sem. Leningrad. Otdel. Mat. Inst. Steklov. (LOMI) **21** (1971), 112–158; English transl. in J. Soviet Math. **1** (1973), no. 4.

12. ______, *On operators admitting regularization*, Zap. Naučn. Sem. Leningrad. Otdel. Mat. Inst. Steklov. (LOMI) **21** (1971), 159–163; English transl. in J. Soviet Math. **1** (1973), no. 4.

13. O. A. Ladyženskaja and V. A. Solonnikov, *Solution of some nonstationary problems of magnetohydrodynamics for a viscous incompressible fluid*, Trudy Mat. Inst. Steklov. **59** (1960), 115–173. (Russian)

14. V. I. Smirnov, *Course in higher mathematics*. Vol. 5, rev. ed., "Nauka", Moscow, 1959; English transl., Pergamon Press, Oxford, and Addison-Wesley, Reading, Mass., 1964.

15. L. N. Slobodeckiĭ, *Estimates in L_p of solutions of elliptic systems*, Dokl. Akad. Nauk SSSR **123** (1958), 616–619. (Russian)

16. G. G. Branover and A. B. Cinober, *Magnetohydrodynamics of incompressible media*, "Nauka", Moscow, 1970. (Russian)

17. S. Agmon, A. Douglis and L. Nirenberg, *Estimates near the boundary for solutions of elliptic partial differential equations satisfying general boundary conditions*. I, Comm. Pure Appl. Math. **12** (1959), 623–727.

18. L. V. Kantorovič and G. P. Akilov, *Functional analysis*, 2nd rev. ed., "Nauka", Moscow, 1977; English transl. of 1st ed., *Functional analysis in normed spaces*, Macmillan, 1964.

Translated by J. F. HEYDA

Труды Матем. Инст. Стеклов.
147(1980)

Proceedings of the Steklov
Institute of Mathematics
1981, Issue 2

ON THE TRACE FORMULA IN THE MULTICHANNEL FRIEDRICHS MODEL

UDC 517.43 + 517.948.35

D. R. JAFAEV

ABSTRACT. Within the framework of the multichannel Friedrichs model an explicit expression is obtained for the "jump" of the trace of the connected part of the resolvent on passing across the continuous spectrum in terms of the scattering matrices for the whole system and its subsystems.
Bibliography: 5 titles.

1. In multichannel problems a trace formula* is an expression for the quantity $\Omega(E + i0)$, $\Omega(z) = \operatorname{Im} \operatorname{Sp} R_c(z)$, where $R_c(z)$ is the so-called connected part of the resolvent (for precise definitions see below), in terms of the scattering matrices of the entire system ($S(E)$) and its various subsystems ($s_\alpha(E)$). Within the framework of the Friedrichs model, in [2] a connection was established between the quantity $\Omega(E + i0)$ and the derivative with respect to E of a certain expression formed from $\ln S$ and $\ln s_\alpha$. In the present work for the same model an expression is obtained (see (1)) for $\Omega(E + i0)$ in terms of a certain combination of the quantities S^*, S', s_α^* and s_α'. This version of the trace formula was suggested in [2], where a system of three particles is considered.

In the multichannel Friedrichs model the original Hilbert space is the space $\mathfrak{H} = L_2(\mathbb{R}^2; N)$ formed from functions $f(\lambda_1, \lambda_2)$ with values in an "auxiliary" Hilbert space N. We set $H = H_0 + V_1 + V_2$ and $H_\alpha = H_0 + V_\alpha$, $\alpha = 1, 2$, where H_0 is the operator of multiplication by $\lambda_1 + \lambda_2$, and V_α is an "integral" operator with kernel $q_\alpha(\lambda_\alpha, \lambda_\alpha')\delta(\lambda_\beta - \lambda_\beta')$, $\alpha \neq \beta$, and $\delta(\cdot)$ is the delta function. We assume that $q_\alpha(\lambda, \lambda')$ are nuclear operators in N and that $q_\alpha^*(\lambda, \lambda') = q_\alpha(\lambda', \lambda)$. It is moreover assumed that in the nuclear norm the functions $q_\alpha(\lambda, \lambda')$ have a certain number of derivatives with respect to λ and λ' and decay together with their derivatives sufficiently fast as $|\lambda| + |\lambda'| \to \infty$.

In the space $\mathfrak{h} = L_2(\mathbb{R}; N)$ we also consider the operators $h_\alpha = h_0 + v_\alpha$, where h_0 is the operator of multiplication by λ, and v_α is an integral operator with kernel $q_\alpha(\lambda, \lambda')$. For simplicity, we assume that the operators h_α and H have no eigenvalues. We denote by $R(z)$, $R_\alpha(z)$, $R_0(z)$, and $r_\alpha(z)$ respectively the resolvents of the operators H, H_α, H_0, and h_α; $R_c = R - R_0 - (R_1 - R_0) - (R_2 - R_0)$, and $S(E)$, $S_\alpha(E)$, and $s_\alpha(E)$ are the scattering matrices for the pairs (H_0, H), (H_0, H_α), and (h_0, h_α) at the point E of the spectral axis.

1980 *Mathematics Subject Classification*. Primary 47A40; Secondary 47A55.

* *Editor's note*. In the Russian literature, Sp A denotes the trace of A.

The object of this paper is to prove the equality

$$2i\Omega(E+i0)=\mathrm{Sp}_{\mathfrak{H}}\Big[S^*(E)S'(E)-\sum_{\alpha=1}^{2}S_\alpha^*(E)S_\alpha'(E)\Big]$$
$$+(4\pi i)^{-1}\sum_{\alpha\neq\beta}\int d\lambda\,\mathrm{sp}_{\mathcal{N}}[s_\alpha^*(\lambda)s_\alpha'(\lambda)s_\beta'(E-\lambda)s_\beta^*(E-\lambda)]. \quad (1)$$

Here and below an integral with no limits indicated is understood to go over $\mathbb{R}$, the indices α and β everywhere take the values 1 and 2, and we often omit the variables over which summation is taken.

In [1] the expression for $2i\Omega(E + i0)$ also decomposes into two terms. The first term is equal to

$$\frac{d}{dE}\mathrm{Sp}_{\mathfrak{H}}[\ln S(E)-\sum\ln S_\alpha(E)], \quad (2)$$

while the second term is formed from the operators $\ln s_\alpha$ in correspondence with the Campbell-Hausdorff formula. In spite of formal equality, the first terms on the right sides of (1) and (2) are different. Thus, the relation (1) and the trace formula of [1] are of distinct character.

2. We set $T(z) = V - VR(z)V$, $T_\alpha(z) = V_\alpha - V_\alpha R_\alpha(z)V_\alpha$, and $t_\alpha(z) = v_\alpha - v_\alpha r_\alpha(z)v_\alpha$. We assume everywhere that $z = E + i\varepsilon$, $\varepsilon > 0$. Integral operators in the space $\mathfrak{h}$ with kernels that have the properties of smoothness and decay of the functions q_α enumerated above are called *operators with smooth kernels*. This definition is carried over to operators depending on the spectral parameter z (or $\bar{z}$); here we assume that the kernels of the operators together with a certain number of derivatives with respect to E are Hölder continuous in E and ε for $\varepsilon > 0$. An analogous definition is introduced for integral operators in $\mathfrak{H}$. It follows from the results of [3] and [4] that the operators $t_\alpha(z)$ have smooth kernels. The kernel of $T_\alpha(z)$ is equal to $t_\alpha(\lambda_\alpha, \lambda_\alpha'; z-\lambda_\beta)\delta(\lambda_\beta-\lambda_\beta')$.

Let $P(z)$ be an integral operator in $\mathfrak{H}$ with smooth kernel $P(\lambda_1, \lambda_2, \lambda_1', \lambda_2'; z)$. We denote by $P'(z)$ the operator with kernel

$$\Big[\frac{1}{2}\Big(\frac{\partial}{\partial\lambda_1}+\frac{\partial}{\partial\lambda_2}+\frac{\partial}{\partial\lambda_2'}+\frac{\partial}{\partial\lambda_1'}\Big)+\frac{\partial}{\partial E}\Big]P(\lambda_1, \lambda_2, \lambda_1', \lambda_2'; z).$$

We apply this definition also to the operators $T_\alpha(z)$ with kernels which contain delta functions. The operator $T_\alpha'(z)$ is given by the kernel $t_\alpha'(\lambda_\alpha, \lambda_\alpha'; z-\lambda_\beta)\cdot\delta(\lambda_\beta-\lambda_\beta')$, where

$$t_\alpha'(\lambda, \lambda'; z)=\frac{1}{2}\Big(\frac{\partial}{\partial\lambda}+\frac{\partial}{\partial\lambda'}+\frac{\partial}{\partial E}\Big)t_\alpha(\lambda, \lambda'; z). \quad (3)$$

It is easy to demonstrate the validity of the formula

$$(T_\alpha R_0 T_\beta)'=T_\alpha' R_0 T_\beta+T_\alpha R_0 T_\beta', \quad \alpha\neq\beta. \quad (4)$$

We denote by $P^{\mathfrak{h}}(E)$ the operator in $\mathfrak{h}$ with kernel

$$2\pi i P\Big(\frac{E}{2}+\mu, \frac{E}{2}-\mu, \frac{E}{2}+\mu', \frac{E}{2}-\mu'; E+i0\Big);$$

then

$$(P')^{\mathfrak{h}}(E) = \frac{d}{dE} P^{\mathfrak{h}}(E).$$

This definition also carries over to operators with kernels that have singularities. We recall that the scattering matrices are introduced by the relations

$$S(E) = I - T^{\mathfrak{h}}(E), \quad S_\alpha(E) = I - T_\alpha^{\mathfrak{h}}(E), \tag{5}$$

and $s_\alpha(E) = I - 2\pi i t_\alpha(E, E; E + i0)$ is an operator in N. Below we often denote the function $t_\alpha(\lambda, \lambda; \lambda + i0)$ by $t_\alpha(\lambda)$; similarly, we set $t'_\alpha(\lambda, \lambda; \lambda + i0) \equiv t'(\lambda)$.

Following [1], we say that an operator F with kernel $f(\mu, \mu')$ in $\mathfrak{h}$ *has a trace* if the function $(\mathrm{sp}_N f)(\mu, \mu')$ is smooth, and we define

$$\mathrm{Sp}_{\mathfrak{h}} F = \int (\mathrm{sp}_N f)(\mu, \mu)\, d\mu. \tag{6}$$

In the derivation of (1) an important role is played by the so-called preliminary trace formula established for a system of three particles in [2]. We set $\Delta_z = 2i\varepsilon R_0(z) R_0(\bar{z}) = R_0(z) - R_0(\bar{z})$, and

$$K_z(P) = -P'(z) - P^*(z)\Delta_z P'(z). \tag{7}$$

In this problem the preliminary trace formula has the form

$$2i\Omega(z) = \mathrm{Sp}\, \Delta_z [K_z(T) - \sum K_z(T_\alpha)]. \tag{8}$$

We shall also need the Hilbert identity

$$T_\alpha(z) - T_\alpha^*(z) = -T_\alpha(z)\Delta_z T_\alpha^*(z) = -T_\alpha^*(z)\Delta_z T_\alpha(z). \tag{9}$$

We distinguish three types of operators in $\mathfrak{H}$ depending on the parameter z (or $\bar{z}$). The class $\mathfrak{A}_1^{(\alpha)}$ consists of operators containing the factor $\delta(\lambda_\alpha - \lambda'_\alpha)$ in their kernels; $\mathfrak{A}_1 = \mathfrak{A}_1^{(1)} \cup \mathfrak{A}_1^{(2)}$. The class $\mathfrak{A}_2$ contains operators with kernels having singular denominators of the type $(\lambda'_\alpha + \lambda_\beta - E \pm i\varepsilon)^{-1}$; for example, $T_\alpha R_0 T_\alpha \in \mathfrak{A}_2$. With the exception of the singularities enumerated, the kernels of operators of the classes $\mathfrak{A}_1$ and $\mathfrak{A}_2$ are assumed smooth. The class $\mathfrak{A}_3$ consists of operators with smooth kernels. We set $\mathfrak{A} = \mathfrak{A}_1 \cup \mathfrak{A}_2 \cup \mathfrak{A}_3$. We often omit the dependence of the operators on z.

PROPOSITION 1. 1) *If $P_\gamma \in \mathfrak{A}_1^{(\gamma)}$, $\gamma = \alpha, \beta$, then for $\alpha \neq \beta$ the operators $P_\alpha R_0 P_\beta R_0 P_\alpha$ belong to the class $\mathfrak{A}_3$.*

2) *If $P_\alpha \in \mathfrak{A}$ and $P \in \mathfrak{A}_3$, then $P_\alpha R_0 P \in \mathfrak{A}_3$ and $P R_0 P_\alpha \in \mathfrak{A}_3$.*

3) *If $P_i \in \mathfrak{A}_1^{(\alpha)}$, $i = 1, 2$, then $P_1 R_0 P_2 \in \mathfrak{A}_1^{(\alpha)}$.*

PROPOSITION 2. *Set*

$$T_0 = T_1 + T_2 - T_1 R_0 T_2 - T_2 R_0 T_1, \quad Q = T - T_0. \tag{10}$$

Then $Q \in \mathfrak{A}_3$.

Propositions 1 and 2 are easily proved by the technique of L. D. Faddeev [5].

PROPOSITION 3. *If $P \in \mathfrak{A}_3$, then*

$$\lim_{\varepsilon\to 0} \operatorname{Sp} \Delta_\varepsilon P(z) = \operatorname{Sp}_{\mathfrak{h}} P^{\mathfrak{h}}(E). \tag{11}$$

This assertion is well known; it is a consequence of the relation

$$(x-E-i0)^{-1}-(x-E+i0)^{-1}=2\pi i\delta(x-E). \tag{12}$$

PROPOSITION 4. *Let $P_i \in \mathfrak{A}$ $(i = 1, 2)$; then*

$$(P_1\Delta P_2)^{\mathfrak{h}}(E)=P_1^{\mathfrak{h}}(E)\,P_2^{\mathfrak{h}}(E). \tag{13}$$

The proof of (13) follows directly from the definitions and (12).

PROPOSITION 5. *Let F and G be operators in $\mathfrak{h}$, and suppose that F is an integral operator with kernel $f(\mu, \mu')(\mu' - \mu \pm i0)^{-1}$, while G is the operator of multiplication by $g(\mu)$, where f and g are smooth functions. Then the trace of the commutator $[F, G]$ exists, and*

$$\operatorname{Sp}_{\mathfrak{h}}[F, G]=\int \operatorname{sp}_N(f(\mu, \mu)\,g'(\mu))\,d\mu. \tag{14}$$

Indeed, the kernel of $[F, G]$ is

$$(\mu'-\mu\pm i0)^{-1}[f(\mu, \mu')\,g(\mu')-g(\mu)\,f(\mu, \mu')].$$

Transposing the operators under the sign sp_N, we find that the function

$$(\operatorname{sp}_N[F, G])(\mu, \mu')=(\mu'-\mu\pm i0)^{-1}\operatorname{sp}_N[f(\mu, \mu')(g(\mu')-g(\mu))]$$

is smooth. Its value on the diagonal $\mu = \mu'$ is found by l'Hôpital's rule. The proof of (14) is complete.

From (5), (7), and (13) we obtain

$$\operatorname{Sp}_{\mathfrak{h}}[S^*(E)\,S'(E)-\sum S_\alpha^*(E)\,S_\alpha'(E)]=\operatorname{Sp}_{\mathfrak{h}}[K_{E+i0}(T)^{\mathfrak{h}}-\sum K_{E+i0}(T_\alpha)^{\mathfrak{h}}]. \tag{15}$$

3. We proceed to the proof of (1). We first verify that the limit

$$\lim_{\varepsilon\to 0}\operatorname{Sp}\Delta_\varepsilon[K_\varepsilon(T)-K_\varepsilon(T_0)]=\operatorname{Sp}_{\mathfrak{h}}[K_{E+i0}(T)^{\mathfrak{h}}-K_{E+i0}(T_0)^{\mathfrak{h}}] \tag{16}$$

exists. According to the definitions (7) and (10), $K(T) - K(T_0) = -Q' - Q^*\Delta Q' - T_0^*\Delta Q' - Q^*\Delta T_0'$. From Proposition 2 and assertion 2) of Proposition 1 it follows that the operator $K(T) - K(T_0)$ belongs to the class $\mathfrak{A}_3$. Therefore, by Proposition 3 the left side of (16) exists and is equal to

$$-\operatorname{Sp}_{\mathfrak{h}}[Q'+Q^*\Delta Q'+T_0^*\Delta Q'+Q^*\Delta T_0']^{\mathfrak{h}}.$$

The relation (16) now follows from Proposition 4.

According to (8), (15), and (16), to prove (1) it suffices to show that

$$\lim_{\varepsilon\to 0}\operatorname{Sp}\Delta_\varepsilon[K_\varepsilon(T_0)-\sum K_\varepsilon(T_\alpha)]=\operatorname{Sp}_{\mathfrak{h}}[K_{E+i0}(T_0)^{\mathfrak{h}}-\sum K_{E+i0}(T_\alpha)^{\mathfrak{h}}]$$
$$+(4\pi i)^{-1}\sum_{\alpha\neq\beta}\int d\lambda\,\operatorname{sp}_N[s_\alpha^*(\lambda)\,s_\alpha'(\lambda)\,s_\beta'(E-\lambda)\,s_\beta^*(E-\lambda)]. \tag{17}$$

We describe the plan for proving (17). Comparing the definitions (7) and (10), we find that the operator $\Delta[K(T_0) - \sum K(T_\alpha)]$ can be represented as a sum of monomials in T_α, T_β, T_α^*, T_β^*, T_α' and T_β' each of which has degree i, where $i = 2, 3, 4$. We denote the sum of the monomials of degree i by A_i. The trace of an individual monomial, in general, has no limit as $\varepsilon \to 0$. However, using (9) and combining certain monomials into groups, it is possible to represent $\Delta[K(T_0) - \sum K(T_\alpha)]$ as a sum of a few terms each of which has a trace with a limit as $\varepsilon \to 0$. Formula (13) makes it possible to carry through analogous transformations with the operator $K_{E+i0}(T_0)^{\mathfrak{h}} - \sum K_{E+i0}(T_\alpha)^{\mathfrak{h}}$; here to the terms on the left side of (17) with a trace which has a limit as $\varepsilon \to 0$ there correspond terms with a well-defined trace in $\mathfrak{h}$. The relation of the left side of (17) to the first term of the right side is established on the basis of Proposition 5.

Explicit expressions for the operators A_i follow from (7) and (10):

$$A_2 = \sum_{\alpha \neq \beta} \Delta[(T_\alpha R_0 T_\beta)' - T_\alpha^* \Delta T_\beta'], \tag{18}$$

$$A_3 = \sum_{\alpha \neq \beta} \Delta (T_\alpha R_0 T_\beta)^* \Delta \sum T_\alpha' + \sum \Delta T_\alpha^* \Delta \sum_{\alpha \neq \beta} (T_\alpha R_0 T_\beta)', \tag{19}$$

$$A_4 = - \sum_{\alpha \neq \beta} \Delta (T_\alpha R_0 T_\beta)^* \Delta \sum_{\alpha \neq \beta} (T_\alpha R_0 T_\beta)'. \tag{20}$$

We set $B_i = (\Delta^{-1} A_i)^{\mathfrak{h}}$; then

$$K_{E+i0}(T_0)^{\mathfrak{h}} - \sum K_{E+i0}(T_\alpha)^{\mathfrak{h}} = B_2 + B_3 + B_4.$$

4. We first consider the operators A_2 and B_2. It follows from (4), (9), and (18) that

$$A_2 = \sum_{\alpha \neq \beta} \Delta (T_\alpha' R_0 T_\beta + T_\beta R_0^* T_\alpha') - \sum_{\alpha \neq \beta} \Delta T_\beta^* \Delta T_\beta \Delta T_\alpha'. \tag{21}$$

We shall establish that the trace of the first term on the right side of (21) has a limit as $\varepsilon \to 0$. The kernel of the operator $T_\alpha' R_0 T_\beta$ is equal to

$$(\lambda_\alpha' + \lambda_\beta - z)^{-1} t_\alpha' (\lambda_\alpha, \lambda_\alpha'; z - \lambda_\beta) t_\beta (\lambda_\beta, \lambda_\beta'; z - \lambda_\alpha'). \tag{22}$$

Transposing the operators under the sign of the trace in $\mathfrak{H}$, we find that

$$\operatorname{Sp}[\Delta (T_\alpha' R_0 T_\beta + T_\beta R_0^* T_\alpha')] = \operatorname{Sp}(R_0 T_\alpha' R_0 T_\beta - R_0^* T_\alpha' R_0^* T_\beta)$$

$$= \int \int d\lambda_\alpha d\lambda_\beta [(\lambda_\alpha + \lambda_\beta - z)^{-2} + (\lambda_\alpha + \lambda_\beta - \bar{z})^{-2}] \operatorname{sp}_N [t_\alpha' (\lambda_\alpha, \lambda_\alpha; z - \lambda_\beta)$$

$$\times t_\beta (\lambda_\beta, \lambda_\beta; z - \lambda_\alpha)]. \tag{23}$$

It now follows from the well-known relation $(x - E - i0)^{-2} - (x - E + i0)^{-2} = -2\pi i \delta'(x - E)$ that the limits of (23) as $\varepsilon \to 0$ is equal to

$$2\pi i \frac{\partial}{\partial E} \int d\lambda \operatorname{sp}_N [t_\alpha' (\lambda) t_\beta (E - \lambda)] - 2\pi i \int \int d\lambda_\alpha d\lambda_\beta \delta (\lambda_\alpha + \lambda_\beta - E)$$

$$\times \frac{\partial}{\partial E} \operatorname{sp}_N [t_\alpha' (\lambda_\alpha, \lambda_\alpha; E + i0 - \lambda_\beta) t_\beta (\lambda_\beta, \lambda_\beta; E + i0 - \lambda_\alpha)]. \tag{24}$$

According to (13), to (21) there corresponds the following expression for B_2:

$$B_2 = \sum_{\alpha \neq \beta} [(T'_\alpha R_0 T_\beta)^\flat + (T_\beta R_0^* T'_\alpha)^\flat] - \sum_{\alpha \neq \beta} T_\beta^{*\flat} T_\beta^\flat T'^\flat_\alpha.$$

We shall compute the trace of one of the terms in the first sum. To be specific, we assume that $\alpha = 1$ and $\beta = 2$. According to the expression (22) for the kernel of the operator $T'_\alpha R_0 T_\beta$, the kernels of the operators $(T'_1 R_0 T_2)^\flat$ and $(T_2 R_0^* T'_1)^\flat$ are equal respectively to

$$2\pi i (\mu' - \mu - i0)^{-1} t'_1\left(\frac{E}{2} + \mu, \frac{E}{2} + \mu'; \frac{E}{2} + \mu + i0\right) \times t_2\left(\frac{E}{2} - \mu, \frac{E}{2} - \mu'; \frac{E}{2} - \mu' + i0\right)$$

and

$$2\pi i (\mu - \mu' + i0)^{-1} t_2\left(\frac{E}{2} - \mu, \frac{E}{2} - \mu'; \frac{E}{2} - \mu + i0\right) \times t'_1\left(\frac{E}{2} + \mu, \frac{E}{2} + \mu'; \frac{E}{2} + \mu' + i0\right). \quad (25)$$

Thus $[\mathrm{sp}_N(T'_1 R_0 T_2 + T_2 R_0^* T'_1)^\flat](\mu, \mu')$ is equal to $F(\mu, \mu'; E)(\mu' - \mu - i0)^{-1}$, where

$$F(\mu, \mu'; E) = 2\pi i \,\mathrm{sp}_N\Big[t'_1\left(\frac{E}{2} + \mu, \frac{E}{2} + \mu'; \frac{E}{2} + \mu + i0\right) \times t_2\left(\frac{E}{2} - \mu, \frac{E}{2} - \mu'; \frac{E}{2} - \mu' + i0\right) - t'_1\left(\frac{E}{2} + \mu, \frac{E}{2} + \mu'; \frac{E}{2} + \mu' + i0\right) t_2\left(\frac{E}{2} - \mu, \frac{E}{2} - \mu'; \frac{E}{2} - \mu + i0\right)\Big].$$

We have here made use of the possibility of permuting operators under the sign of sp_N. Since $F(\mu, \mu; E) = 0$, it follows that $[\mathrm{sp}_N(T'_1 R_0 T_2 + T_2 R_0^* T'_1)^\flat](\mu, \mu')$ is a smooth function of μ and μ', and hence, according to (6), $\mathrm{Sp}_\flat(T'_1 R_0 T_2 + T_2 R_0^* T'_1)^\flat$ exists. To find $[\mathrm{sp}_N(T'_1 R_0 T_2 + T_2 R_0^* T'_1)^\flat](\mu, \mu)$ we apply l'Hôpital's rule:

$$\mathrm{Sp}_\flat (T'_1 R_0 T_2 + T_2 R_0^* T'_1)^\flat = \int \frac{\partial}{\partial \mu'} F(\mu, \mu'; E)\big|_{\mu=\mu'} d\mu. \quad (26)$$

The integral (26) is equal to the second term of (24). Therefore, the difference of (24) and (26) coincides with the first term of (24), which by (3) can be rewritten in the form

$$4\pi i \int d\lambda \,\mathrm{sp}_N [t'_\alpha(\lambda) t'_\beta(E - \lambda)]. \quad (27)$$

5. We now consider monomials of third degree. According to (19), (4), and (9), the operator $A_3 - \sum_{\alpha\neq\beta} \Delta T^*_\beta \Delta T_\beta \Delta T'_\alpha$ can be represented as the sum of four terms:

$$\sum_{\alpha\neq\beta} \Delta\,(T^*_\beta \Delta T'_\alpha R_0 T_\beta + T_\beta \Delta T^*_\beta R^*_0 T'_\alpha), \tag{28}$$

$$\sum_{\alpha\neq\beta} \Delta\,(T_\beta R^*_0 T^*_\alpha \Delta T'_\alpha + T^*_\alpha \Delta T'_\alpha R_0 T_\beta), \tag{29}$$

$$\sum_{\alpha\neq\beta} \Delta\,(T^*_\alpha R_0 T_\beta R_0 T'_\alpha - T^*_\alpha R^*_0 T^*_\beta R^*_0 T'_\alpha), \tag{30}$$

$$\sum_{\alpha\neq\beta} \Delta\,(T_\beta \Delta T^*_\beta R^*_0 T^*_\alpha \Delta T'_\alpha + T^*_\alpha R^*_0 T^*_\beta \Delta T_\beta R_0 T'_\alpha). \tag{31}$$

We carry out a similar transformation with the operator $B_3 - \sum_{\alpha\neq\beta} T^{*\mathfrak{h}}_\beta T^{\mathfrak{h}}_\beta T'^{\mathfrak{h}}_\alpha$.

We shall explain the decomposition of terms of third degree into the sum of (28)–(31). In (28) and (29) terms are combined which have the same cyclic order of the position of the operators T'_α, T_β, T^*_β and T'_α, T_β, T^*_α. The trace of each of the terms contained in (30) has a limit as $\varepsilon \to 0$. Finally, the sum (31) consists of "remainder" monomials of fourth degree.

We first consider (28). Permuting the operators under the sign of the trace in $\mathfrak{H}$ and using Propositions 1 and 3, we find that

$$\begin{aligned}\mathrm{Sp}\,(\Delta T^*_\beta \Delta T'_\alpha R_0 T_\beta + \Delta T_\beta \Delta T^*_\beta R^*_0 T'_\alpha) = \mathrm{Sp}\,\Delta\,(T^*_\beta R_0 T'_\alpha R_0 T_\beta \\ - T^*_\beta R^*_0 T'_\alpha R^*_0 T_\beta) \to \mathrm{Sp}_{\mathfrak{h}}\,(T^*_\beta R_0 T'_\alpha R_0 T_\beta - T^*_\beta R^*_0 T'_\alpha R^*_0 T_\beta)^{\mathfrak{h}}\end{aligned} \tag{32}$$

as $\varepsilon \to 0$. Further, in correspondence with the plan of the work it is necessary to establish the existence of a trace in $\mathfrak{h}$ of the operator

$$T^{*\mathfrak{h}}_\beta (T'_\alpha R_0 T_\beta)^{\mathfrak{h}} + T^{\mathfrak{h}}_\beta (T^*_\beta R^*_0 T'_\alpha)^{\mathfrak{h}} \tag{33}$$

and compare its trace with the right side of (32).

Applying Proposition 4, we rewrite the operator under the sign $\mathrm{Sp}_{\mathfrak{h}}$ on the right side of (32) in the form $T^{*\mathfrak{h}}_\beta (T'_\alpha R_0 T_\beta)^{\mathfrak{h}} + (T^*_\beta R^*_0 T'_\alpha)^{\mathfrak{h}} T^{\mathfrak{h}}_\beta$. To prove the existence of a trace for the operator (33) it suffices to show that the commutator $[(T^*_\beta R^*_0 T'_\alpha)^{\mathfrak{h}}, T^{\mathfrak{h}}_\beta]$ has a well-defined trace in $\mathfrak{h}$. The difference of the right side of (32) and the trace of the operator (33) is equal to $\mathrm{Sp}_{\mathfrak{h}}[(T^*_\beta R^*_0 T'_\alpha)^{\mathfrak{h}}, T^{\mathfrak{h}}_\beta]$. Taking, for example, $\alpha = 1$ and $\beta = 2$, we find that in correspondence with (25) the operator $(T^*_\beta R^*_0 T'_\alpha)^{\mathfrak{h}}$ is given by the kernel $F(\mu, \mu'; E)(\mu' - \mu - i0)^{-1}$, where $F(\mu, \mu'; E)$ is a smooth function and

$$F(\mu, \mu; E) = -2\pi i t^*_2(E/2 - \mu)\, t'_1(E/2 + \mu).$$

The operator $T^{\mathfrak{h}}_\beta$ is the operator of multiplication by the function $2\pi i t_2(E/2 - \mu)$. It follows from Proposition 5 that the trace

$$\mathrm{Sp}_{\mathfrak{h}}\left[(T^*_\beta R^*_0 T'_\alpha)^{\mathfrak{h}},\ T^{\mathfrak{h}}_\beta\right] = 2\,(2\pi i)^2 \int d\lambda\, \mathrm{sp}_{\mathcal{N}} t'_\alpha(\lambda)\, t'_\beta(E - \lambda)\, t^*_\beta(E - \lambda) \tag{34}$$

exists.

The operator (29) is treated in a manner similar to (28). Since we can permute operators under the Sp sign, it follows that

$$\operatorname{Sp}\Delta(T_\beta R_0^* T_\alpha^* \Delta T_\alpha' + T_\alpha^* \Delta T_\alpha' R_0 T_\beta) = \operatorname{Sp}\Delta(T_\alpha' R_0 T_\beta R_0 T_\alpha^* - T_\alpha' R_0^* T_\beta R_0^* T_\alpha^*).$$

It follows from Propositions 1, 3, and 4 that this expression has a limit as $\varepsilon \to 0$ equal to

$$\operatorname{Sp}_{\mathfrak{h}} \left(T_\alpha'^{\mathfrak{h}} (T_\beta R_0^* T_\alpha^*)^{\mathfrak{h}} + (T_\alpha' R_0 T_\beta)^{\mathfrak{h}} T_\alpha^{*\mathfrak{h}} \right). \tag{35}$$

Further, it is necessary to establish the existence of

$$\operatorname{Sp}_{\mathfrak{h}} \left((T_\beta R_0^* T_\alpha^*)^{\mathfrak{h}} T_\alpha'^{\mathfrak{h}} + T_\alpha^{*\mathfrak{h}} (T_\alpha' R_0 T_\beta)^{\mathfrak{h}} \right) \tag{36}$$

and to compute the difference of (35) and (36). This difference can be written in the form

$$\operatorname{Sp}_{\mathfrak{h}} \left[T_\alpha'^{\mathfrak{h}}, (T_\beta R_0^* T_\alpha^*)^{\mathfrak{h}} \right] + \operatorname{Sp}_{\mathfrak{h}} \left[(T_\alpha' R_0 T_\beta)^{\mathfrak{h}}, T_\alpha^{*\mathfrak{h}} \right]. \tag{37}$$

Applying Proposition 5, we find that each of the terms in (37) is well defined, and the entire expression (37) is equal to

$$2(2\pi i)^2 \int d\lambda \operatorname{sp}_{\mathcal{N}} [t_\alpha^*(\lambda) t_\alpha''(\lambda) + t_\alpha^{*\prime}(\lambda) t_\alpha'(\lambda)] t_\beta(E-\lambda)$$
$$= 2(2\pi i)^2 \int d\lambda \operatorname{sp}_{\mathcal{N}} t_\alpha^*(\lambda) t_\alpha'(\lambda) t_\beta'(E-\lambda). \tag{38}$$

The equality (38) is established by integration by parts. It follows from the existence of the traces in (35) and (37) that (36) is meaningful.

According to Propositions 1 and 3, the trace of each of the operators under the summation sign in (30) has a limit which coincides with the trace of the corresponding operator in $\mathfrak{h}$.

6. We proceed to consider the terms of fourth degree. The sum of monomials of fourth degree is equal to the combination of (20) and (31). Using (9), we can write this sum in the form

$$\sum_{\alpha \neq \beta} [(\Delta T_\beta \Delta T_\beta^* R_0^* T_\alpha^* \Delta T_\alpha' - \Delta T_\beta^* R_0^* T_\alpha^* \Delta T_\alpha' R_0 T_\beta)$$
$$- \Delta T_\alpha^* R_0^* T_\beta^* \Delta T_\alpha' R_0 T_\beta - \Delta T_\beta^* R_0^* T_\alpha^* \Delta T_\beta R_0 T_\alpha']. \tag{39}$$

We start by considering the first term in (39). Permuting the operators under the Sp sign and applying Propositions 1 and 3, we find that

$$\operatorname{Sp}(\Delta T_\beta \Delta T_\beta^* R_0^* T_\alpha^* \Delta T_\alpha' - \Delta T_\beta^* R_0^* T_\alpha^* \Delta T_\alpha' R_0 T_\beta) = \operatorname{Sp}\Delta T_\beta^* R_0^* T_\alpha^* (R_0^* - R_0) T_\alpha' R_0^* T_\beta$$
$$\to \operatorname{Sp}_{\mathfrak{h}} (T_\beta^* R_0^* T_\alpha^* (R_0^* - R_0) T_\alpha' R_0^* T_\beta)^{\mathfrak{h}} \tag{40}$$

as $\varepsilon \to 0$. To this operator there corresponds the operator

$$T_\beta^{\mathfrak{h}} (T_\beta^* R_0^* T_\alpha^*)^{\mathfrak{h}} T_\alpha'^{\mathfrak{h}} - (T_\beta^* R_0^* T_\alpha^*)^{\mathfrak{h}} (T_\alpha' R_0 T_\beta)^{\mathfrak{h}}, \tag{41}$$

acting in $\mathfrak{h}$. According to Proposition 4, the operator under the sign $\mathrm{Sp}_{\mathfrak{h}}$ on the right side of (40) can be written in the form

$$(T_\beta^* R_0^* T_\alpha^*)^{\mathfrak{h}}\, T_\alpha'^{\mathfrak{h}} T_\beta^{\mathfrak{h}} - (T_\beta^* R_0^* T_\alpha^*)^{\mathfrak{h}} (T_\alpha' R_0 T_\beta)^{\mathfrak{h}}. \tag{42}$$

The difference of (42) and (41) is equal to $[(T_\beta^* R_0^* T_\alpha^*)^{\mathfrak{h}}, T'^{\mathfrak{h}}, T_\beta^{\mathfrak{h}}]$. It follows from Proposition 5 that the trace of this commutator is well defined and equal to

$$2\,(2\pi i)^3 \int d\lambda\, \mathrm{sp}_N t_\alpha^*(\lambda)\, t_\alpha'(\lambda)\, t_\beta'(E-\lambda)\, t_\beta^*(E-\lambda). \tag{43}$$

By Propositions 1 and 3 the trace of the second and third terms under the summation sign in (39) has a limit as $\varepsilon \to 0$, and this limit coincides with the trace of the corresponding operator in $\mathfrak{h}$.

We return to (17). From our arguments it follows that all the terms it contains are well defined. To prove (17) it remains to note that the sum of (27), (34), (38), and (43) is equal to

$$(4\pi i)^{-1} \int d\lambda\, \mathrm{sp}_N\, [s_\alpha^*(\lambda)\, s_\alpha'(\lambda)\, s_\beta'(E-\lambda)\, s_\beta^*(E-\lambda)].$$

The proof of (1) is complete.

Bibliography

1. V. S. Buslaev, *On trace formulas for multichannel problems*, Zap. Naučn. Sem. Leningrad. Otdel. Mat. Inst. Steklov. (LOMI) **27** (1972), 47–66; English transl. in J. Soviet Math. **3** (1975), no. 4.

2. V. S. Buslaev and S. P. Merkur'ev, *On the relation between the third virial coefficient and the S-matrix*, Teoret. i. Mat. Fiz. **5** (1970), 372–387; English transl. in Theoret. and Math. Phys. **5** (1970).

3. L. D. Faddeev, *On the Friedrichs model in the theory of perturbations of a continuous spectrum*, Trudy Mat. Inst. Steklov. **73** (1964), 292–313; English transl. in Amer. Math. Soc. Transl. (2) **62** (1967).

4. V. S. Buslaev, *Spectral identities and the trace formula in the Friedrichs model*, Problemy Mat. Fiz., vyp. 4, Izdat. Leningrad. Gos. Univ., Leningrad, 1970, pp. 48–60; English transl. in Topics in Math. Phys., no. 4, Plenum Press, 1971.

5. L. D. Faddeev, *Mathematical questions in the quantum theory of scattering for a system of three particles*, Trudy Mat. Inst. Steklov. **69** (1963); English transl., *Mathematical aspects of the three-body problem in the quantum scattering theory*, Israel Program Sci. Transls., Jerusalem, and Davey, New York, 1965.

Translated by J. R. SCHULENBERGER

ABCDEFGHIJ–AMS–8987654321